THE SMITH CONJECTURE

Paul A. Smith

THE SMITH CONJECTURE

Edited by   JOHN W. MORGAN
            and
            HYMAN BASS

Department of Mathematics
Columbia University
New York, New York

1984

ACADEMIC PRESS, INC.

(Harcourt Brace Jovanovich, Publishers)

Orlando San Diego San Francisco New York London
Toronto Montreal Sydney Tokyo São Paulo

This is a volume in
PURE AND APPLIED MATHEMATICS

A Series of Monographs and Textbooks

Editors: SAMUEL EILENBERG AND HYMAN BASS

A list of recent titles in this series appears at the end of this volume.

COPYRIGHT © 1984, BY ACADEMIC PRESS, INC.
ALL RIGHTS RESERVED.
NO PART OF THIS PUBLICATION MAY BE REPRODUCED OR
TRANSMITTED IN ANY FORM OR BY ANY MEANS, ELECTRONIC
OR MECHANICAL, INCLUDING PHOTOCOPY, RECORDING, OR ANY
INFORMATION STORAGE AND RETRIEVAL SYSTEM, WITHOUT
PERMISSION IN WRITING FROM THE PUBLISHER.

ACADEMIC PRESS, INC.
Orlando, Florida 32887

United Kingdom Edition published by
ACADEMIC PRESS, INC. (LONDON) LTD.
24/28 Oval Road, London NW1 7DX

Library of Congress Cataloging in Publication Data
Main entry under title:

The Smith conjecture.

  (Pure and applied mathematics)
  Includes index.
  1. Three-manifolds (Topology) 2. Finite groups.
I. Morgan, John W., Date. II. Bass, Hyman,
Date. III. Series: Pure and applied mathematics
(Academic Press)
QA3.P8 [QA613]   510s [514'.3]   83-15846
ISBN 0–12–506980–4 (alk. paper)

PRINTED IN THE UNITED STATES OF AMERICA

84 85 86 87   9 8 7 6 5 4 3 2 1

CONTENTS

Contributors ix
Preface xi
Acknowledgments xiii
List of Notation xv

PART A. INTRODUCTION

I. The Smith Conjecture   3

John W. Morgan

1. Formulations   3
2. Generalizations   5
3. Some Consequences Relating to the Poincaré Conjecture   5
4. Additional Remarks   6

II. History of the Smith Conjecture and Early Progress   7

John W. Morgan

1. History of the Smith Conjecture   7
2. Early Progress   8

III. An Outline of the Proof — 11
John W. Morgan

1. Preliminaries — 11
2. First Reductions — 13
3. The Argument in Brief — 13

References for Part A — 17

PART B. THE CASE OF NO INCOMPRESSIBLE SURFACE

IV. The Proof in the Case of No Incompressible Surface — 21
Peter B. Shalen

Introduction — 21
1. The Algebraic Approach to the Smith Conjecture — 22
2. Hyperbolic Geometry and Algebraic Integers — 23
3. The Existence of Hyperbolic Structures and the Torus Theorem — 28
4. $PSL_2(\mathbf{C})$ and Incompressible Surfaces — 31
5. History — 35
References — 36

V. On Thurston's Uniformization Theorem for Three-Dimensional Manifolds — 37
John W. Morgan

Introduction — 37
1. An Introduction to Hyperbolic Geometry — 43
2. Kleinian Groups — 47
3. Statement of the Main Theorem—The Case of Finite Volume — 51
4. Hierarchies and Pared Manifolds — 55
5. Statement of the Main Theorem—The General Case — 60
6. Convex Hyperbolic Structures of Finite Volume — 61
7. The Gluing Theorem—Statement and First Reduction — 70
8. Combination Theorems — 74
9. Deformation Theory — 78
10. The Fixed Point Theorem — 84
11. The First Step in the Proof of the Bounded Image Theorem — 87
12. Completion of the Proof of the Bounded Image Theorem — 95
13. Special Cases — 106
14. Kleinian Groups with Torsion — 107
15. Patterns of Circles — 117
16. The Inductive Step in the Proof of Theorems A' and B' — 119
References — 124

VI. Finitely Generated Subgroups of GL_2 — 127

Hyman Bass

1. The GL_2-Subgroup Theorem — 127
2. Arboreal Group Theory — 130
3. The Tree of SL_2 over a Local Field — 132
4. Proof of the GL_2-Subgroup Theorem — 134
   References — 136

PART C. THE CASE OF AN INCOMPRESSIBLE SURFACE

VII. Incompressible Surfaces in Branched Coverings — 139

C. McA. Gordon and R. A. Litherland

1. Introduction — 139
2. Terminology and Statement of Results — 140
3. Proofs of Theorems 1 and 2 — 142
4. The Equivalent Loop Theorem for Involutions — 147
   References — 151

VIII. The Equivariant Loop Theorem for Three-Dimensional Manifolds and a Review of the Existence Theorems for Minimal Surfaces — 153

Shing-Tung Yau and William H. Meeks, III

1. Morrey's Solution for the Plateau Problem in a General Riemannian Manifold — 154
2. The Existence Theorem for Manifolds with Boundary — 159
3. Existence of Closed Minimal Surfaces — 160
4. Existence of the Free Boundary Value Problem for Minimal Surfaces — 162
   References — 163

PART D. GENERALIZATIONS

IX. Group Actions on R^3 — 167

William H. Meeks, III and Shing-Tung Yau

References — 179

X. Finite Group Actions on Homotopy 3-Spheres   181
Michael W. Davis and John W. Morgan

1.	Orbifolds	183
2.	Two-Dimensional Orbifolds	187
3.	Three-Dimensional Orbifolds	189
4.	Seifert-Fibered Orbifolds	192
5.	Seifert-Fibered Orbifolds and Linear Actions	196
6.	Statement of the Main Results	206
7.	A Special Case	207
8.	Completion of the Proof	213
	Appendix	216
	References	225

XI. A Survey of Results in Higher Dimensions   227
Michael W. Davis

1.	The Montgomery–Samuelson Example	230
2.	G-Complexes	230
3.	The Brieskorn Examples	231
4.	Oliver's Example	232
5.	Local Properties: Groups of Homeomorphisms versus Groups of Diffeomorphisms	234
6.	Work of Lowell Jones	235
7.	Actions on Disks	236
8.	Actions on Spheres	238
9.	Actions on Euclidean Spaces	238
	References	239

Index   241

CONTRIBUTORS

Numbers in parentheses indicate the pages on which the authors' contributions begin.

HYMAN BASS (127), Department of Mathematics, Columbia University, New York, New York 10027

MICHAEL W. DAVIS (181, 227), Department of Mathematics, Ohio State University, Columbus, Ohio 43210

C. McA. GORDON (139), Department of Mathematics, University of Texas at Austin, Austin, Texas 78712

R. A. LITHERLAND[†] (139), Department of Pure Mathematics and Mathematical Statistics, University of Cambridge, Cambridge, England

WILLIAM H. MEEKS, III (153, 167), Instituto de Matemàtica Pura e Aplicada, Estrada Dona Castorina, 110, Rio de Janiero, Brazil 22460

JOHN W. MORGAN (3, 7, 11, 37, 181), Department of Mathematics, Columbia University, New York, New York 10027

PETER B. SHALEN (21), Department of Mathematics, Rice University, Houston, Texas 77001

SHING-TUNG YAU[‡] (153, 167), Department of Mathematics, Stanford University, Stanford, California 94305

[†] Present address: Department of Mathematics, Louisiana State University, Baton Rouge, Louisiana 70803.
[‡] Present address: Institute for Advanced Study, School of Mathematics, Princeton University, Princeton, New Jersey 08540.

PREFACE

In the 1940s Paul A. Smith asked whether or not the fixed point set of a periodic, orientation-preserving homeomorphism of S^3 to itself was always an unknotted circle. In the fall of 1978 this question was answered in the affirmative for diffeomorphisms. The proof rests on the work of mathematicians in diverse areas of mathematics. In the spring of 1979 a symposium was held at Columbia University on the solution to Smith's question. It brought together the principal actors in the drama to present various pieces of the proof. In addition to written versions of the presentations, this volume includes an introduction which explains how the pieces fit together. (See especially Chapter III, Section 3.) There are also two papers (Chapters IX and X) containing generalizations of the results on the Smith conjecture. The last article (Chapter XI) is a discussion of the situation in dimensions greater than three.

It seemed entirely appropriate to have such a symposium at Columbia University. Paul Smith spent most of his mathematical life at Columbia. In the spring of 1979 he was Professor Emeritus at Columbia and still lived in the neighborhood. He was one of the most attentive members of the audience as the resolution of his 38-year-old question unfolded.

It was also at Columbia that a significant step in the solution of the conjecture occurred. During a conversation with Bass, Thurston learned of Bass's result (Chapter VI). He saw, in the light of his own work (Chapter V), the relevance to the Smith conjecture. He also saw the need to treat the cases covered in Part C. What was needed to deal with these missing cases came clearly into focus during conversations between Thurston and Gordon (the latter being motivated by his earlier work with Litherland; see Chapter VII). At about the same time, Meeks and Yau had established exactly the required result (Chapter VIII). However, there was a gap in communication between Thurston and Gordon, on the one hand, and Meeks and Yau, on the other. This gap was bridged by Gordon when he learned of the existence of the work of Meeks and Yau. With that, the proof was complete.

There was a purpose beyond the purely mathematical in holding the symposium. That was to honor Paul Smith. His work has had tremendous influence on topology, and the symposium provided a look at one direction in which this influence has led the field. During the symposium Smith said that out of consideration for the younger mathematicians he would be sure to make his next question easier to solve than this one. He was jesting of course, for he knew full well that mathematics needs deep and hard problems and that the younger mathematicians assembled at the symposium owed him a debt of gratitude for making his questions hard and fertile ones on which to work.

Sadly, nothing marks the passage of time between the symposium and the publication of this volume more clearly than Paul Smith's death. All who knew him are saddened and made poorer by his passing. He was a fine man as well as a first-rate mathematician.

ACKNOWLEDGMENTS

Thanks are due to many. Both Joan Birman and Michael Davis gave generously of their time and expertise in helping to prepare this volume. The symposium was made possible by support from the National Science Foundation and the J. F. Ritt Lecture Fund. The NSF also bore the cost of preparing this volume. We express our thanks also to Columbia University for hosting the symposium. We are grateful to Kate March for her help in preparing the manuscripts. Finally, we thank the Academic Press staff for their friendly and efficient assistance in the preparation of this volume for publication.

LIST OF NOTATION

$\tilde{\Sigma}$	Homotopy 3-sphere
$h: \tilde{\Sigma} \to \tilde{\Sigma}$	Periodic diffeomorphism
Σ	The quotient space of $\tilde{\Sigma}$ by the action generated by h
$\tilde{K}$	The fixed set of h
$p: \tilde{\Sigma} \to \Sigma$	Natural projection
$K = p(\tilde{K})$	
$\Gamma = \pi_1(\Sigma - K)$	
$\tilde{\Gamma} = \pi_1(\tilde{\Sigma} - \tilde{K})$	
$M = \Sigma - K$	
$F \subset M$	Closed surface
$\tilde{F} = p^{-1}(F)$	
D	2-Disk
N	A closed tubular neighborhood of K
$\tilde{N} = (p^{-1}(N))$	
μ	Meridian on N
B	3-Ball

PART **A**

INTRODUCTION

CHAPTER I

The Smith Conjecture

John W. Morgan†

Department of Mathematics
Columbia University
New York, New York

1. Formulations

Let S^3 be the unit sphere in $\mathbf{R}^4$. Smith proved [Sm 1] that any periodic, orientation-preserving homeomorphism of S^3 to itself with fixed points has a fixed point set that is homeomorphic to a circle. He then asked [Ei 1] if the fixed point set must be unknotted. (Unknotted here means that there is a homeomorphism of S^3 to itself that throws the given simple closed curve onto a geometric circle, i.e., onto S^3 intersected with a two-dimensional linear subspace of $\mathbf{R}^4$.) As Smith realized, an affirmative answer to this question is equivalent to the statement that every such homeomorphism of S^3 to itself is standard, that is, topologically equivalent to a linear (i.e., orthogonal) action [Moi 1, Sm 4]. Montgomery and Zippin [Mon–Z] showed that in this generality the answer to the question is no. They gave examples of periodic homeomorphisms of S^3 whose fixed point sets are wildly embedded

---

† With assistance from Joan Birman and Michael W. Davis.

simple closed curves (wild in the sense that there are not even *local* homeomorphisms that throw the fixed point set onto a standard arc). This type of pathology is ruled out if we require the homeomorphism to be a diffeomorphism (or piecewise linear (PL) homeomorphism). One formulates the differentiable version of Smith's question thus:

If $h: S^3 \to S^3$ is an orientation-preserving, periodic diffeomorphism with nonempty fixed point set, then is this fixed set an unknotted circle?

What became known as the Smith conjecture was the conjecture that the answer to this question is yes. Another way to frame the question is to ask

Is every orientation-preserving, periodic diffeomorphism $h: S^3 \to S^3$, with fixed points, conjugate (by a diffeomorphism of S^3 to itself) to an orthogonal diffeomorphism?

It is the purpose of this volume to present the recent arguments that answer this question affirmatively.

It turns out that one makes no essential use of the fact that the space being acted upon is S^3. The arguments apply more generally to homotopy 3-spheres. *Henceforth, $\tilde{\Sigma}$ denotes a homotopy 3-sphere.* The main theorem proved in this volume is the following:

THEOREM (Solution of the Smith Conjecture). *Let $h: \tilde{\Sigma} \to \tilde{\Sigma}$ be an orientation-preserving, periodic diffeomorphism (different from the identity) with fixed points. Then the fixed point set of h is an unknotted circle[1] in $\tilde{\Sigma}$.*

Remarks. (1) The arguments proving the Smith conjecture can easily be adapted for the PL case, or for the topological case, provided that in the topological case one assumes that the fixed point set $\tilde{K} \subset \tilde{\Sigma}$ is locally flat.
(2) All known examples in which the fixed point set $\tilde{K}$ is wild have the property that $\tilde{K}$ bounds a topologically embedded disk.

There is a reformulation of this result in terms of group actions. Suppose that we are given an effective, orthogonal action of a finite group $G \times S^3 \to S^3$ and a homotopy 3-cell H. Choose a ball $B \subset S^3$ that is disjoint from all its translates under nontrivial elements of G. Remove B and all its translates from S^3 and sew a copy of H into each hole. The restricted G-action on $S^3 - (\bigcup_{g \in G} gB)$ extends in an obvious way to an action on the resulting homotopy 3-sphere. Any action constructed in this manner is called *essentially linear.*

---

[1] An unknotted circle is one that bounds a differentiably embedded 2-disk $D \subset \tilde{\Sigma}$. If $\tilde{\Sigma} = S^3$, then this notion agrees with the other definition of unknotted.

I. The Smith Conjecture

THEOREM (Reformulation of the Solution to the Smith Conjecture). *Let $G \times \tilde{\Sigma} \to \tilde{\Sigma}$ be a finite cyclic group action generated by an orientation-preserving diffeomorphism with a nonempty fixed point set. This action is equivariantly diffeomorphic to an essentially linear action.*

2. Generalizations

The techniques used to establish the Smith conjecture can be used to prove various generalizations. Two such generalizations are treated in Chapters IX and X of this volume. In Chapter IX Meeks and Yau prove the following:

THEOREM *If G is a finite group of orientation-preserving diffeomorphisms of $\mathbf{R}^3$, then the action is equivariantly diffeomorphic to a linear action, except possibly when G is isomorphic to A_5, the alternating group on 5 letters.*[2]

In Chapter X Davis and Morgan treat another generalization:

THEOREM. *If G is a finite group of orientation-preserving diffeomorphisms of $\tilde{\Sigma}$ so that all isotropy groups are cyclic and one isotropy group has order divisible by a prime > 5, then G is equivariantly diffeomorphic to an essentially linear action.*[3]

These results lead one to the following:

QUESTION. *Is every nonfree action of a finite group on a homotopy 3-sphere or a contractible 3-manifold essentially linear?*

The results quoted above show that in certain special cases the answer is yes. There is no known example of a nonlinear, differentiable action of a finite group on S^3 or $\mathbf{R}^3$.

3. Some Consequences Relating to the Poincaré Conjecture

The theorem on the solution of the Smith conjecture affirms several special cases of the Poincaré conjecture: Suppose that $h: \tilde{\Sigma} \to \tilde{\Sigma}$ is an orientation-preserving periodic diffeomorphism of period n with fixed points. Let Σ be the quotient of $\tilde{\Sigma}$ by the cyclic group generated by h. It is easy to see that Σ is a homotopy 3-sphere. It follows from the theorem on the reformulation

---

[2] See footnote 1 in Chapter X for the case $G \approx A_5$.
[3] This result has been expanded to cover more cases (see Feighn [Fe]).

of the solution of the Smith conjecture that $\tilde{\Sigma}$ is diffeomorphic to the connected sum of n copies of Σ. In particular, $\tilde{\Sigma}$ is not an irreducible homotopy 3-sphere. Also, if the quotient space Σ is diffeomorphic to S^3, then so is $\tilde{\Sigma}$. This last result can be phrased another way: A cyclic covering of S^3 branched along a smooth knot is never a counterexample to the Poincaré conjecture. On the other hand, not all 3-manifolds (or even all homology 3-spheres) are cyclic branched covers of S^3 [My].

Birman and Hilden [Bir–H] have shown that every 3-manifold with a Heegaard splitting of genus 2 is a two-sheeted branched cyclic cover of S^3. As a result, there is no counterexample to the Poincaré conjecture with a Heegaard splitting of genus 2.

4. Additional Remarks

The proof of the Smith conjecture represents a culmination of the efforts of many mathematicians. Of course, the work of Smith himself on cyclic group actions was seminal. Almost all of the apparatus of three-dimensional topology, as it has developed from the foundation laid by Kneser, Papakyriakopoulos, and Haken, to its contemporary form in the work of Waldhausen, Stallings, Epstein, Jaco, Shalen, and Johannson, among others, is needed as well. In this volume, we regard this apparatus as "classical," as forming the mathematical environment in which the proof resides. The broad outlines of this proof were first brought into focus by Thurston. The proof, as it finally emerged, represents a confluence of ideas and methods from diverse areas of mathematics (minimal surface theory, hyperbolic geometry, and kleinian groups, and the algebra of 2×2 matrices). The applications of these ideas and methods to three-dimensional topology are due to Thurston, Meeks, and Yau, with help from Bass, Shalen, Gordon, and Litherland. The most surprising feature is how well these new techniques mesh with the purely topological techniques already available. Together, they should have a profound influence on three-dimensional topology, the solution of the Smith conjecture being but a beginning.

CHAPTER II

History of the Smith Conjecture and Early Progress

John W. Morgan†

Department of Mathematics
Columbia University
New York, New York

1. History of the Smith Conjecture

The study of periodic diffeomorphisms of the disk and the sphere began with the work of Brouwer [Bro] and Kerekjarto [Ke] in the 1920s. They proved that an orientation-preserving periodic diffeomorphism of the 2-disk or the 2-sphere was conjugate by a diffeomorphism to a rotation. These original proofs were incomplete and the gap was filled later by Eilenberg [Ei 2].

It was in this context that Smith's work was done. In a series of papers produced during the 1930s and 1940s [Sm 1–3] he studied homeomorphisms of the n-disk and the n-sphere periodic of prime power period p^r. He showed that the $\mathbf{Z}/p$-homology of the fixed point set is the same as that of a smaller dimensional disk or sphere. Furthermore, if the homeomorphism is orientation-preserving, then the codimension of the fixed point set is even. Now consider $f: S^3 \to S^3$ a periodic homeomorphism that preserves the orientation

† With assistance from Joan Birman and Michael W. Davis.

and that has fixed points. Some power f^t of f is of prime power period. Thus its fixed points $F(f^t)$ must have the $\mathbf{Z}/p$-homology of a circle. This means that the fixed point set of f^t is precisely a circle in S^3. The homeomorphism f acts on $F(f^t)$ as an orientation-preserving homeomorphism of finite order. Since f has fixed points, $f|F(f^t)$ contains fixed points. But the only orientation-preserving periodic homeomorphism of the circle with fixed points is the identity. Hence $f|F(f^t)$ is the identity, and the fixed points of f form a circle $F(f) \subset S^3$.

Clearly, it was this result that led Smith to ask which circles in S^3 arise as fixed points of periodic mappings. Of course, any geodesic circle in the homogeneous metric is the fixed point set of a linear periodic isometry. Conjugating these by homeomorphism of S^3 produces examples where the fixed point set is any unknotted circle in S^3. Smith then asked [Ei 1], Are there any other circles which arise as fixed points of periodic homeomorphisms?

2. Early Progress

We are studying a periodic homeomorphism $h: \tilde{\Sigma} \to \tilde{\Sigma}$ with nonempty fixed point set $\tilde{K} \subset \tilde{\Sigma}$. The quotient of $\tilde{\Sigma}$ by the group generated by h is denoted Σ, and $K \subset \Sigma$ is the image of $\tilde{K}$.

One of the earliest results on the Smith conjecture was the construction by Montgomery and Zippen [Mon-Z] in 1954 of a topological involution of the 3-sphere with the fixed point set a wild knot. This led naturally to a modification of the Smith conjecture by the addition of the hypothesis that h be PL (implying that $\tilde{K}$ is tame) or, equivalently, that h be a diffeomorphism. There followed a series of papers that ruled out various classes of tame knots as counterexamples. In 1955, Montgomery and Samelson [Mon-Sa] established that a two-strand cable knot could not occur as the fixed point set of a PL involution h. In 1958, Kinoshita [Ki] and Fox [F 1] found conditions on the Alexander polynomials of tame $\tilde{K}$ and K that allowed them to rule out various combinations of knot type and period as counterexamples. These were summarized in Fox [Fo 2]. Most of these conditions arise from the fact that if $\tilde{K}$ is a counterexample, then the knot group $\tilde{\Gamma}$ is a normal subgroup of the knot group Γ with quotient group $\mathbf{Z}/n\mathbf{Z}$. Other special cases which were settled include: $\tilde{K}$ cannot be a torus knot (Giffen [G 1], also Fox [Fo 3]), K cannot be a torus knot (Fox [Fo 3]), K cannot be a doubled knot (Giffen [G 3]), and $\tilde{K}$ cannot be a two-bridge knot (Cappell and Shaneson [Ca-Shan]) or a cabled knot or cabled braid or doubled knot (Myers [My]).

---

[1] A knot K in S^3 is tame if it is equivalent to a polygonal knot; otherwise it is wild.

II. History of the Smith Conjecture and Early Progress

Major progress of a more general nature was made by Waldausen [W] in 1969. He showed that the Smith conjecture for PL homeomorphism of even period was true. His methods were entirely those of PL topology and have not been generalized to odd periods.

Moise [Moi 1] showed in 1961 that if h is PL and $\tilde{K}$ is unknotted, then h is conjugate to an orthogonal rotation. This result was reproved by Smith [Sm 4] in 1965 using simpler techniques. In 1964, Bing [Bin] produced for each integer $n > 1$ uncountably many inequivalent nonorthogonal rotations of E^3 about a wild line of period n. Even more, he showed that if X is any closed subset of E^1 without isolated points, then for each $n > 1$ there is a rotation h of E^3 of period n with the wild subset of $\tilde{K}$ topologically equivalent to X. Bing asked whether or not a (not necessarily PL) periodic map h of E^3 would be equivalent to a standard PL map if $F(h)$ were the z-axis, a question which was settled affirmatively by Moise [Moi 2] in 1978.

The analogue of the Smith conjecture in dimension $m > 3$ was investigated by Giffen [G 2], who showed that for each $n > 1, m > 4$ there exists a smooth m-sphere pair $(S^m, \tilde{K})$ that is inequivalent to the standard pair (S^m, S^{m-2}), such that $\tilde{K}$ is fixed pointwise by a smooth transformation of period n of S^m. He produced similar counterexamples for $m = 4$ and odd n for piecewise linear maps.

CHAPTER III

An Outline of the Proof†

John W. Morgan‡

Department of Mathematics
Columbia University
New York, New York

1. Preliminaries

We are given a homotopy 3-sphere and an orientation-preserving, periodic diffeomorphism $h: \tilde{\Sigma} \to \tilde{\Sigma}$ of period $n > 1$. Smith's results [Sm 2] imply that the fixed point set of h is either empty or diffeomorphic to S^1. We are concerned with the latter case. We denote the fixed points of h by $\tilde{K} \subset \tilde{\Sigma}$. Our task is to show that $\tilde{K}$ is unknotted in $\tilde{\Sigma}$. For the purpose of the argument it is often more convenient to work with the quotient space of the cyclic group action generated by h, $\Sigma = \tilde{\Sigma}/\langle h \rangle$. Notice that if $0 < l < n$, then the fixed point set of h^l contains $\tilde{K}$ and is diffeomorphic to a circle. Hence this fixed point set is exactly $\tilde{K}$. This means that the cyclic group generated by h acts freely on $\tilde{\Sigma} - \tilde{K}$. By the *equivariant tubular neighborhood theorem* [Bre] there is an invariant neighborhood $\tilde{N}$ of $\tilde{K}$ in $\tilde{\Sigma}$ of the form $K \times D^2$, where D^2

---

† A glossary of technical terms used in this chapter and throughout this volume can be found in the List of Notation that precedes Chapter I.
‡ With assistance from Joan Birman and Michael W. Davis.

is the unit disk in **C**. Furthermore, we can choose this product structure so that the action of h on N is given by

$$h(\alpha, z) = (\alpha, \zeta \cdot z), \quad \alpha \in \tilde{K}, \quad \text{and} \quad z \in D^2$$

where $\zeta = e^{2\pi i r/n}$ for some r relatively prime to n. In view of this the quotient space Σ naturally inherits the structure of a smooth manifold, so that the projection map $p: \tilde{\Sigma} \to \Sigma$ is an n-fold cyclic covering branched over $K = p(\tilde{K})$.

Let $\Gamma = \pi_1(\Sigma - K)$ and $\tilde{\Gamma} = \pi_1(\tilde{\Sigma} - \tilde{K})$. The map $p: \tilde{\Sigma} - \tilde{K} \to \Sigma - K$ is a cyclic (unbranched) covering. Thus we have an exact sequence

$$1 \to \tilde{\Gamma} \xrightarrow{p_*} \Gamma \to \mathbf{Z}/n\mathbf{Z} \to 1.$$

Let $\tilde{u} \in \pi_1(\tilde{\Sigma} - \tilde{K})$ and $u \in \pi_1(\Sigma - K)$ be the classes of the meridians of the knots. Clearly, $p_*\tilde{u}$ is conjugate to u^n. The fact that $\tilde{\Sigma}$ is simply connected is equivalent to the fact that $\tilde{u}$ is a normal generator for $\tilde{\Gamma}$. It follows immediately that u is a normal generator for Γ and hence that Σ is simply connected, i.e., Σ is also a homotopy sphere (see Crowell and Fox [Cr-F]).

The action of $\langle h \rangle$ on $\tilde{\Sigma}$ is identified with the group of covering transformations of this cyclic branched cover. Thus, the Smith conjecture can be rephrased: *If a cyclic branched cover of a homotopy 3-sphere Σ branched over $K \subset \Sigma$ is simply connected, then does it follow that $K \subset \Sigma$ unknotted?*

The space $N = p(\tilde{N})$ is a tubular neighborhood of $K \subset \Sigma$ of the form $K \times D^2$. Thus the complement of K, $\Sigma - K$, is diffeomorphic to the interior of a compact manifold with boundary, $\Sigma - \text{int } N$. For many technical reasons it is often desirable to work with this compact manifold rather than $\Sigma - K$.

This brings us to the first basic result that we need.

THEOREM. *The following conditions are equivalent:*

(1) $K \subset \Sigma$ *is unknotted.*
(2) $\Gamma \cong \mathbf{Z}$.
($\tilde{1}$) $\tilde{K} \subset \tilde{\Sigma}$ *is unknotted.*
($\tilde{2}$) $\tilde{\Gamma} \cong \mathbf{Z}$.
(3) *The action of the cyclic group $G = \langle h \rangle$ on $\tilde{\Sigma}$ is essentially linear.*

The solution of the Smith conjecture presented here uses the implication (2) $\Rightarrow$ ($\tilde{1}$). The equivalences (1) $\Leftrightarrow$ (2) $\Leftrightarrow$ ($\tilde{1}$) $\Leftrightarrow$ ($\tilde{2}$) follow from results based on the classical work of Papakyriakopoulous [P]. Proofs are presented in Chapter IV. We sketch here a proof that (1) $\Leftrightarrow$ (3). Certainly, if $G \times S^3$ is a linear action of a cyclic group with the generator having a fixed circle, then that circle is unknotted. This is visibly true also for an essentially linear action. Conversely, if $K \subset \Sigma$ is unknotted, then choose a $D^2 \subset \Sigma$ whose boundary is K. A small neighborhood of D^2 in Σ will be a 3-ball containing

III. An Outline of the Proof

K as an unknotted circle in its interior. The complement of this is a homotopy 3-cell in Σ. Replacing this homotopy cell by the 3-ball produces the standard unknotted pair $S^1 \subset S^3$, up to diffeomorphism. The cyclic branched covering of this pair yields a linear action. It follows that the branched covering of the original pair yields an essentially linear action.

2. First Reductions

We are studying knots K in homotopy spheres Σ, so that the n-fold cyclic covering of Σ branched along K is simply connected. We wish to show that K is unknotted in Σ. The argument is by contradiction. The first reduction is to show that if there is a counterexample, then there is one when

(a) $\Sigma - K$ is irreducible,[1] and
(b) the knot $K \subset \Sigma$ is prime.[2]

To establish (a), notice that we can write $\Sigma = \Sigma_1 \# \Sigma_2$, where $K \subset \Sigma_1$ and where $\Sigma_1 - K$ is irreducible. This is a consequence of the finiteness of connected sum decompositions for 3-manifolds [Kn]. Clearly, $K \subset \Sigma$ is knotted if and only if $K \subset \Sigma_1$ is, and a cyclic cover of Σ_1 branched along K is simply connected if and only if the corresponding covering of Σ is. Thus, if a cyclic branched cover of $K \subset \Sigma$ yields a counterexample to the Smith conjecture, then the corresponding branched cover of $K \subset \Sigma_1$ yields a counterexample with $\Sigma_1 - K$ irreducible.

If $K \subset \Sigma$ is not prime, then, by definition, there is $S^2 \subset \Sigma$ meeting K in exactly two points and decomposing the pair (Σ, K) into $(\Sigma_1, K_1) \# (\Sigma_2, K_2)$, where $(K_i - \text{int } D^1) \subset (\Sigma_i - \text{int } B^3) \subset \Sigma$ and each K_i is knotted in Σ_i. The preimage $\tilde{S} = p^{-1}(S^2)$ is the n-fold cyclic cover of S^2 branched at two points. Thus $\tilde{S}$ is itself a 2-sphere. Thus $\tilde{S}$ gives a decomposition of $\tilde{\Sigma}$ as $\tilde{\Sigma}_1 \# \tilde{\Sigma}_2$, where $\tilde{\Sigma}_i$ is the cyclic cover of Σ_i branched over K_i. If $\tilde{\Sigma}$ is a homotopy 3-sphere, then each $\tilde{\Sigma}_i$ is also. Since all knots have a finite prime factorization [Kn], we can use this argument to replace any given counterexample with one for which the knot is prime. Notice that this simplification process does not destroy the fact that $\Sigma - K$ is irreducible.

3. The Argument in Brief

Having made these first reductions we proceed here to sketch the argument that completes the proof. The outline below is to serve as a guide to the various chapters included in this volume. There are two cases to consider.

---

[1] That is, every 2-sphere in $\Sigma - K$ bounds a 3-ball.
[2] That is, any 2-sphere in Σ meeting K transversely in two points bounds a ball meeting K in an unknotted arc.

Case (i). $\Sigma - K$ *contains an incompressible surface F which is nonperipheral (nonperipheral meaning that F is not the boundary of a tubular neighborhood of K in Σ).*

Case (ii). Every incompressible surface in $\Sigma - K$ is peripheral.

Case (i) will be shown to be impossible by Gordon and Litherland (Chapter VII). The techniques are those of classical PL topology, plus an essential new fact, namely, the equivariant version of Dehn's lemma and the loop theorem. This result is due to Meeks and Yau and will be discussed by Yau in Chapter VIII. Basically, one finds a minimal surface solution to Dehn's lemma and the loop theorem, and this solution automatically satisfies the equivariant version of the theorem when there is a finite group action via isometries. Using this, in Chapter VII Gordon employs standard "cut and paste" techniques and two-dimensional Smith theory to establish that Case (i) cannot arise.

The argument dealing with Case (ii) is due to Thurston and is discussed in Chapter IV by Shalen. Central to this argument is the notion of a peripheral subgroup of $\pi_1(\Sigma - K) = \Gamma$, defined as follows. Let $\pi_1(\infty)$ be

$$\text{Im}(\pi_1(N - K) \to \Gamma),$$

where N is a tubular neighborhood of K in Σ. (Note that $N - K \cong K \times (D^2 - \{0\})$, and hence $\pi_1(N - K) \cong \mathbf{Z} \times \mathbf{Z}$.) Call a subgroup of Γ *peripheral* if it is conjugate to a subgroup of $\pi_1(\infty)$. Using results of Jaco and Shalen [Ja-Shal] and Johannson [Jo], Shalen will show that in Case (ii), either

(iia) $\Sigma - K$ *is a Seifert fiber space* or
(iib)   *every subgroup of Γ isomorphic to $\mathbf{Z} \times \mathbf{Z}$ is peripheral.*

Actually, Cases (iia) and (iib) are not mutually exclusive. There is one knot complement that satisfies them both—$S^1 \times \mathbf{R}^2$. This allows us to replace (iib) by

(iib') *Every subgroup of Γ isomorphic to $\mathbf{Z} \times \mathbf{Z}$ is peripheral and $\Sigma - K$ is not diffeomorphic to $S^1 \times \mathbf{R}^2$.*

By classical arguments, sketched in Chapter IV, one can show that in Case (iia) the complement must be diffeomorphic to $S^1 \times \mathbf{R}^2$. This is the case in which $K \subset \Sigma$ is unknotted (see (2) $\Rightarrow$ (1) of the theorem in Section 1).

Finally, suppose that we are in Case (iib'). The hypotheses of this case are exactly those of Thurston's uniformization theorem (see Chapter V), which tells us that $\Sigma - K$ admits a hyperbolic structure. This structure is equivalent to a discrete and faithful representation $\rho: \Gamma \to \text{PSL}_2(\mathbf{C})$, which is well-defined up to conjugation. This representation is automatically irreducible. The next step in the argument is to conjugate ρ until its image lies in a "nice"

III. An Outline of the Proof

subgroup of $PSL_2(\mathbf{C})$. This is done by applying Bass's theorem (see Chapter VI) to the irreducible representation ρ. This result presents three possibilities for an irreducible representation. Using classical results in 3-manifold topology of Epstein and Stallings [Ep], one shows that under hypothesis (iib′) the only possibility on Bass's list that can occur is that ρ is conjugate to a representation into $PSL_2(A)$, where A is the ring of integers in an appropriate algebraic number field inside $\mathbf{C}$.

At this point the argument becomes purely algebraic. We reduce modulo a prime ideal $\mathfrak{p} \subset A$, which contains n (the integer that is the order of the cyclic covering in question). By using the resulting representation of Γ in $PSL_2(A/\mathfrak{p})$, one argues that no cyclic n-fold cover, as in Case (ii) + (iib′), can ever be simply connected. This shows that Case (ii) + (iib′) cannot occur and completes the proof of the Smith conjecture.

The flow chart shown in Fig. 1 gives a pictorial representation of the logical structure of the argument that we have just sketched. The chapter numbers or references cited beside an arrow refer to the place where the implications are discussed.

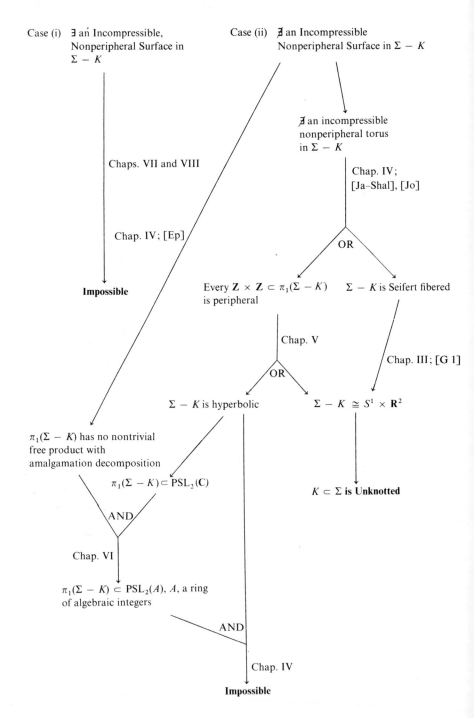

Figure 1 Flow Chart. Initial assumptions: $\Sigma - K$ irreducible, K prime (see Chap. II).

References for Part A

[Bin] Bing, R. H., Inequivalent families of periodic homeomorphisms E^3, *Ann. of Math.* (2) **80** (1964), 78–93.

[Bir–H] Birman, J. S., and Hilden, H., Heegaard splittings of branched coverings of S^3, *Trans. Amer. Math. Soc.* **213** (1975), 315–352.

[Bre] Bredon, G., "Introduction to Compact Transformation Groups." Academic Press, New York, 1972.

[Bro] Brouwer, L. E. J., Über die periodischen transformationen der Kugel, *Math. Ann.* **80** (1919), 262–280.

[Ca–Shan] Cappell, S., and Shaneson, J., A note on the Smith conjecture, to appear.

[Cr–F] Crowell, R., and Fox, R. H., "An Introduction to Knot Theory," 2d ed. Springer-Verlag, Berlin and New York, 1977.

[Ei 1] Eilenberg, S., On the problems of topology, *Ann. of Math.* (2) **50** (1949), 247–260.

[Ei 2] Eilenberg, S., Sur les transformations périodiques de la surface du sphère, *Fund. Math.* **22** (1934), 28–41.

[Ep] Epstein, D. B. A., Free products with amalgamations and 3-manifolds, *Proc. Amer. Math. Soc.* **12** (1961), 669–670.

[Fe] Feighn, M., On the generalized Smith conjecture, Ph.D. thesis, Columbia Univ., New York, 1981.

[Fo 1] Fox, R. H., On knots whose points are fixed under a periodic transformation of the 3-sphere, *Osaka. J. Math.* **10** (1958), 31–35.

[Fo 2] Fox, R. H., Knots and periodic transformations, in "Topology of 3-manifolds" (M. K. Fort, Jr. ed.) pp. 177–182. Prentice-Hall, Englewood Cliffs, New Jersey, 1962.

[Fo 3] Fox, R. H., "Two theorems about periodic transformations of the 3-sphere," *Michigan Math. J.* **14** (1967), 331–334.

[G 1] Giffen, C. H., Fibered knots and periodic transformations, Ph.D. thesis, Princeton Univ., Princeton, New Jersey, 1964.

[G 2] Giffen, C. H., The generalized Smith conjecture, *Amer. J. Math.* **88** (1966), 187–198.

[G 3] Giffen, C. H., Cyclic branched coverings of doubled curves in 3-manifolds, *Illinois J. Math.* **11** (1967), 644–646.

References for Part A

[Ja-Shal] Jaco, W., and Shalen, P., *Seifert fibered spaces in 3-manifolds*, Mem. Amer. Math. Soc., No. 220 (1979).

[Jo] Johannson, K., "Homotopy Equivalences of 3-manifolds with Boundary," Lecture Notes in Mathematics, Vol. 761. Springer-Verlag, Berlin and New York, 1979.

[Ke] Kerekjarto, B., Über die periodischen Transformationen der Kreisschube und Kugelflache, *Math. Ann.* **80** (1919), 36–38.

[Ki] Kinoshita, Shin'ichi, On knots and periodic transformations, *Osaka J. Math.*, **10** (1958), 43–52.

[Kn] Kneser, H., Geschlossen flachen in dreidimensionalen Mannigfaltigkeiten, *Jahresber. Deutsch. Math. Verein.* **38** (1929), 248–260.

[Moi 1] Moise, E., Periodic homeomorphisms of the 3-sphere, *Illinois J. Math.* **6** (1962), 206–225.

[Moi 2] Moise, E., Statically tame periodic homeomorphisms of compact, connected 3-manifolds I: Homeomorphisms conjugate to rotations of the 3-sphere, preprint.

[Mon-Sa] Montgomery, D., and Samelson, H., A theorem on fixed points of involutions in S^3, *Canad. J. Math.* **7** (1955), 208–220.

[Mon-Z] Montgomery, D., and Zippen, L., Examples of transformation groups, *Proc. Amer. Math. Soc.* **5** (1954), 460–465.

[My] Myers, R., Companionship of knots and the Smith conjecture. Ph.D. thesis. Rice Univ., Houston, Texas, 1977.

[P] Papakyriakopoulos, C. D., On Dehn's lemma and the asphericity of knots, *Ann. of Math.* (2) **66** (1957), 1–26.

[Sm 1] Smith, P. A., Transformations of finite period, *Ann. of Math.* (2) **39** (1938), 127–164.

[Sm 2] Smith, P. A., Transformations of finite period II, *Ann. of Math.* (2) **40** (1939), 690–711.

[Sm 3] Smith, P. A., Transformations of finite period III, *Ann. of Math.* (2) **42** (1941), 446–458.

[Sm 4] Smith, P. A., Periodic transformations of 3-manifolds, *Illinois J. Math.* **9** (1965), 343–348.

[W] Waldhausen, F., Über Involutionen der 3-Sphäre, *Topology* **8** (1969), 81–91.

PART **B**

**THE PROOF IN THE CASE
OF NO INCOMPRESSIBLE SURFACE**

CHAPTER IV

The Proof in the Case of No Incompressible Surface

Peter Shalen

Department of Mathematics
Rice University
Houston, Texas

Introduction

The following "half" of the Smith conjecture was announced by William Thurston in 1977. (It was referred to in Chapter III as Case (ii).)

THEOREM. *Let h be a periodic diffeomorphism of a smooth homotopy 3-sphere $\tilde{\Sigma}$ with one-dimensional fixed point set $\tilde{K}$. Set $\Sigma = \tilde{\Sigma}/h$, let $p: \tilde{\Sigma} \to \Sigma$ denote the projection map, and set $K = p(\tilde{K})$. Suppose that every incompressible, closed, orientable surface in $\Sigma - K$ is isotopic in $\Sigma - K$ to the boundary of the tubular neighborhood of K. Then K is unknotted* (and hence so is $\tilde{K}$).

Recall that a smooth, closed, orientable surface F in the interior of a 3-manifold M is *incompressible* if $\pi_1(F) \to \pi_1(M)$ is a monomorphism, and if F is not a boundary of a 3-ball in M. By the work of Papakyriakopoulos, this is equivalent to a geometric definition (see Hempel [H, Chapter 4]).

The hypothesis on incompressible surfaces implies in particular that M is *irreducible* (or "prime") in the sense that every smooth 2-sphere in M bounds a 3-ball. As was explained in Chapter III, the assumption of irreducibility is not a real restriction.

Note that the conclusion of the above theorem implies that the boundary of the tubular neighborhood of K is not, in fact, incompressible. Thus, if the hypothesis about incompressible surfaces is true, it is "vacuously" true.

The proof of the theorem depends on a fundamental result of Thurston's about the existence of *hyperbolic structures* on 3-manifolds, and on an algebraic result of Bass's; indirectly, it also depends on our ideas. In Sections 1–4 we shall present the proof of the above theorem, referring to Thurston's and Bass's theorems (the proofs of which are discussed in the next two chapters in this volume) and to older results in the literature. In Section 5, we shall discuss some of the immediate history of the proof, including our own role in it.

We shall work in the smooth category. Many of the results we quote are proved in the piecewise linear category, but there is no difficulty in translating the statements. Unlabeled homomorphisms will be understood to be induced by inclusion. Basepoints will be suppressed whenever the choice of them does not affect the truth of a statement.

1. The Algebraic Approach to the Smith Conjecture

Let h be a periodic diffeomorphism of a closed 3-manifold $\tilde{\Sigma}$ having a one-dimensional fixed point set $\tilde{K}$. Let n denote the period of h. Set $\Sigma = \tilde{\Sigma}/h$, let $p: \tilde{\Sigma} \to \Sigma$ denote the quotient projection, and set $K = p(\tilde{K})$. Let N be a tubular neighborhood of K in Σ, and set $\tilde{N} = p^{-1}(N)$. We may identify N by a diffeomorphism with $D^2 \times S^1$; set $\mu = (\partial D^2) \times \{x\}$ for some $x \in S^1$ (μ is a *meridian curve* of K), and set $\tilde{\mu} = p^{-1}(\mu)$.

Set $\Gamma = \pi_1(\Sigma - \mathring{N})$, $\tilde{\Gamma} = \pi_1(\tilde{\Sigma} - \mathring{\tilde{N}})$. Clearly, $\tilde{\Sigma} - \tilde{N}$ is an n-fold cyclic covering space of $\Sigma - \mathring{N}$, so that $\tilde{\Gamma}$ may be identified with a normal subgroup of Γ, with $\Gamma/\tilde{\Gamma} \approx \mathbf{Z}/n\mathbf{Z}$. Choosing orientations of μ and $\tilde{\mu}$, we may consider the elements $[\tilde{\mu}] \in \tilde{\Gamma}$, $[\mu] \in \tilde{\Gamma} \subset \Gamma$; if the orientations are chosen compatibly, we have $[\tilde{\mu}] = [\mu]^n$. On the other hand, we may identify $\tilde{N}$ with $D^2 \times S^1$ so that $\tilde{\mu} = \partial D^2 \times \{x\}$ for some $x \in S^1$. Hence, by the Seifert–van Kampen theorem, $\pi_1(\tilde{\Sigma}) \approx \tilde{\Gamma}/([\tilde{\mu}])$, the quotient of $\tilde{\Gamma}$ by the normal closure of $[\tilde{\mu}]$.

Now suppose that $\tilde{\Sigma}$ is simply connected. Then, by the above observation, $\tilde{\Gamma}/([\tilde{\mu}]) = 1$. This suggests an algebraic approach to the Smith conjecture: we assume that K is knotted in Σ, and we try to produce a nontrivial representation ρ of $\tilde{\Gamma}$ in some group H such that $\rho([\mu]^n) = 1$. If this can be done,

IV. The Proof in the Case of No Incompressible Surface

then ρ will induce a nontrivial representation of $\tilde{\Gamma}/([\tilde{\mu}])$ in H, contradicting the triviality of the group $\tilde{\Gamma}/([\tilde{\mu}])$.

One very special case of the Smith conjecture that can be handled rather directly by the above method is the case in which $\Sigma - \mathring{N}$ is a Seifert-fibered manifold [H, Chapter 12]. This case (which was discussed in Chapter III as Case (iia)) must be handled separately in order to prove the theorem stated in the introduction to this chapter. A proof of the conjecture when $\Sigma - \mathring{N}$ is Seifert-fibered is essentially contained in [G]; however, for the reader's convenience, we shall include a proof in Section 2.

The method of proof of the theorem stated in the introduction, in the generic case in which $\Sigma - K$ is not Seifert-fibered (Case (iib') of Chapter III), is to introduce a geometric structure on $\Sigma - K$ that gives rise to a faithful representation of $\Gamma \approx \pi_1(\Sigma - K)$ in a matrix group; the latter in turn defines a representation of Γ in a group of matrices over a finite field, to which the above method can be applied. In this case, the argument, while carried out in an algebraic context, involves very deep considerations in the topology and geometry of 3-manifolds.

The notations and conventions introduced in this section will be used throughout the chapter.

2. Hyperbolic Geometry and Algebraic Integers

The geometric structure on $\Sigma - K$ that was mentioned at the end of Section 1 is a *hyperbolic structure* of a special type. In this section we give a rudimentary discussion of hyperbolic 3-manifolds and show their relevance to the Smith conjecture; as a by-product, we obtain a proof of the conjecture in the case where $\Sigma - \mathring{N}$ is Seifert-fibered.

The following discussion of hyperbolic 3-manifolds is based on Fatou [F], which contains a good introduction to the subject for topologists and gives proofs of the results stated here.

Let R be any integral domain. The group of all 2×2 matrices of determinant 1 with entries in R is the *special linear group* $SL_2(R)$. The center of $SL_2(R)$ is $\{I, -I\}$; the quotient $SL_2(R)/\{\pm I\}$ is the *projective special linear group* $PSL_2(R)$. We denote the image of $(\begin{smallmatrix} a & b \\ c & d \end{smallmatrix})$ in $PSL_2(R)$ by $\pm(\begin{smallmatrix} a & b \\ c & d \end{smallmatrix})$. Recall that for $R = \mathbf{C}$, $PSL_2(\mathbf{C})$ acts on the Riemann sphere $\mathbf{C}^+$ (the one-point compactification of $\mathbf{C}$) by Möbius transformations; the element $\pm(\begin{smallmatrix} a & b \\ c & d \end{smallmatrix})$ maps z to $(az + b)/(cz + d)$.

If we identify $\mathbf{C}$ with $\mathbf{R}^2 \times \{0\} \subset \mathbf{R}^3$, then this action of $PSL_2(\mathbf{C})$ on $\mathbf{C}^+$ may be extended in a very useful way to an action on $(\mathbf{R}^3)^+$. This action leaves invariant the half space $\mathbb{H}^3 = \mathbf{R}^2 \times (0, \infty)$, and one can define a noneuclidean geometry on $\mathbb{H}^3$ (hereafter called *hyperbolic 3-space*) in such a way that this action of $PSL_2(\mathbf{C})$ gives the full group of orientation-preserving isometries of $\mathbb{H}^3$. Moreover, every discrete, torsion-free subgroup Γ of $PSL_2(\mathbf{C})$ acts freely and properly discontinuously on $\mathbb{H}^3$, in the sense that each point $x \in \mathbb{H}^3$ has a neighborhood U such that $\gamma U \cap U = \emptyset$ for all $\gamma \in \Gamma - \{1\}$. Thus $\mathbb{H}^3/\Gamma$ is a 3-manifold whose universal covering and fundamental group may be identified with $\mathbb{H}^3$ and Γ, respectively. An orientable 3-manifold defined in this way will be called *hyperbolic*.[1]

Hyperbolicity imposes certain topological restrictions on a 3-manifold. Let P be a rank-two free abelian subgroup of Γ. Then P is in particular a rank-two free abelian, discrete subgroup of $PSL_2(\mathbf{C})$; it is easy to show that any such subgroup is conjugate to a group of elements of the form $\pm \begin{pmatrix} 1 & \lambda \\ 0 & 1 \end{pmatrix}$, $\lambda \in \mathbf{C}$. Moreover, a very elegant argument (see Fatou [F]) shows that there is a 3-manifold-with-boundary $E \subset M$, closed as a subset of M and diffeomorphic to $S^1 \times S^1 \times [0, \infty)$, such that P is contained in a conjugate of $\text{Im}(\pi_1(E) \to \pi_1(M))$.

In Section 3 we shall state a weak form of a fundamental result of Thurston's, that a huge class of 3-manifolds have hyperbolic structures. The importance of this fact for the Smith conjecture comes from the following result, which is also due to Thurston.

PROPOSITION 1. *Let h be a periodic diffeomorphism of a homotopy 3-sphere $\tilde{\Sigma}$ with one-dimensional fixed point set $\tilde{K}$. Set $\Sigma = \tilde{\Sigma}/h$, let $p: \tilde{\Sigma} \to \Sigma$ denote the quotient projection, and set $K = p(\tilde{K})$. Suppose that $\Sigma - K$ is diffeomorphic to $\mathbb{H}^3/\Gamma$, where Γ is a discrete, torsion-free subgroup of $PSL_2(\mathbf{C})$, such that the entries (defined up to sign) of each element of Γ are algebraic integers in $\mathbf{C}$. Then K is unknotted.*

Proof. According to the remarks in Section 1 (and with the notation of that section), it is enough to show that if K is knotted, then there is a nontrivial representation ρ of $\tilde{\Gamma}$ in some group H such that $\rho([\mu]^n) = 1$. We begin by showing that the knottedness of K implies that (a) the trace of $\mu \in PSL_2(\mathbf{C})$ is ± 2, and (b) $\tilde{\Gamma}$ is not conjugate in $PSL_2(\mathbf{C})$ to a group of upper triangular elements.

The proofs of both assertions involve classical results of Papakyriakopoulos, the Dehn lemma, and the loop theorem [H, Chapter 4]. These results show, first, that if K is knotted, then $\pi_1(\partial N) \to \pi_1(\Sigma - \mathring{N})$ is injective (cf.

---

[1] N.B.: Here we are using the word *hyperbolic* to mean simply a quotient of $\mathbb{H}^3$ by any discrete torsion-free subgroup of $PSL_2(\mathbf{C})$. To put it another way, hyperbolicity means the existence of a complete riemannian metric all of whose sectional curvature are -1. Often one uses the word to mean in addition that the volume of the manifold is finite. If M is the interior of a compact manifold whose boundary is a union of tori, then any hyperbolic structure on M in our sense will automatically have finite volume unless M is homeomorphic to (a) $S^1 \times \mathbf{R}^2$ or (b) $T^2 \times \mathbf{R}$.

IV. The Proof in the Case of No Incompressible Surface

Neuwirth [N, pp. 56–57]; the observation there is made for $\Sigma = S^3$, but the same argument works in any homology 3-sphere). Thus

$$P = \operatorname{Im}(\pi_1(\partial N) \to \pi_1(\Sigma - K))$$

is a rank-two free abelian subgroup of Γ. By a remark made earlier in this section, P is conjugate to a group of elements $\pm \begin{pmatrix} 1 & \lambda \\ 0 & 1 \end{pmatrix}$, $\lambda \in \mathbf{C}$; in particular, $[\mu] \in P$ has trace ± 2, which is assertion (a). To prove (b), we use the fact if K is knotted, then $\pi_1(\Sigma - K)$ contains a nonabelian free group ([N, Chapter 4, Section 3]. If $T \subset \Sigma$ is a minimal Seifert surface for k, i.e., a smooth, orientable surface bounded by K, and if it has minimal genus among all Seifert surfaces, then the Dehn lemma and loop theorem imply that $\pi_1(T) \to \pi_1(\Sigma - K)$ is injective. But T has positive genus if K is unknotted, and so $\pi_1(T)$ is a nonabelian free group. Again, the argument works in any homology 3-sphere.) Now since $\tilde{\Gamma}$ has finite index in Γ, $\tilde{\Gamma}$ also contains a nonabelian free group; hence $\tilde{\Gamma}$ is not solvable. But the group of all upper triangular elements of $\mathrm{PSL}_2(\mathbf{C})$ is solvable; in fact it has a homomorphism onto the multiplicative group $(\mathbf{C} - \{0\})/\{\pm 1\}$ whose kernel consists of elements of the form $\pm \begin{pmatrix} 1 & \lambda \\ 0 & 1 \end{pmatrix}$, $\lambda \in \mathbf{C}$. This proves (b).

The observations (a) and (b) reduce the proof of Proposition 1 to the following purely algebraic result. (We take $g = [\mu]$, and note that $n > 1$ since h has a one-dimensional fixed point set.)

LEMMA. *Let Γ be a finitely generated subgroup of $\mathrm{PSL}_2(\mathfrak{A})$, where $\mathfrak{A}$ denotes the ring of all algebraic integers in $\mathbf{C}$. Let $\tilde{\Gamma}$ be a subgroup of Γ that is not conjugate in $\mathrm{PSL}_2(\mathbf{C})$ to a group of upper triangular elements, and let $g \in \Gamma$ be such that trace $g = \pm 2$ and $g^n \in \tilde{\Gamma}$, where n is an integer > 1. Then there is a nontrivial homomorphism $\rho: \tilde{\Gamma} \to \mathrm{PSL}_2(\Lambda)$, for some field Λ, such that $\rho(g^n) = I$.*

Proof. Let $L \subset \mathbf{C}$ be the extension of $\mathbf{Q}$ generated by the entries of the generators of Γ. Then L is a finite extension of $\mathbf{Q}$ and $\Gamma \subset \mathrm{PSL}_2(A)$, where A denotes the ring of algebraic integers in L. Note that $n \in \mathbf{Z} \subset A$; moreover, $n > 1$ implies that $1/n \notin A$, since $A \cap \mathbf{Q} \subset \mathbf{Z}$ [Z–Sa, p. 261, Example 1]. Thus n is a noninvertible element of A and is therefore contained in a prime ideal $\mathfrak{p}$. Let $\mathcal{O} = A_{(\mathfrak{p})}$ be the ring obtained by localizing A at $\mathfrak{p}$ ([Z–Sa, p. 228]). Then, by the results of Zariski and Samuel [Z–Sa], Chap. 5], $\mathcal{O}$ is a discrete valuation ring, i.e., it is a principal ideal domain (PID) having a unique maximal ideal (π), and every ideal of $\mathcal{O}$ is generated by a power of π. It follows easily that every nonzero element of $\mathcal{O}$ may be written uniquely in the form $\pi^k e$, where $k \geq 0$ is an integer and $e \in \mathcal{O}$ is a unit.

We have $g \in \Gamma \subset \mathrm{PSL}_2(A) \subset \mathrm{PSL}_2(\mathcal{O})$, and $\operatorname{tr} g = \pm 2$. Now it is easy to show that if $\mathcal{O}$ is *any* PID, then any matrix in $\mathrm{SL}_2(\mathcal{O})$ with trace 2 is conjugate in $SL_2(\mathcal{O})$ to a matrix $\begin{pmatrix} 1 & \lambda \\ 0 & 1 \end{pmatrix}$ for some $\lambda \in \mathcal{O}$. (In fact, the given matrix has characteristic polynomial $X^2 - 2X + 1$, and hence is the identity on some one-dimensional subspace $V \subset L^2$, where L is the quotient field of $\mathcal{O}$. Since $\mathcal{O}$ is a unique factorization domain (UFD), V contains a vector (a, b), where a and b are relatively prime elements of $\mathcal{O}$; and since $\mathcal{O}$ is a PID, the vector (a, b) can be extended to a basis for $\mathcal{O}^2$. In this basis the matrix takes the form $\begin{pmatrix} 1 & \lambda \\ 0 & 1 \end{pmatrix}$.) Projecting into $PSL_2(\mathcal{O})$, we see that any element of trace ± 2 is conjugate in $PSL_2(\mathcal{O})$ to $\pm \begin{pmatrix} 1 & \lambda \\ 0 & 1 \end{pmatrix}$ for some $\lambda \in \mathcal{O}$.

Hence, by restricting some inner automorphism of $PSL_2(\mathcal{O})$ to Γ, we obtain a faithful representation $\rho_0: \Gamma \to PSL_2(\mathcal{O})$ such that $\rho_0(g) = \pm \begin{pmatrix} 1 & \lambda \\ 0 & 1 \end{pmatrix}$, $\lambda \in \mathcal{O}$. Note that $\rho_0(g^n) = \pm \begin{pmatrix} 1 & n\lambda \\ 0 & 1 \end{pmatrix}$ and that $n\lambda \in (\pi)$, since $n \in \mathfrak{p} \subset (\pi)$. Now, if we set $\Lambda = \mathcal{O}/(\pi)$ and let $q: PSL_2(\mathcal{O}) \to PSL_2(\Lambda)$ be the homomorphism that reduces the entries mod π, then Λ is a field, since (π) is a maximal ideal, and $q\rho_0|\tilde{\Gamma}: \tilde{\Gamma} \to PSL_2(\Lambda)$ is a homomorphism that maps g^n to the identity; however, this homomorphism may be trivial.

To correct this, we invoke the hypothesis that $\tilde{\Gamma}$ is not conjugate in $PSL_2(\mathbf{C})$ to a group of upper triangular elements. This means that there are matrices in $\rho_0(\tilde{\Gamma})$ whose lower left-hand entry is nonzero. Let l be the least nonnegative integer for which there is an element in $\rho_0(\tilde{\Gamma})$ of the form

$$\pm \begin{pmatrix} a & b \\ \pi^l e & d \end{pmatrix}$$

with e a unit in $\mathcal{O}$. Then we can define a faithful representation

$$\rho_1: \tilde{\Gamma} \to PSL_2(\mathbf{C})$$

by

$$\rho_1(\gamma) = \begin{pmatrix} \pi^l & 0 \\ 0 & 1 \end{pmatrix} \rho_0(\gamma) \begin{pmatrix} \pi^{-l} & 0 \\ 0 & 1 \end{pmatrix}.$$

If

$$\rho_0(\gamma) = \pm \begin{pmatrix} a & b \\ c & d \end{pmatrix},$$

then

$$\rho_1(\gamma) = \pm \begin{pmatrix} a & \pi^l b \\ \pi^{-l} c & d \end{pmatrix}.$$

From the definition of l, it now follows that $\rho_1(\tilde{\Gamma}) \subset PSL_2(\mathcal{O})$, and that for some $\gamma_1 \in \tilde{\Gamma}$, $\rho_1(\gamma_1)$ has the form

$$\pm \begin{pmatrix} a_1 & b_1 \\ c_1 & d_1 \end{pmatrix},$$

where c_1 is a unit in $\mathcal{O}$, i.e., $c_1 \notin (\pi)$.

IV. The Proof in the Case of No Incompressible Surface

On the other hand,

$$\rho_1(g^n) = \pm \begin{pmatrix} 1 & \pi^l \alpha \\ 0 & 1 \end{pmatrix};$$

since $\alpha \in (\pi)$ and $l \geq 0$, we have $\pi^l \alpha \in (\pi)$. Therefore, if we set $\rho = q\rho_1 : \tilde{\Gamma} \to \text{PSL}_2(F)$, we have $\rho(g^n) = \pm I$, but $\rho(\gamma_1) \neq \pm I$, since $c_1 \notin (\pi)$. ∎

It may be shown that the field Λ used in the above proof is in fact a finite field. It is isomorphic to $A/\mathfrak{p}$.

As a by-product of this lemma we can prove another special case of the Smith conjecture.

PROPOSITION 2. *Let h be a periodic homeomorphism of a homotopy 3-sphere $\tilde{\Sigma}$ with one-dimensional fixed point set $\tilde{K}$. Let $p : \tilde{\Sigma} \to \Sigma$ be the quotient projection, set $K = p(\tilde{K})$, and let N be a regular neighborhood of K in Σ. Suppose that $\Sigma - \overset{\circ}{N}$ is a Seifert-fibered manifold. Then K is unknotted in Σ.*

Proof. By the discussion in Section 1, it is enough to show that if K is knotted, then there is a nontrivial representation ρ of $\tilde{\Gamma}$ in a group H such that $\rho([\mu]^n) = 1$, where $n > 1$ is the period of H. By the above lemma, this can be done provided that we can find a representation ρ^* of Γ in $\text{PSL}_2(\mathfrak{A})$, where $\mathfrak{A}$ denotes the ring of all algebraic integers in $\mathbf{C}$, such that

(a) $\text{tr } \rho^*([\mu]) = \pm 2$ and
(b) $\rho^*(\Gamma)$ is not conjugate in $\text{PSL}_2(\mathbf{C})$ to a group of upper triangular matrices.

Presentations for the fundamental groups of arbitrary bounded Seifert-fibered 3-manifolds are given in [H, Chapter 12]. Since Γ is the group of a knot in a homology 3-sphere, its commutator quotient is infinite cyclic; by examining the presentations in [H], it follows that Γ has the form

$$\langle c_1, \ldots, c_q, t : [c_i, t] = 1, c_i^{r_i} = t^{s_i} \rangle,$$

where the r_i are distinct integers >1. Moreover, the method of deriving this presentation shows that $\text{Im}(\pi_1(\partial N) \to \pi_1(\Sigma - \overset{\circ}{\partial N}))$ is generated by the central element t and a conjugate of $c_1 \cdots c_q$. Hence $[\mu]$ is conjugate to $(c_1 \cdots c_q)^p t^m$ for some $p, m \in \mathbf{Z}$.

Note also that $q > 1$; otherwise, by the way the above presentation is derived, we could conclude that $\Sigma - \overset{\circ}{N}$ was Seifert-fibered over a disk with one singular fiber, and was therefore a solid torus; this would imply that K was unknotted in Σ.

To define the representation ρ^*, we set

$$\rho^*(t) = \pm I, \qquad \rho^*(c_1) = \pm \begin{pmatrix} \theta & 1 \\ 0 & \theta^{-1} \end{pmatrix},$$

where θ is a primitive $(2r_1)$th root of unity;

$$\rho^*(c_2) = \pm \begin{pmatrix} \omega & 0 \\ \lambda & \omega^{-1} \end{pmatrix},$$

where ω is a primitive $(2r_2)$th root of unity and λ is an algebraic integer that we shall define presently; and $\rho^*(c_i) = \pm I$, $i > 2$. Since each of the matrices $\rho^*(c_i)$ ($i = 1, 2$) has two distinct characteristic roots, it is conjugate to a diagonal matrix that must have the same characteristic roots; it follows that $\rho^*(c_i)^{r_i} = \pm I$ in $\mathrm{PSL}_2(\mathfrak{A})$, so that ρ^* is a well-defined representation.

The representation ρ will satisfy (a) provided that $\mathrm{tr}\, \rho^*(c_1 c_2) = \pm 2$. But we have $\mathrm{tr}\, \rho^*(c_1 c_2) = \theta\omega + \theta^{-1}\omega^{-1} + \lambda$. Hence there are *two* values of λ for which (a) holds. Note also that both these values of λ are $\neq 0$; for otherwise we would have $\theta\omega = \pm 1$, which would imply $r_1 = r_2$.

On the other hand, $\rho^*(\Gamma)$ will satisfy (b) unless $\rho^*(c_1)$ and $\rho^*(c_2)$ have a common eigenvector; i.e., unless, regarded as Möbius transformations, they have a common fixed point on $\mathbf{C}^+$. But, by a direct calculation, the fixed points of $\rho^*(c_1)$ are ∞ and $1/(\theta - \theta^{-1})$, while those of $\rho^*(c_2)$ are 0 and $(\omega - \omega^{-1})/\lambda$. Since ω and θ are $\neq \pm 1$, there is *only one* value of $\lambda \neq 0$ (namely, $(\theta - \theta^{-1})(\omega - \omega^{-1})$) for which (b) fails to hold. Hence there is *at least one* value of λ for which (a) and (b) both hold.

This proves that the required representation ρ^* exists and completes the proof in the case that $\Sigma - K$ is Seifert-fibered. ∎

3. The Existence of Hyperbolic Structures and the Torus Theorem

It is clear from Proposition 1 of Section 2 that producing hyperbolic structures on knot complements in irreducible 3-manifolds is useful for the solution of the Smith conjecture. In this section we discuss a fundamental result of Thurston's that gives necessary and sufficient conditions for such a structure to exist, and results of Waldhausen, Johannson, and Jaco and Shalen that permit one to reformulate Thurston's theorem in a more applicable form.

If K is a smooth 1-sphere in a closed 3-manifold Σ and N is a tubular neighborhood of K, then $\Sigma - K$ is diffeomorphic to the interior of the compact 3-manifold $\Sigma - \overset{\circ}{N}$. In general one may ask, given a compact 3-manifold M, when does $\overset{\circ}{M}$ admit a hyperbolic structure? An obvious necessary condition is that the universal covering of $\overset{\circ}{M}$ be diffeomorphic

IV. The Proof in the Case of No Incompressible Surface

to $\mathbf{R}^3$. This in turn implies that M is irreducible. (In fact, it is a classical theorem of Alexander's (see Moise [Mo, Chapter 17, Theorem 12]) that $\mathbf{R}^3$ is irreducible, and an elementary argument (see Milnor [Mi]) shows that a 3-manifold with an irreducible covering space is irreducible.)

A more interesting necessary condition for the hyperbolicity of $\overset{\circ}{M}$ was mentioned in Section 2: every rank-two free abelian subgroup of $\pi_1(\overset{\circ}{M})$ must be contained in a conjugate of $\text{Im}(\pi_1(E) \to \pi_1(\overset{\circ}{M}))$, for some manifold-with-boundary $E \subset \overset{\circ}{M}$ that is closed as a subset of $\overset{\circ}{M}$ and is diffeomorphic to $S^1 \times S^1 \times [0, \infty)$. In terms of M this means that any rank-two free abelian subgroup of $\pi_1(M)$ is *peripheral*; a subgroup of $\pi_1(M)$, where M is a 3-manifold, is said to be peripheral if it is contained in a conjugate of

$$\text{Im}(\pi_1(B) \to \pi_1(M)),$$

where B is a component of ∂M.

Thurston's theorem says that the necessary conditions stated above are often sufficient. In particular, this is so if $\partial M \neq \varnothing$, and this is the case in which we shall quote the result.

THURSTON'S THEOREM (Weak Form) *Let M be a compact, orientable, irreducible 3-manifold with nonempty boundary. Suppose that every rank-two free abelian subgroup of $\pi_1(M)$ is peripheral. Then either M is diffeomorphic to the twisted I-bundle over the Klein bottle, or $\overset{\circ}{M}$ is diffeomorphic to a hyperbolic manifold.*

The version of Thurston's theorem presented in Chapter V is stronger than the above version in two respects. First, the manifold M is allowed to be closed, provided that it contains some closed, orientable, incompressible surface. Second, if one assumes that each boundary component of M is a torus and excludes a few degenerate cases, then, with the metric defined by its hyperbolic structure, $\overset{\circ}{M}$ has *finite volume*. These refinements will not be needed in what follows.

In order to apply the above results one needs a way of recognizing whether $\pi_1(M)$ contains nonperipheral rank-two free abelian subgroups. There is one eminently "visible" type of rank-two free abelian subgroup of $\pi_1(M)$, namely, $\text{Im}(\pi_1(T) \to \pi_1(M))$, where T is an incompressible torus in $\overset{\circ}{M}$. Moreover, a theorem of Stallings and Brown and Crowell [B–C] implies that this subgroup can be peripheral only if T is boundary-parallel (i.e., if it is the frontier of a submanifold of M diffeomorphic to $S^1 \times S^1 \times I$).

However, many rank-two free abelian subgroups of 3-manifold groups are not contained in subgroups of this form. The most obvious example is for $M \approx F \times S^1$, where F is a surface. In this case $\pi_1(M) \approx \pi_1(F) \times \langle t \rangle$, and for any $\alpha \in \pi_1(F) - \{1\}$, α and t generate a rank-two free abelian subgroup.

It is not hard to show that the latter subgroup cannot be conjugate to a subgroup of $\text{Im}(\pi_1(T) \to \pi_1(M))$, T an incompressible torus in $\dot{M}$, unless α is a power of an element of $\pi_1(T)$ defined up to conjugacy by a simple closed curve.

More generally, it is not hard to show that if M is a bounded Seifert-fibered space, then, with a few degenerate exceptions, $\pi_1(M)$ is very rich in rank-two free abelian subgroups that are not defined by incompressible tori. (In fact, every bounded Seifert-fibered space has a finite-sheeted covering diffeomorphic to $T \times S^1$ for some surface T, and the above remarks can be applied.)

It turns out that, in a sense, all rank-two free abelian subgroups of fundamental groups of bounded, orientable, irreducible 3-manifolds are accounted for by the above examples. This is made precise in the statement of the following result, which is a weak version of a theorem that was proved independently by Johannson [Jo] and by Jaco and Shalen [Ja–Sh], and which refines an earlier theorem of Waldhausen [W].

TORUS THEOREM (Weak Form) *Let M be a compact, orientable, irreducible 3-manifold with nonempty boundary. Then for any rank-two free abelian subgroup H of $\pi_1(M)$ there is a Seifert-fibered space $C \subset \overset{\circ}{M}$, whose boundary components are incompressible tori in M, such that H is conjugate to a subgroup of $\text{Im}(\pi_1(C) \to \pi_1(M))$.*

The theorem may be interpreted, via the sphere theorem of Papakyriakopoulos ([H, Chapter 4]), as asserting that any map of a torus into M is homotopic to a map whose image is contained in such a submanifold C. In the stronger version of the torus theorem, M is again allowed to be closed, provided that it contains an incompressible surface. Moreover, the Seifert-fibered submanifold C may be chosen *independently* of the given map of the torus into M, provided that we allow C to be disconnected; in fact, there is a *canonical* choice of C (which may be empty) that depends only on the manifold M. It turns out that the components of $\overset{\circ}{M} - C$ may be shown, using Thurston's theorem, to have hyperbolic structures. In this way a large part of the classification problem for 3-manifolds can be usefully reformulated in terms of hyperbolic geometry.

What we need here is the following consequence of Thurston's theorem and the torus theorem.

PROPOSITION 3. *Let M be a compact, irreducible, orientable 3-manifold with $\partial M \neq \emptyset$. Suppose that every incompressible torus $T \subset \overset{\circ}{M}$ is boundary-parallel in M. Then either $\overset{\circ}{M}$ is diffeomorphic to a hyperbolic 3-manifold, or M is Seifert-fibered.*

IV. The Proof in the Case of No Incompressible Surface

Proof. Assume that $\overset{\circ}{M}$ is *not* diffeomorphic to a hyperbolic 3-manifold. Then by Thurston's theorem, $\pi_1(M)$ contains a nonperipheral rank-two free abelian subgroup H. By the torus theorem, H is conjugate to a subgroup of $\mathrm{Im}(\pi_1(C) \to \pi_1(M))$ for some Seifert-fibered space $C \subset M$; and the components of ∂C are incompressible tori. By the hypothesis, each component of ∂C is boundary-parallel. Since M is connected, it follows that either C is contained in a regular neighborhood of some torus component of ∂M, or M is a regular neighborhood of C. But the first alternative would imply that H, which is contained in $\mathrm{Im}(\pi_1(C) \to \pi_1(M))$, was peripheral—a contradiction. Hence M is a regular neighborhood of C, and is therefore Seifert-fibered. ∎

Observe that under the hypotheses of the theorem stated in the introduction to this chapter, every closed, incompressible surface in $\Sigma - \overset{\circ}{N}$, and in particular every incompressible torus, is boundary-parallel. Thus by Proposition 3, either $\Sigma - \overset{\circ}{N}$ is Seifert-fibered and Proposition 2 may be applied or $\Sigma - K$ has a hyperbolic structure. Such a structure provides an isomorphism of Γ with a subgroup of $\mathrm{PSL}_2(\mathbf{C})$, and if the entries of the elements of this subgroup are algebraic integers, we can apply Proposition 1. We shall see in the next section that this is in fact the case under the hypotheses of the theorem; this requires using the fact that $\Sigma - K$ contains no closed incompressible surfaces of higher genus.

4. $\mathrm{PSL}_2(\mathbf{C})$ and Incompressible Surfaces

The most general question raised by the discussion at the end of the last section is this: given a compact, orientable, irreducible 3-manifold M, what does the existence of non-boundary-parallel, closed, incompressible surfaces in M mean in terms of $\pi_1(M)$? This has been understood for a good many years, and we shall begin this section with an account of it.

If $F \subset \overset{\circ}{M}$ is a closed incompressible surface which is orientable but *nonseparating*, then

(∗) there is an epimorphism $\lambda: \pi_1(M) \to \mathbf{Z}$ such that $\lambda(P) = 0$ for every peripheral subgroup P of $\pi_1(M)$.

In fact we can define λ by $\lambda([\gamma]) = \gamma \cdot F$, where the dot denotes the intersection pairing. To see that λ is surjective we need only observe that, since F is nonseparating, there is a simple closed curve that intersects F transversally at a single point; and since $F \subset \overset{\circ}{M}$, it is clear that $\lambda(P) = 0$ for P peripheral.

On the other hand, if $F \subset \overset{\circ}{M}$ is a closed, *separating*, incompressible surface, and if A_1 and A_2 are the closures of the components of $M - F$, we can write $\pi_1(M) = \pi_1(A_0) *_{\pi_1(F)} \pi_1(A_1)$, a free product with amalgamation (see

Kurosh [K, p. 29ff]; the homomorphisms $\pi_1(F) \to \pi_1(A_i)$ are injective, since F is incompressible). Moreover, it follows from a result due to Stallings and Brown and Crowell [B–C] that neither homomorphism $\pi_1(F) \to \pi_1(A_i)$ is surjective unless F is boundary-parallel. Since each component of ∂M is contained in A_0 or A_1, one concludes in this case that

(∗∗) $\pi_1(M)$ is a free product with amalgamation $\Gamma_0 *_\Lambda \Gamma_1$, with $\Gamma_0 \neq \Lambda \neq \Gamma_1$, in such a way that each peripheral subgroup of $\pi_1(M)$ is conjugate to a subgroup of Γ_0 or Γ_1.

Thus if M contains a closed, orientable, incompressible surface that is not boundary-parallel, then either (∗) or (∗∗) holds. What we really need here is a converse to this fact. The following result, essentially due to Stallings, Epstein, and Waldhausen, is very similar to results that have appeared in the literature, but the proof is so simple and elegant that we shall include it.

PROPOSITION 4. *Let M be a compact, orientable, irreducible 3-manifold for which at least one of the conditions* (∗), (∗∗) *holds. Then M contains a closed, incompressible, orientable surface that is not boundary-parallel.*

Proof. First suppose that (∗) holds. Identify $\pi_1(S^1)$ with $\mathbf{Z}$, so that λ is an epimorphism from $\pi_1(M)$ to $\pi_1(S^1)$. Since S^1 is aspherical (i.e., has trivial higher homotopy groups), λ is induced by a map $f: M \to S^1$. Since $\lambda(P) = 0$ for each peripheral subgroup P, we can again invoke the asphericity of S^1 to homotope f to a map g such that $g(B)$ is a single point for each component B of ∂M. We may take g to be smooth. Then if y is a generic point of S^1, g is transversal to $\{y\}$ and $y \notin g(\partial M)$; thus $g^{-1}(y)$ is a smooth, closed 2-manifold that is locally separating in M and hence orientable. It is an important observation due to Stallings that g is in turn homotopic to a map j that is still transversal to $\{y\}$, such that the components of $j^{-1}(y)$ are not only closed and orientable but are incompressible as well. This is included, for example, in Lemma 6.5 of Hempel [H], which shows more generally that if g is a map of a compact, orientable 3-manifold M into a k-manifold X, and if $Y \subset X$ is a locally separating, aspherical $(k-1)$-manifold such that $\pi_1(Y) \to \pi_1(X)$ is a monomorphism and each component of $X - Y$ is aspherical, then G is homotopic to a map j transversal to Y and such that $j^{-1}(Y)$ is incompressible. Moreover, the proof shows that if g is transversal to Y and $g^{-1}(Y)$ is closed, we may take $j^{-1}(Y)$ to be closed.

Now if each component of $F = j^{-1}(y)$ were boundary-parallel, then some component of $M - F$ would be a deformation retract of M, and hence j (and f) would be homotopic in S^1 to a map of M into $S^1 - \{x\}$. This contradicts the hypothesis that λ is an epimorphism.

IV. The Proof in the Case of No Incompressible Surface

If (**) holds, the proof is similar except that we must map M into a more complicated space. Let Y be a $K(\Lambda, 1)$ (i.e., an aspherical space with fundamental group Λ); let W_i be $K(\Gamma_i, 1)$ for $i = 0, 1$; let $\alpha_i: Y \to W_i$ be a map that induces the "inclusion" of Λ in Γ_i; and let Z_i denote the mapping cylinder of α_i. Then there is a natural identification of Y with a subspace of Z_i ($i = 0, 1$), and we set $X = Z_0 \cup_Y Z_1$. Clearly, $\pi_1(X) \approx \Gamma_0 *_\Lambda \Gamma_1$.

Moreover, X is easily seen to be aspherical. (In fact, if $\tilde{X}$ is the universal covering of X and p is the projection map, then since $\pi_1(Y)$ and the $\pi_1(Z_i)$ map monomorphically into $\pi_1(X)$, each component of $p^{-1}(Y)$ or $p^{-1}(Z_i)$ is a copy of the universal covering of Y or Z_i. Now Z_i (which is homotopy equivalent to W_i) and Y are aspherical, and hence the components of $p^{-1}(Y)$ and $p^{-1}(Z_i)$ are contractible. The Mayer–Vietoris theorem now shows that $H_i(\tilde{X}) = 0$ for $i > 1$; since $\pi_1(X) = 0$, the Hurewicz theorem now shows that $\tilde{X}$ is contractible, as required.)

Thus X is a $K(\Gamma_0 *_\Lambda \Gamma_1, 1)$, and so there is a map $f: M \to X$ inducing the natural isomorphism of fundamental groups. Since each peripheral subgroup of $\pi_1(M)$ is conjugate to a subgroup of Γ_0 or Γ_1, f is homotopic to a map g such that $g(\partial M) \cap Y = \emptyset$. Now X and Y are in general not manifolds, but Y has a collar neighborhood in X, and so we can assume that h is transversal to Y. Then $g^{-1}(Y)$ is an orientable surface, which is closed since

$$g(\partial M) \cap Y = \emptyset.$$

Moreover, since Y and the components of $X - Y$ are aspherical, the proof of Lemma 6.5 of Hempel [H] still shows that g is homotopic to a map j such that each component of $F = j^{-1}(Y)$ is a closed, orientable, incompressible surface.

Finally, if each component of F is boundary-parallel in M, we can show as above that j is homotopic to a map of M into $X - Y$. Since j induces an isomorphism of fundamental groups, it follows that Z_i carries $\pi_1(X)$ for some i, i.e., that $\Gamma_i = \Gamma_0 *_\Lambda \Gamma_1$. But, by a general fact about free products with amalgamation [K, p. 32], $\Gamma_0 \cap \Gamma_1 = \Lambda$. Hence $\Gamma_{1-i} = \Lambda$, contradicting the hypothesis (**). ∎

In the case discussed at the end of Section 3, $\overset{\circ}{M}$ is hyperbolic, so that $\pi_1(M)$ can be identified with a subgroup Γ of $PSL_2(\mathbf{C})$. Moreover, if the components of ∂M are tori whose fundamental groups map monomorphically into $\pi_1(M)$, then the peripheral subgroups are free abelian of rank two. By a remark in Section 2, a rank-two free abelian subgroup of Γ is *unipotent*, i.e., it is conjugate to a group of elements of the form $\pm \begin{pmatrix} 1 & \lambda \\ 0 & 1 \end{pmatrix}$, $\lambda \in \mathbf{C}$. (An element is unipotent if it generates a unipotent subgroup.) Thus (*) (or (**)) is implied by conclusion (a) (or (b)) of the following theorem, whose relevance to the question discussed at the end of Section 3 is now apparent.

BASS'S THEOREM. *Let Γ be a finitely generated subgroup of $\mathrm{PSL}_2(\mathbf{C})$. Then one of the following alternatives holds.*

(a) *There is an epimorphism $\lambda\colon \Gamma \to \mathbf{Z}$ such that $\lambda(u) = 0$ for all unipotent elements $u \in \Gamma$.*

(b) *Γ is an amalgamated free product $\Gamma_0 *_\Lambda \Gamma_1$ with $\Gamma_0 \neq \Lambda \neq \Gamma_1$, and every finitely generated unipotent subgroup of Γ is contained in a conjugate of Γ_0 or Γ_1.*

(c) *Γ is conjugate to a group of triangular matrices $\left(\begin{smallmatrix} a & b \\ 0 & a^{-1} \end{smallmatrix}\right)$ with a a root of unity.*

(d) *Γ is conjugate to a subgroup of $\mathrm{PSL}_2(A)$ where A is a ring of algebraic integers.*

A nearly identical theorem is proved by Bass in Chapter VI. The only significant difference is that Bass's statement involves $\mathrm{GL}_2(\mathbf{C})$ rather than $\mathrm{PSL}_2(\mathbf{C})$. If $\Gamma \subset \mathrm{PSL}_2(\mathbf{C})$, and if we apply Bass's statement to $q^{-1}(\Gamma) \subset \mathrm{SL}_2(\mathbf{C}) \subset \mathrm{GL}_2(\mathbf{C})$, where $q\colon \mathrm{SL}_2(\mathbf{C}) \to \mathrm{PSL}_2(\mathbf{C})$ is the quotient projection, the above statement follows very easily. (The least trivial point is that if $q^{-1}(\Gamma) = \Gamma_0 *_\Lambda \Gamma_1$, then $\{\pm I\} \subset \Lambda$, so that $\Gamma = q(\Gamma_0) *_{q(\Lambda)} q(\Gamma_1)$; and this is immediate from the fact [Ma–Ka–So, Corollary 4.5] that the center of such an amalgamated free product is always contained in the amalgamated subgroup.)

Combining Bass's theorem with results discussed earlier, we can now establish a striking result about general 3-manifolds.

PROPOSITION 5. *Let M be a compact, orientable, irreducible 3-manifold. Suppose that $\partial M \neq \emptyset$ and that each component of ∂M is a torus. Suppose further that every closed, orientable, incompressible surface in M is boundary parallel. Then either* (i) *M is Seifert-fibered or* (ii) *there is a ring A of aglebraic integers and a torsion-free group $\Gamma \subset \mathrm{PSL}_2(A)$, that is discrete as a subgroup of $\mathrm{PSL}_2(\mathbf{C})$, and is such that $\overset{\circ}{M}$ is diffeomorphic to $\mathbb{H}^2/\Gamma$.*

Proof. Since, in particular, every incompressible torus in M is boundary-parallel, Proposition 3 asserts that either (i) holds or $\overset{\circ}{M}$ is diffeomorphic to a hyperbolic 3-manifold. We may therefore assume that $\overset{\circ}{M} \approx \mathbb{H}^3/\Gamma$, where $\Gamma \subset \mathrm{PSL}_2(\mathbf{C})$ is discrete and torsion free. Since M is compact, Γ is finitely generated; we shall apply Bass's theorem to it.

We may assume that for each component B of ∂M, $\pi_1(B) \to \pi_1(M)$ is a monomorphism. For otherwise, by the Dehn lemma and loop theorem [H, Chapter 4], there is a disk $D \subset M$ such that $\partial D = D \cap \partial M$ is a non-contractible curve in B. Since B is a torus, the manifold $\overline{M - N}$, where N is tubular neighborhood of D, has a 2-sphere boundary component, and by

IV. The Proof in the Case of No Incompressible Surface

the irreducibility of M, $\overline{M - N}$ is a 3-ball. Since M is orientable we must now have $M \approx D^2 \times S^1$, and so (i) holds.

This last assumption implies that the peripheral subgroups of $\Gamma = \pi_1(M)$ are free abelian of rank two. Applying the discussion preceding the statement of Bass's theorem, we see that they are unipotent. Hence if Γ satisfies conclusion (a) or (b) of Bass's theorem, M satisfies one of the conditions $(*)$, $(**)$ preceding the statement of Proposition 4. But the conclusion of Proposition 4 contradicts the hypothesis of the present Proposition. Hence (a) and (b) are ruled out.

Suppose that (c) holds. An element of $PSL_2(\mathbf{C})$ of the form $\begin{pmatrix} a & b \\ 0 & a^{-1} \end{pmatrix}$, where $a \neq \pm 1$ is a root of unity, has finite order > 1. (It has distinct characteristic roots and can therefore be diagonalized to $\begin{pmatrix} a & 0 \\ 0 & a^{-1} \end{pmatrix}$.) Thus a torsion-free subgroup of $PSL_2(\mathbf{C})$ that satisfies (c) must be unipotent and hence abelian. But an orientable, bounded, irreducible 3-manifold M with torus boundary components and abelian fundamental group is diffeomorphic to $S^1 \times D^2$, or $S^1 \times S^1 \times I$, and hence satisfies (i).

Finally, if Γ satisfies alternative (d) of Bass's theorem, then M obviously satisfies alternative (ii) of the proposition. ∎

The proof of the theorem stated in the introduction to this chapter is now immediate. We apply Proposition 5 to $M = \Sigma - \mathring{N}$. If (i) holds, the conclusion follows from Proposition 2; if (ii) holds, it follows from Proposition 1.

5. History

In the spring of 1977, Thurston had announced his theorem on the existence of hyperbolic structures (with a special case missing). He was aware of the connection with the Smith conjecture and had observed certain partial results, presumably including Proposition 1 of this chapter.

We had been aware for several years, thanks to discussions with W. Abikoff, B. Maskit, C. Miller, W. Jaco, and R. Kulkarni, that the existence of a hyperbolic structure was perhaps not a hopelessly restrictive condition on a 3-manifold; and that such a structure provides faithful representations of the fundamental group in $PSL_2(\mathbf{C})$. We were also aware, thanks to discussions with Bass, of the theory of Serre and Bass [Se] on subgroups of $SL_2(\mathbf{C})$ (or $PSL_2(\mathbf{C})$) and knew that it often provided amalgamated free product decompositions. Of course we knew, as a three-dimensional topologist, that by the work of Stallings, Epstein, and Waldhausen in the 1960s (cf. Proposition 4) such decompositions lead to incompressible surfaces. It was clear that these ideas should interact. Having in mind a particular application, and aware of Thurston's work on hyperbolic structures (but not of Proposition 1), we formulated a conjecture which we persuaded Bass to prove (cf. his tactful remarks in Chapter VI).

Our original conjecture (cf. Chapter VI) did not involve algebraic integers, but in proving it Bass was led to the theorem of his stated in Section 4. As we had mentioned to Bass that our conjecture should interact with Thurston's work, he talked with Thurston, who immediately saw that Bass's theorem implies "one-half" of the Smith conjecture.

The interaction among Thurston's theorem, the Serre–Bass theory, and the Stallings–Epstein–Waldhausen construction has had a number of applications in three-dimensional topology, of which the Smith conjecture is the most striking. (See Shalen [Sh] for another example.) We are convinced that the most remarkable applications are yet to come.

References

[B–C]	Brown, E. M., and Crowell, R. H., Deformation retractions of 3-manifolds into their boundaries, *Ann. of Math.* **82** (1965) 445–458.
[F]	Fatou, P., Fonctions automorphes, *in* "Théorie des fonctions algébriques" (P. E. Appel and E. Goursat, eds.), Vol. 2, pp. 158–160. Gauthier-Villars, Paris, 1930.
[G]	Giffen, C. H., Fibered knots and periodic transformations, Ph.D. thesis. Princeton Univ., Princeton, New Jersey, 1964.
[H]	Hempel, J., "3-manifolds," Ann. of Math. Studies, Vol. 86. Princeton Univ. Press, Princeton, New Jersey, 1976.
[Jo]	Johannson, K., "Homotopy Equivalence of 3-manifolds with Boundary," Lecture Notes in Mathematics, Vol. 761. Springer-Verlag, Berlin and New York, 1979.
[Ja–Sh]	Jaco, W., and Shalen, P., Seifert-fibered spaces in 3-manifolds, *Mem. Amer. Math. Soc.* **21**, (1979).
[Ku]	Kurosh, A. G. "The Theory of Groups," Vol. 2. Chelsea, Bronx, New York, 1960.
[Mi]	Milnor, J., A unique factorization theorem for 3-manifolds, *Amer. J. Math.* **84**, 1–7 (1962).
[Ma–Ka–So]	Magnus, W., Karrass, A., and Solitar, D., "Combinatorial Group Theory." Wiley (Interscience), New York, 1966.
[Mo]	Moise, E. E., "Geometric Topology in Dimensions 2 and 3," Graduate Text in Mathematics, Vol. 47, Springer–Verlag, Berlin and New York, 1977.
[N]	Neuwirth, L. P., "Knot groups," *Ann. of Math. Studies*, Vol. 56. Princeton, New Jersey, 1965.
[Se]	Serre, J.-P., "Trees." Springer-Verlag, Berlin and New York, 1980.
[Sh]	Shalen, P., Separating, incompressible surfaces in 3-manifolds, *Invent. Math.* **52**, 105–126 (1979).
[W]	Waldhausen, F., On the determination of some bounded 3-manifolds by their fundamental group alone, *Proc. Internat. Sympos. Topology 1968*, Hercy-Novi, Yugoslavia; pp. 331–332, 1969.
[Z–Sa]	Zariski, O., and Samuel, P., "Commutative Algebra," Vol. I. Van Nostrand–Reinhold, New York, 1958.

CHAPTER V

On Thurston's Uniformization Theorem for Three-Dimensional Manifolds

John W. Morgan

Department of Mathematics
Columbia University
New York, New York

Introduction

This chapter contains an outline of Thurston's existence theorem for hyperbolic structures on three-dimensional manifolds—the three-dimensional uniformization theorem. It is based on lectures given by Thurston at the symposium on the Smith conjecture at Columbia University in the spring of 1979 and at the Summer School at Bowdoin College in August 1980. Working from notes of these lectures as well as innumerable private conversations with Thurston, we produced the manuscript presented here.

This chapter by no means gives a complete proof of the uniformization theorem. However, it does present a fairly detailed description of the structure of the proof and breaks the argument into smaller pieces. Some of these pieces are treated here in great detail, others are dismissed with nothing but a statement of the result. In general, our criterion for deciding how to treat the various pieces was simple—the parts of the argument that are more formal, rely on the Bers–Ahlfors–Teichmüller theory, or rely on three-dimensional topology are treated in detail; those parts that rely on a more detailed study of geodesics, pleated surfaces, and length–area estimates in the 3-manifold are not treated at all. One finds no mention of these objects here.

Although the general outlines and the grand themes presented in this chapter are due entirely to Thurston, the detailed logical structure and explicit formulations of the intermediate results are often our own. They are our attempt at imposing a logical structure, suitable to us, on what we understood Thurston to be saying. As such, the responsibility for the correctness of this detailed matter falls on our shoulders.

The first two sections constitute a general introduction to the theory of complete hyperbolic structures on surfaces and 3-manifolds, and the corresponding theory of kleinian groups. Section 3 contains the statements of the main theorem concerning the existence of hyperbolic structures in the case when the manifolds have finite volume. It is Theorem B of Section 3, applied to knot complements, that is used in the proof of the Smith conjecture. As anyone familiar with the theory of 3-manifolds is aware, statements such as Theorems A and B are proved by inductively cutting the manifold in question into simpler pieces. Our cutting is performed along *superincompressible surfaces*. These are defined in Section 4. Also in this section one finds all the topological results necessary to carry out the inductive process of cutting apart the 3-manifolds. It turns out that as we apply this process of cutting to manifolds that, according to Theorems A and B, should have hyperbolic structures, the new manifolds we obtain satisfy nontrivial topological conditions. We formalize these conditions under the rubric of *pared manifolds* (Definition 4.8).

The general existence theorems for hyperbolic structures are stated in Section 5 (Theorems A' and B'). The structures that we must consider are, in general, of infinite volume. Their defining property is that the complete hyperbolic manifolds contain convex submanifolds of finite volume that carry the fundamental group, i.e., the structures are *geometrically finite*. The main existence theorems say that any pared manifold that is sufficiently large (this is a technical term) has a geometrically finite hyperbolic structure. These results generalize Theorems A and B on the existence of structures of finite volume. They are the ones that are proved by induction. In a nutshell, the proof of Theorems A' and B' goes as follows: Begin with a pared manifold; cut it open; assume by induction that the resulting simpler pared manifold has a geometrically finite structure; deform this structure until the ends that need to be glued together have matching structures; then glue up to form a geometrically finite structure on the original pared manifold. The rest of the chapter is devoted to carrying out such a program.

V. Uniformization Theorem for Three-Dimensional Manifolds

In Section 6, we study geometrically finite structures in their own right. Using results due to Margulis [11] concerning the nature of short, nonnull-homotopic loops, we study the cusps of these manifolds. As a consequence, we are able to show that the underlying topological manifold of a geometrically finite hyperbolic manifold is a pared manifold (though not necessarily a sufficiently large one).

As we indicate above, the inductive step in the proof of the existence theorem is of the following type. One has a convex hyperbolic 3-manifold of finite volume M, disjoint proper, superincompressible surfaces Z_0 and Z_1 in ∂M, and a homeomorphism $\varphi: Z_0 \to Z_1$. The problem is, When does the manifold obtained from M by gluing Z_0 to Z_1 via φ have a geometrically finite hyperbolic structure? The simplest case of this gluing problem is when, in addition to the above, $\partial M = Z_0 \amalg Z_1$. In this case, if the glued-up manifold has a hyperbolic structure, then this structure is of finite volume. It is this simplified version of the gluing problem that we study in Sections 7–12.

The main result along these lines is the *gluing theorem* stated in Section 7: Suppose that M has a convex hyperbolic structure of finite volume, that $\partial M = Z_0 \amalg Z_1$ with both Z_is being superincompressible, and that the result M' of gluing Z_0 to Z_1 by a homeomorphism φ is a pared manifold. Then this pared manifold has a complete hyperbolic structure of finite volume.

There is also a technical result (Corollary 7.3) proved in Section 7. It concerns the representations $\rho: \pi_1(M) \to \mathrm{PGL}_2(\mathbf{C})$ determined by the geometrically finite hyperbolic structures on M. It says that, with the assumptions above, $\rho|\pi_1(Z_i): \pi_1(Z_i) \to \mathrm{PGL}_2(\mathbf{C})$ is a faithful representation onto a quasi-fuchsian group for $i = 0$ and 1.

In Section 8 we review Maskit's combination theorems. These deal with gluing kleinian groups together along common quasi-fuchsian subgroups. The results, applied to our setup, say that a convex structure on M "glues together" to give a complete structure of finite volume on M' if and only if there is an element in $\mathrm{PGL}_2(\mathbf{C})$ which conjugates $\rho(\pi_1(Z_0))$ to $\rho(\pi_1(Z_1))$ and induces $\varphi_\#: \pi_1(Z_0) \xrightarrow{\cong} \pi_1(Z_1)$. This result is stated as Theorem 8.1.

Thus the gluing problem is replaced by a deformation problem: Given a geometrically finite hyperbolic structure on M, deform it until the quasi-fuchsian groups associated to the subgroups $\pi_1(Z_0)$ and $\pi_1(Z_1)$ of $\pi_1(M)$ become conjugate (in $\mathrm{PGL}_2(\mathbf{C})$). Section 9 reviews the general Bers's deformation theory. The principal result is that if M has a geometrically finite structure, then the space of all such structures, denoted $Q(M)$, is naturally homeomorphic to the Teichmüller space $\mathcal{T}(\partial M)$ of ∂M. Taking covering spaces corresponding to the components of ∂M leads to a map called the *skinning map*: Let $\alpha \in \mathcal{T}(\partial M) = Q(M)$. For each component of ∂M, form the associated covering space (or, equivalently, restrict the representation

$\rho_\alpha: \pi_1(M) \to \mathrm{PGL}_2(\mathbf{C})$ determined by α to $\pi_1(Z_0)$ and $\pi_1(Z_1)$). This determines a quasi-fuchsian group for each component of ∂M. By Bers's theory such groups are parametrized by $\mathcal{T}(\partial M) \times \mathcal{T}(\partial M)$. Thus we have a map defined by taking the covering spaces $\mathcal{T}(\partial M) \to \mathcal{T}(\partial M) \times \mathcal{T}(\partial M)$. This map is of the form $\alpha \to (\alpha, \sigma_M(\alpha))$. The map $\sigma_M: \mathcal{T}(\partial M) \to \mathcal{T}(\partial M)$ is the skinning map.

Section 9 finishes with Theorem 9.4, which reformulates the existence of a geometrically finite hyperbolic structure on M that glues up via φ to form one for $M' = M/\varphi$ as equivalent to the existence of a fixed point for the mapping $\tau \circ \sigma_M: \mathcal{T}(\partial M) \to \mathcal{T}(\partial M)$. Here τ is the involution on $\mathcal{T}(\partial M)$ induced by the topological involution on $\partial M = Z_0 \amalg Z_1$ given by

$$\varphi \amalg \varphi^{-1}: Z_0 \amalg Z_1 \to Z_1 \amalg Z_0 = Z_0 \amalg Z_1.$$

Theorem 10.1 (*the fixed point theorem*) states the result that guarantees the existence of a fixed point for $\tau \circ \sigma_M: \mathcal{T}(\partial M) \to \mathcal{T}(\partial M)$. For any $\alpha \in \mathcal{T}(\partial M)$, the sequence $\{(\tau \circ \sigma_M)^n(\alpha)\}_{n=1}^\infty$ converges to a point in $\mathcal{T}(\partial M)$ that is the unique fixed point for $\tau \circ \sigma_M$. The space $\mathcal{T}(\partial M)$ is a complete metric space homeomorphic to a euclidean space. The proof of Theorem 10.1 is divided into two parts. In Section 10 we show that $\tau \circ \sigma_M$ is a strictly distance-decreasing function. This is a fairly easy consequence of classical results of Teichmüller on the so-called Teichmüller mapping. As we also show in Section 10, we could then apply contracting mapping arguments to prove that $\tau \circ \sigma_M$ has a fixed point if we knew that for every point (or in fact for one point) $\alpha \in \mathcal{T}(\partial M)$ the sequence $\{(\tau \circ \sigma_M)^n(\alpha)\}_{n=1}^\infty$ were *bounded* in $\mathcal{T}(\partial M)$.

This is the heart of the matter in the proof of the gluing theorem. One must show that sequences as above are bounded. The result which says that they are indeed bounded is called the *bounded image theorem*. It is the subject of Sections 11 and 12.

Up to this point we have given a complete presentation of the argument, liberally using deep but published results from the fields of topology and kleinian groups. However, it is when we approach the bounded image theorem that this article becomes only an outline of the argument. The proof of this result requires a detailed geometric study of certain surfaces, pleated surfaces, in a hyperbolic 3-manifold. It also requires delicate length–area estimates. This is entirely new material of Thurston's which, unfortunately, is not all available.

The outline of the proof of the bounded image theorem occupies Sections 11 and 12. The set of geometrically finite structures on M, $Q(M)$, sits as an open subset inside $\mathrm{AH}(M)$, the set of all conjugacy classes of discrete and faithful representations of $\pi_1(M)$ in $\mathrm{PGL}_2(\mathbf{C})$. The latter sits as a closed subset in the complex affine variety of all $\mathrm{PGL}_2(\mathbf{C})$-characters of $\pi_1(M)$. By Bers's theory, $Q(M)$ is identified with $\mathcal{T}(\partial M)$. In what follows we use this

V. Uniformization Theorem for Three-Dimensional Manifolds

identification without mention. We consider sequences in $Q(M)$ of the form $\{(\tau \circ \sigma_M)^n(\alpha)\}_{n=1}^\infty$ for $\alpha \in Q(M)$. The main result of Section 11 (Theorem 11.7) is that all such sequences are bounded in $AH(M)$. This means that for any $\alpha \in Q(M)$ and any $g \in \pi_1(M)$, the sequence of complex numbers $\{\text{tr}[((\tau \circ \sigma_M)^n(\alpha))(g)]\}_{n=1}^\infty$ is bounded. (Here, we are viewing $(\tau \circ \sigma_M)^n(\alpha)$ as a conjugacy class of representations of $\pi_1(M)$ in $\text{PGL}_2(\mathbf{C})$.) It has as an immediate consequence the fact that any subsequence of $\{(\tau \circ \sigma_M)^n(\alpha)\}$ has a further subsequence that converges in $AH(M)$, i.e., the conjugacy classes of representations converge. The limit representation is discrete and faithful (since it is a point in $AH(M)$), but it may not be geometrically finite, i.e., it may not sit in the open subspace $Q(M) \subset AH(M)$.

To prove Theorem 11.7, we introduce the characteristic submanifold W of the 3-manifold M. In our case this is a maximal I-bundle over a surface that meets ∂M in the ∂I-subbundle. The frontier of W in M is a collection of essential annuli in M. The first result (Theorem 11.2) is that the sequence $\{(\tau \circ \sigma_M)^n(\alpha)\}_{n=1}^\infty$ is bounded on the cores of these annuli. We state this result without proof. Its proof involves area growth estimates for a certain branch surface, as well as the ergodicity of the geodesic flow on a surface. There is no reference for this result of Thurston's.

Once we have this, there is a general boundedness result (Theorem 11.3) which shows that $\{(\tau \circ \sigma_M)^n(\alpha)\}_{n=1}^\infty$ is bounded on any element of $\pi_1(M)$ whose free homotopy class is represented by a loop in $M - W$. This is a relative version of a result proved in Thurston [18].

The next result that we need (Theorem 11.6) is that associated to any sequence of quasi-fuchsian representations of a surface group, $\rho_i: \pi_1(S) \to \text{PGL}_2(\mathbf{C})$, there is a subsurface $B \subset S$, unique up to isotopy, called the *maximal bounded subsurface of S*. This subsurface includes, up to free homotopy, every loop whose homotopy class remains bounded under the ρ_i. This result is obvious for fuchsian representations. It is not too much more difficult to prove for quasi-fuchsian ones. We do not prove this result of Thurston's; there is no reference for it.

Section 11 finishes with the proof that, given Theorems 11.2, 11.3, and 11.6, the sequence $\{(\tau \circ \sigma_M)^n(\alpha)\}$ is bounded on all of $\pi_1(M)$.

The purpose of Section 12 is to show that the limit points of subsequences of $\{(\tau \circ \sigma_M)^n(\alpha)\}$, which were proved to exist in $AH(M)$ in Section 11, are actually points in $Q(M) \subset AH(M)$. As an immediate consequence, we have the bounded image theorem. The basic result we need for this is Theorem 12.1. This theorem describes how a manifold which is a limit of quasi-fuchsian manifolds can be a covering space of another hyperbolic manifold. This result is stated without proof. Its proof depends on an analysis of pleated surfaces. Much of the proof can be extracted from Sections 8 and 9 of [19] (one should consult especially 8.8.5, 8.9.2, 8.10.5, 8.10.9, 8.11.1, 9.2, and 9.2.2).

Using this result, we give a detailed argument to complete the proof of the bounded image theorem.

There is one case in which the fixed point theorem and the bounded image theorem are not true. That is the case in which M has a finite covering of the form $S \times I$ for some surface S. In this case, however, $M' = M/\varphi$ has a finite sheeted covering which fibers over the circle. Except for this case, we have now completed the proof of the gluing theorem. This special case is handled by a different (and much simpler and more direct) argument. An outline of this case appears in Section 13. A more detailed outline can be found in [16].

At this point, we have completed the discussion of the gluing result in the case where $\partial M = Z_0 \amalg Z_1$. We take up the more general gluing problem where Z_0 and Z_1 have boundary inside ∂M. Thurston's idea, a beautifully simple one, is to create a finer hyperbolic structure on M in which $\partial M - (Z_0 \amalg Z_1)$ becomes a union of totally geodesic faces meeting at angles of the form π/n. Such a structure corresponds to a geometrically finite kleinian group with torsion. Such structures and their associated kleinian groups are discussed in detail in Section 14. There is the notion of the boundary of such a structure. It is represented by $Z_0 \amalg Z_1$ together with an induced polygonal structure (and angles of incidence) on $\partial(Z_0 \amalg Z_1)$.

Theorem 14.2 gives a generalization of the gluing theorem to this situation. This result states that if M has a structure as described above and if the gluing homeomorphism $\varphi: Z_0 \to Z_1$ preserves the polygonal structures (with angles) on ∂Z_0 and ∂Z_1, then there is a deformation of the given structure to another of the same type which glues together to form a structure on M. The basic idea is to pass to a torsion-free, normal subgroup of finite index (using Selberg's theorem). Here we find ourselves confronted with exactly the type of problem we solved in Sections 7–12. Using the results there, we glue together. Then by Mostow's theorem the finite group of topological homeomorphisms that acts on the "lifted" manifolds is actually realized as a group of isometries. The quotient by this group solves the original gluing problem.

We are left then with the problem of enhancing a convex hyperbolic structure on M to one which on $\partial M - (Z_0 \amalg Z_1)$ is piecewise geodesic with angles of the form π/n. This is accomplished in Sections 15 and 16. In Section 15 we show how to do this on $S \times I$, for S a surface, so that one end is geodesic and the other realizes a predetermined piecewise geodesic pattern. This result is a generalization of Andre'ev's theorem [3]. The proof appears in the revised version of Thurston's Princeton notes [19]. We do not say anything about the proof here.

Finally, in Section 16 (Theorem 16.5) we show how to glue the structures created on $\partial M \times I$ in Section 15 to the convex hyperbolic structure assumed by

V. Uniformization Theorem for Three-Dimensional Manifolds

induction to exist on M. This will produce a new structure on M that has the requisite piecewise geodesic structure on $\partial M - (Z_0 \amalg Z_1)$. Combining this with the results of Sections 7 through 15 (excluding Section 13) completes the proof of the inductive step in the proof of the existence theorem. Since the initial step in the induction requires only that we find a convex hyperbolic structure on the ball, this completes the proof of the existence Theorems A, B, A', and B' stated in Sections 3 and 5.

1. An Introduction to Hyperbolic Geometry

The domain of three-dimensional manifolds is a fertile meeting ground for topology and geometry. Unlike dimension two, for which the richness is all on the side of the geometry, and unlike the higher dimensions, where the topology and geometry diverge significantly, dimension three exhibits a deep interaction between the geometry and topology. This interaction can be used to study both the topological and geometric structures. The concern of this chapter is to explain how geometry is a powerful tool in the study of the topology of three-dimensional manifolds in general and knot complements in particular.

There are several kinds of homogeneous three-dimensional geometries, but the most interesting and most important kind is *hyperbolic geometry*, also called *noneuclidean geometry* or *lobachevskian geometry*. The following is a model for this geometry.

(1) *Points* are the points of the interior of the unit 3-ball in $\mathbf{R}^3$.
(2) *Lines* are the intersection of the interior of the unit 3-ball with circles (or lines) in 3-space that meet the unit 2-sphere perpendicularly.
(3) *Planes* are the intersections of the interior of the unit 3-ball with 2-spheres (or planes) in 3-space that meet the unit 2-sphere perpendicularly.

This geometry is actually the geometry associated to a homogeneous riemannian geometry. To see this, one introduces the *Poincaré metric*

$$ds^2 = \frac{dx_1^2 + dx_2^2 + dx_3^2}{4(1-r^2)^2}, \quad \text{where} \quad r = \sqrt{x_1^2 + x_2^2 + x_3^2},$$

on the interior of the unit ball. This is a complete metric that looks the same in all directions from all points, i.e., given points x and y and directions τ_x and τ_y at x and y, there is an isometry carrying x to y and τ_x to τ_y. In this metric the geodesics are the arcs described in (2) and the totally geodesic planes are planes described in (3). It is a metric with all sectional curvatures at all points equal to minus one. This riemannian manifold is denoted by $\mathbb{H}^3$.

Of course, the Poincaré metric and the usual riemannian metric on the unit ball are very different. Another way to say this is that our intuitive euclidean sense of size is a distorted picture of hyperbolic size. Actually, at the origin of the ball the distortion is light and the euclidean intuition is reasonable. But as an object is moved closer to the boundary, the distortions grow. Something of constant hyperbolic size becomes small exponentially fast in the euclidean metric as it is moved to the boundary. But the Poincaré metric and the euclidean metric are conformally equivalent.

To study this geometry in detail we must first study its group of isometries. It turns out to be a very nice group. It is just the group of conformal transformations (i.e., angle preserving transformations) of the *closed* unit ball to itself. This group in turn consists of the conformal transformations of the boundary of the unit ball to itself. Said another way, every conformal isomorphism of S^2 to itself has a unique conformal extension to the closed unit ball, and this conformal extension is an isometry of the Poincaré metric on the interior. Furthermore, every isometry of hyperbolic space arises in this fashion. The group of orientation-preserving, conformal isomorphisms of S^2 has another description. To see this, recall that stereographic projection from the point $(0, 0, 1) \in S^2$ to the (x_1, x_2)-plane is a conformal mapping (see Fig. 1.1). If we identify the (x_1, x_2)-plane with the complex plane by $z = x_1 + ix_2$, then we can extend stereographic projection to a conformal diffeomorphism from the unit 2-sphere to the Riemann sphere $= \mathbf{C} \cup \{\infty\}$. We can also think of the Riemann sphere as being the projective space of $\mathbf{C}^2$ with homogeneous coordinates $[z_1, z_2]$ (where $(z_1/z_2) = z$). This allows us to identify the orientation-preserving conformal transformations of the unit 2-sphere with those of the Riemann 2-sphere.

A conformal map of the Riemann sphere is determined by a 2×2 matrix $\begin{pmatrix} a & b \\ c & d \end{pmatrix}$ whose determinant is nonzero. The equation of the mapping is $[z_1, z_2] \to [az_1 + bz_2, cz_1 + dz_2]$, or in nonhomogeneous coordinates,

$$z \to \frac{az + b}{cz + d}.$$

If we replace a matrix A by λA, $\lambda \in \mathbf{C}^*$, then the map of the Riemann sphere is unchanged. Furthermore, all orientation-preserving conformal self-mappings

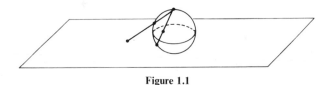

Figure 1.1

V. Uniformization Theorem for Three-Dimensional Manifolds

of the Riemann sphere arise in this manner. This tells us that the group of orientation-preserving conformal transformations of the unit 2-sphere (or the Riemann sphere) is identified with the projective general linear group

$$PGL_2(\mathbf{C}) = GL_2(\mathbf{C})/\mathbf{C}^* = SL_2(\mathbf{C})/\{\pm \text{id}\}.$$

Having described the full group of isometries of hyperbolic geometry, we discuss now the dynamic properties of various types of elements in the group. Every mapping of the closed unit ball to itself must have a fixed point. Any orientation-preserving conformal mapping of S^2 with three fixed points is the identity. From these two facts one sees that every orientation-preserving isomorphism of the closed unit ball to itself is one of three types:

(1) a fixed point in the interior, and hence an entire fixed line in $\mathbb{H}^3$ — *elliptic*;
(2) two fixed points on the boundary and none in the interior — *hyperbolic*;
(3) a fixed point on the boundary and none in the interior — *parabolic*.

The group of orientation-preserving isometries that fix a given point in the interior (i.e., a point in hyperbolic space) is isomorphic to $SO(3)$. The reason is that all directions look the same from this point, and hence one can find an isometry fixing the point and inducing an arbitrary rotation on the tangent plane of the point.

An isometry with two fixed points on the sphere at infinity and none in the interior corresponds to an element $\alpha \in PGL_2(\mathbf{C})$ with two fixed points on the projective line. This means that if we lift α to an element of $GL_2(\mathbf{C})$, then it has two invariant lines in $\mathbf{C}^2$, i.e., two independent eigenvectors. The isometry of $\mathbb{H}^3$ must leave invariant the unique geodesic whose end points are the fixed points on the sphere at infinity. This geodesic is the *axis* of the isometry. The full transformation is a "screw motion" about this axis. Points are moved along the axis toward the fixed point corresponding to the larger of the two eigenvalues at a rate that is the log of the absolute value of the ratio of the two eigenvalues. (If the two eigenvalues have the same absolute value, then the entire axis is fixed and the element in question is elliptic rather than hyperbolic.) The angle of rotation about the axis is the argument of the ratio of the eigenvalues. Under iteration of a hyperbolic element $\alpha \in PGL_2(\mathbf{C})$, all points of the closed unit ball except one tend to the attracting fixed point of α in the 2-sphere at infinity. The one that does not tend to this point is the other fixed point of α.

A parabolic element corresponds to $\alpha \in PGL_2(\mathbf{C})$ which, when lifted to a matrix in $GL_2(\mathbf{C})$, has only one eigenvalue. To study such an element it is

Figure 1.2

convenient to take a different model for hyperbolic 3-space—one analogous to the upper half-plane model for hyperbolic 2-space. It is the upper half-space $\{(x_1, x_2, x_3) | x_3 > 0\}$. The metric is $ds^2 = (dx_1^2 + dx_2^2 + dx_3^2)/x_3^2$. The geodesics are

(1) semicircles meeting the (x_1, x_2)-plane perpendicularly and
(2) vertical straight lines.

We view its boundary as the (x_1, x_2)-plane $\cup \{\infty\}$. We identify this in an obvious way with the Riemann sphere. If we begin with an isometry $\tilde{\alpha}$ whose extension α has a unique fixed point on the boundary, we can arrange that this fixed point be ∞. Thus the isometry gives a fractional linear transformation of the z-plane to itself sending ∞ to ∞. This means that the matrix representation of $\alpha \in \mathrm{PGL}_2(\mathbf{C})$ in the chosen coordinates is $\begin{pmatrix} 1 & b \\ 0 & 1 \end{pmatrix}$ for some $b \in \mathbf{C}$. As a result, the map induced by α on the z-plane is a pure translation. This means that if S is a circle in the z-plane with center z_0 and radius r, then $\alpha(S)$ is a circle in the z-plane with center $\alpha(z_0)$ and radius r. Consequently, $\tilde{\alpha}$ preserves the (euclidean) height of totally geodesic planes meeting the z-plane in circles as well as preserving the point which is highest (see Fig. 1.2). Thus $\tilde{\alpha}$ preserves the euclidean plane $\{x_3 = C\}$ for any $C > 0$. These planes are the *horospheres of* ∞. Regions of the form $\{x_3 \geq C\}$ are the *horoballs of* ∞. In the unit ball model these are pictured as shown in Fig. 1.3. The metric induced on the horospheres is a flat euclidean metric. (This is clear from the formula for the metric.) Notice however that the horospheres are not totally geodesic planes. In fact, each horoball is strictly convex. On each horosphere of ∞, the element $\tilde{\alpha}$ induces a euclidean transformation. Thus the group that

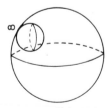

Figure 1.3

consists of all parabolic elements fixing ∞ and of the identity acts on each such horosphere of ∞ by euclidean translations. This group is isomorphic to the group of all such translations, i.e., the additive group of $\mathbf{R}^2$.

These descriptions of the various elements are but a brief introduction to the subject. Much more can be said; geometric properties can be translated into matrix properties and vice versa. Thus the study of hyperbolic geometry becomes the study of "little" questions of the linear algebra of 2×2 matrices. Of course, this is a misleading point of view. The questions about 2×2 matrices are not little; they are deep and hard to grasp. In fact, one often studies them *through* the geometry.

2. Kleinian Groups

A *kleinian group* is a discrete subgroup of $\mathrm{PGL}_2(\mathbf{C})$. Thought of as groups of isometries of $\mathbb{H}^3$, these are exactly the ones which act properly discontinuously on $\mathbb{H}^3$. A kleinian group is torsion-free if and only if its action on hyperbolic space is free. Hence, the study of torsion-free, kleinian groups is (logically, if not psychologically) the study of 3-manifolds whose universal cover is $\mathbb{H}^3$ and whose covering transformations are orientation-preserving isometries. Such manifolds are called *complete hyperbolic manifolds*. We fix an orientation for hyperbolic space. Since any kleinian group consists of orientation-preserving isometries, any hyperbolic 3-manifold has an induced orientation.

It is important to study the boundary values of a kleinian group as well as its action on $\mathbb{H}^3$. That is to say, a kleinian group Γ has a natural representation as a group of conformal transformations of S^2_∞ (the sphere at infinity). This action and the action on $\mathbb{H}^3$ fit together to form an action of Γ on the closed 3-ball B^3. Here quite complicated phenomena appear even for groups acting freely and properly discontinuously on the interior. We define $L_\Gamma \subset S^2_\infty$, the *limit set of* Γ, to be all $x \in S^2_\infty$ for which there is some $y \in \mathrm{int}(B^3)$ and a sequence $\{\gamma_i\}$ of elements in Γ so that $\lim_{i \to \infty} \gamma_i y = x$. The first observation is that if $\lim_{i \to \infty} \gamma_i y = x$, then for any $y' \in \mathrm{int}(B^3)$, $\lim_{i \to \infty} \gamma_i y' = x$. The reason is that as $\gamma_i y$ approaches $x \in S^2_\infty$, the euclidean distance from $\gamma_i y$ to $\gamma_i y'$ must go to zero. The same argument proves that if K is any compact, nonempty subset of $\mathbb{H}^3$, then

$$L_\Gamma = \left(\overline{\bigcup_{\gamma \in \Gamma} \gamma K} \right) \cap S^2_\infty.$$

It follows immediately that $L_\Gamma \subset S^2_\infty$ is a closed Γ-invariant subset. It is also nonempty provided that Γ is not a finite group. The limit set L_Γ can also be described as the closure of the set of all points in S^2_∞ fixed by some element of infinite order in Γ. If L_Γ has more than two points, then it is an uncountable subset of S^2_∞. The limit set is finite exactly when Γ has a subgroup of finite index that is abelian; in this case Γ is called an *elementary group*.

Many properties and proofs break down for elementary groups; the correct properties of elementary groups are almost always easy to work out separately, but to include these special cases in every discussion unnecessarily complicates the exposition. To avoid this difficulty we shall make the assumption that Γ *is not elementary*.

The complement Ω_Γ of L_Γ in S^2_∞ is called the *region of discontinuity of* Γ. This is an open subset of S^2_∞ that is Γ-invariant. If Γ is a kleinian group, then the action of Γ on Ω_Γ and even the action of Γ on $\mathbb{H}^3 \cup \Omega_\Gamma$ inside the closed ball is properly discontinuous.

Let Γ be a torsion-free kleinian group. When Γ is not elementary, L_Γ consists of more than two points. Thus the conformal structure on any component D of Ω_Γ, when lifted to the universal cover $\tilde{D}$ of D, gives an oriented conformal structure equivalent to that of the unit disk. The stabilizer of D in Γ, stab(D), extended by $\pi_1(D)$ becomes a discrete group of orientation-preserving conformal transformations of the unit disk under this identification, i.e., a subgroup of $\mathrm{PGL}_2(\mathbf{R})$. This shows that the natural conformal structure on Ω_Γ/Γ is conformally equivalent to the conformal structure associated to a disjoint union of complete hyperbolic surfaces. These hyperbolic surfaces are called the *surfaces at infinity* of Γ. Topologically they are exactly the boundary components of the 3-manifold $(\mathbb{H}^3 \cup \Omega_\Gamma)/\Gamma$.

Since the metric on $\mathbb{H}^3$ is complete and has all sectional curvatures equal to -1 and since Γ acts on $\mathbb{H}^3$ as a group of isometries, the quotient manifold $\mathbb{H}^3/\Gamma$ has an induced metric that is complete and has all sectional curvatures equal to minus one. The converse is also true. To see that, let M be a 3-manifold equipped with such a metric. By the fundamental lemma of riemannian geometry [15, pp.7–26] sufficiently small open sets in M are isometric with small open sets in $\mathbb{H}^3$. By using this, one can construct a local isometry (preserving orientations) $d: \tilde{M} \to \mathbb{H}^3$.

This map is obtained by beginning with an identification of small open sets and analytically continuing. The map is well defined up to the action of $\mathrm{PGL}_2(\mathbf{C})$ on $\mathbb{H}^3$. It is called the *development map*. Since it is a local isometry and $\tilde{M}$ is complete, it follows that d is a covering projection. Since $\tilde{M}$ is connected and $\mathbb{H}^3$ is simply connected, d is a homeomorphism and hence an isometric isomorphism. Thus d transforms the fundamental group of M, acting as covering transformations on $\tilde{M}$, to a torsion-free kleinian group $\Gamma \subset \mathrm{PGL}_2(\mathbf{C})$:

$$d_*: \pi_1(M) \xrightarrow{\cong} \Gamma \subset \mathrm{PGL}_2(\mathbf{C}).$$

V. Uniformization Theorem for Three-Dimensional Manifolds

(If we change d by an element in $\mathrm{PGL}_2(\mathbf{C})$, then the representation d_* is conjugated by this element.) This map d_* is called the *holomony representation*. The map d induces an isometry $M \to \mathbb{H}^3/\Gamma$.

The above construction shows that complete hyperbolic 3-manifolds, considered up to isometry, are naturally identified with torsion-free kleinian groups considered up to conjugacy. (Of course, there is a result like this in every dimension n in which we replace kleinian groups by discrete subgroups of the isometries of hyperbolic n-space.)

The first questions that one must address in attempting to use this geometry to study three-dimensional manifolds are those of existence and uniqueness of such structures. These are the questions that are addressed in this chapter. Before describing the situation in dimension three, let us begin with the case of surfaces. Surfaces fall into three qualitatively distinct types – those with positive, zero, and negative Euler characteristics. Every surface of negative Euler characteristic has a geometric structure modeled on two-dimensional hyperbolic geometry, i.e., it has a complete riemannian metric of constant curvature equal to -1.

Anticipating the discussion in dimension three, we shall describe briefly how one obtains hyperbolic structures on surfaces. In fact, there is a whole space of hyperbolic structures on a given topological surface, say M_g, the surface of genus g. The space of geometric structures comes equipped with a topology. It is called the *Teichmüller space* $\mathcal{T}(M_g)$. It is homeomorphic to $\mathbf{R}^{6g-6}$. Here is a brief geometric description of the space $\mathcal{T}(M_2)$. We begin with the surface shown in Fig. 2.1 and cut it up into simpler surfaces along

Figure 2.1

closed curves. To do this, take a maximal family of disjoint simple closed curves on M_2, no two of which are homotopic and so that none is homotopically trivial. There must be three curves in any such family for M_2. (See for example, Fig. 2.2.) If we cut M_2 open along these curves, then the result (called two pairs of pants) is as shown in Fig. 2.3.

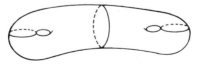

Figure 2.2

Figure 2.3

Our first task is to study hyperbolic structures on a pair of pants, and then to decide when hyperbolic structures on the two pairs of pants can be glued together to form one on the original surface. In studying hyperbolic structures on a pair of pants, we have introduced a new difficulty–namely, our surface now has a boundary. We have to decide how to treat the boundary. Of course, we could just require that the pair of pants be a compact C^∞-manifold with constant curvature -1 on the interior and no condition on the boundary. This is not too reasonable since we wish to glue the resulting structures together. Instead, to simplify the gluing problem, we require that the boundary be geodesic. Thus we pose the problem of classifying hyperbolic structures on a pair of pants whose boundary components are totally geodesic. Associated to any such structure are three positive constants—the lengths of the boundary components. In fact, these lengths are arbitrary positive numbers and the set of three of them characterises the hyperbolic structure on the pair of pants. It takes a little hyperbolic trigonometry to prove this, but basically there is a strong analogy between properties of pairs of pants and properties of triangles. The above result is the analogue of the fact that a triangle is determined by the lengths of its three sides.

Anyway, the space of hyperbolic structures on a pair of pants is $\mathbf{R}^3$. A homeomorphism is given by the logarithms of the lengths of the boundary curves $\{\log(l_1), \log(l_2), \log(l_3)\}$. To form M_2, we must glue two such pairs of pants together. To be able to do that, it is necessary and sufficient that the corresponding boundary curves have the same lengths. If they do have the same lengths, there is one degree of freedom in how one glues: Any two gluings differ by an amount of rotation $\theta \in \mathbf{R}^1$. (The rotation must be taken in $\mathbf{R}$ rather than S^1 since we must keep track of the homotopy equivalence to a given reference surface.) (See Fig. 2.4.) In the end, we have six real-valued invariants for structures arising in this fashion:

$$\{\log(l_1), \log(l_2), \log(l_3), \theta_1, \theta_2, \theta_3\}.$$

If we begin with an arbitrary hyperbolic structure on M_2 and a family of three curves, then the free homotopy classes of the curves are all represented by unique closed geodesics. If the original three curves are disjoint, then the geodesic representatives remain disjoint. We can cut the surface open along these geodesics and get two hyperbolic pairs of pants. The given hyperbolic

V. Uniformization Theorem for Three-Dimensional Manifolds

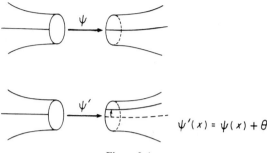

Figure 2.4

structure on M_2 comes from gluing together structures on pairs of pants. Hence, the six real parameters described above give a parametrization of Teichmüller space, i.e., they determine a homeomorphism: $\mathcal{T}(M_2) \overset{\approx}{\to} \mathbf{R}^6$.

For a surface of higher genus g an analogous discussion holds. It takes $(3g-3)$ simple closed curves to cut it into $(2g-2)$ pairs of pants. Associated to each curve we have the logarithm of its length and a rotation parameter telling how we glue the pants back together. Thus we find a homeomorphism

$$\mathcal{T}(M_g) \cong \mathbf{R}^{6g-6} = \mathbf{R}^{-3\chi(M_g)}.$$

3. Statement of the Main Theorem—The Case of Finite Volume

Let us turn our attention to 3-manifolds. We shall begin with the theorem for closed manifolds and introduce complications later.

THEOREM A. *Let M be a closed, irreducible 3-manifold that is Haken and atoroidal. Then M has a riemannian metric which has all sectional curvatures equal to -1. That is, there is a discrete subgroup $\Gamma \subset \mathrm{PGL}_2(\mathbf{C})$ that acts freely (and properly discontinuously) on $\mathbb{H}^3$ and a homeomorphism of M with $\mathbb{H}^3/\Gamma$.*

Remarks (1) A 3-manifold is said to be *irreducible* if every $S^2 \subset M$ bounds a 3-ball in M. A 3-manifold is *atoroidal* if M is irreducible and if $\pi_1(M)$ contains no noncyclic, abelian subgroup (i.e., if every map $\varphi: T^2 \to M$ has a kernel on the level of fundamental groups). The definition of a Haken manifold is given after Proposition 4.5.

(2) A metric on M all of whose sectional curvatures are equal to minus one is called a *hyperbolic metric on M* or a *hyperbolic structure on M*.

(3) If M is closed and has a hyperbolic metric, then there is $\varepsilon > 0$ so that every closed loop on M of length $\leq \varepsilon$ is homotopically trivial. Thus, when we

write M as isometric to $\mathbb{H}^3/\Gamma$, no nontrivial element of Γ can move any point of $\mathbb{H}^3$ a distance less than ε. Parabolic elements do not have this property. Thus $\Gamma - \{e\}$ consists entirely of hyperbolic elements.

(4) If $M \cong \mathbb{H}^3/\Gamma$, then M is irreducible. (Any 2-sphere in M lifts to the universal cover $\tilde{M} \cong \mathbb{H}^3$. There it bounds a ball. In M it bounds the image of this ball, which is also a ball.) The further condition that M be atoroidal is also necessary if M is to be homeomorphic to $\mathbb{H}^3/\Gamma$. To see this, let $G \subset \Gamma$ be an abelian subgroup. Let $\alpha \in G$ be a nontrivial element. As we saw in Remark (3), α is hyperbolic and hence has a unique axis A_α in $\mathbb{H}^3$. Let β be an other element in G. Since β commutes with α, β must fix the axis A_α. Thus A_α is left invariant by β and hence by all of G. The group G then acts freely and properly discontinuously on A_α and hence is cyclic.

(5) The condition that M be Haken is not implied by the existence of a homeomorphism to $\mathbb{H}^3/\Gamma$. It is easy to construct such manifolds which are non-Haken, e.g., most surgeries on the figure eight knot give such examples [19, Section 5]).

(6) A theorem of Mostow [14] implies that any two hyperbolic structures on a closed 3-manifold are isometric. This is very different from the case of surfaces. It allows one to use geometric invariants (e.g., volume and diameter) as topological invariants.

Theorem A does not apply to the case of interest for the Smith conjecture (that of knot complements), because a knot complement is not closed. However, from the point of view of hyperbolic geometry, manifolds like knot complements are almost closed and similar results hold for them. Here is a form of the theorem which does apply to knot complements.

THEOREM B. *Let M be a compact irreducible 3-manifold with nonempty boundary. Suppose that*
  (i) *every boundary component of M is homeomorphic to T^2;*
  (ii) *any abelian, noncyclic subgroup of $\pi_1(M)$ is conjugate to a subgroup of the image of the fundamental group of some boundary component of M (such subgroups are called peripheral);*
  (iii) *M is not $S^1 \times D^2$, $T^2 \times I$, or the interval bundle over the Klein bottle whose total space is orientable.*

Then $\operatorname{int}(M)$ has a complete hyperbolic structure of finite volume.

Remarks. (1) We do not need the hypothesis that M is Haken because M has nonempty boundary.

(2) Condition (ii) on abelian, noncyclic subgroups is the analogue of the atoroidal condition of Theorem A. In its geometric form it says that any map $\varphi \colon T^2 \to M$ that is injective on the fundamental group is homotopic to a map $\varphi' \colon T^2 \to \partial M$.

V. Uniformization Theorem for Three-Dimensional Manifolds

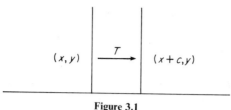

Figure 3.1

(3) The existence of the developing map shows that the hyperbolic structure on int(M) is isometric with $\mathbb{H}^3/\Gamma$ for some torsion-free kleinian group Γ.

(4) The hyperbolic structure on the ends of M is completely understood. Let us examine the case of a surface first. Let E be an end of a complete hyperbolic surface of finite volume. Then $\pi_1(E)$ is cyclic and the representation of $\pi_1(E)$ into $\mathrm{PGL}_2(\mathbf{R})$ coming from the hyperbolic structure sends this group to a group of parabolic transformations. In the upper half-space model with the fixed point at infinity, the generator of $\pi_1(E)$ is sent to a translation T given by $T(x, y) = (x + c, y)$ for some appropriate c (see Fig. 3.1). In fact, in the surface there is a neighborhood of the end, which is isometric to $\{(x, y) | y \geq K\}/(T)$ for some $K > 0$ (see Fig. 3.2). In our case we have M with each end E homeomorphic to $T^2 \times [0, 1)$. Since M is irreducible and is not $S^1 \times D^2$, $\pi_1(E) \to \pi_1(M)$ is an injection. Thus $\pi_1(M) \cong \Gamma$ has a subgroup $\mathbf{Z} \times \mathbf{Z}$ corresponding to the fundamental group of E. As we saw in Remark (4) under Theorem A, this $\mathbf{Z} \times \mathbf{Z}$ must consist of parabolic elements. Since all elements in this subgroup commute, they fix a common point on the boundary. Choose the upper half-space model so that the common fixed point for this subgroup is ∞. Let τ be this parabolic group and let T be $\mathbb{H}^3/\tau$. The hyperbolic manifold T is a covering space of int(M). The function $f : \mathbb{H}^3 \to \mathbf{R}^+$ given by $f(x_1, x_2, x_3) = x_3$ is invariant under the action of τ. The resulting map $f : T \to \mathbf{R}^+$ has $f^{-1}(t) \cong T^2$ for all $t \in \mathbf{R}^+$. Thus T is homeomorphic to $T^2 \times \mathbf{R}^+$. There is an isometric lifting of $E \subset M$ to $\tilde{E} \subset T$. The image $\tilde{E}$ of this lifting must contain one of the ends of T. We claim that it contains the end of T where f approaches $+\infty$. The reason is that Γ is larger than τ (since M is not $T^2 \times I$). Hence Γ either contains hyperbolic elements

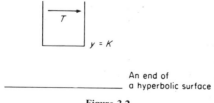

An end of a hyperbolic surface

Figure 3.2

or contains parabolic elements fixing some point other than ∞. Both types of elements identify pairs of points with arbitrarily small x_3-coordinates. Thus no neighborhood of the form $f^{-1}(0, \varepsilon]$ in T can embed in M under the projection mapping $T \to \text{int}(M)$. As a result, E contains a proper subset isometric to

$$E_K = \{(x_1, x_2, x_3) | x_3 \geq K\}/\tau$$

for some $K > 0$. Here τ is a maximal lattice subgroup of $\mathbf{R}^2$, the group of all parabolic motions fixing ∞. Such an end of $\text{int}(M)$ is called a *torus cusp*. A simple computation shows that the volume of any torus cusp is finite. Thus each end of $\text{int}(M)$ has finite volume. It follows immediately that $\text{int}(M)$ itself has finite volume.

Conversely, if N has a complete metric of finite volume, then a theorem of Margulis [11] implies that each end of N is of the form E_K. As a result, N is the interior of a compact manifold M, all of whose boundary components are tori.

(5) From Mostow's theorem and the discussion in Remark (4), it follows that $\text{int}(M)$ has at most one complete hyperbolic metric up to isometry. In particular, the group $\Gamma \subset \text{PGL}_2(\mathbf{C})$ corresponding to the hyperbolic structure on $\text{int}(M)$ is determined up to conjugation in $\text{PGL}_2(\mathbf{C})$ by the topology of M^3.

We now have referred to Mostow's theorem twice. Let us give a precise statement of it.

THEOREM (Mostow's Rigidity Theorem [14]). *Let M and M' have complete hyperbolic metrices of finite volume and dimension greater than two and let $\varphi: M \to M'$ be a homotopy equivalence. There is a unique isometry $I_\varphi: M \to M'$ homotopic to φ.*

COROLLARY 1. *Given M and M' as above and an isomorphism $\varphi_*: \pi_1(M) \to \pi_1(M')$, there is a unique isometry $I_\varphi: M \to M'$, so that I_φ induces φ_* up to inner automorphism.*

Proof. Since the universal covers of M and M' are contractible, any $\varphi_*: \pi_1(M) \to \pi_1(M')$ is realized by a homotopy equivalence $\varphi: M \to M'$. Thus there is an isometry $I_\varphi: M \to M'$ homotopic to φ and, hence, agreeing with φ_* up to inner automorphism. If we have isometries that induce the same homomorphism on the fundamental group, up to inner automorphism, then the isometries are homotopic and, by Mostow's theorem, equal. ∎

COROLLARY 2. *Let M^3 be a complete hyperbolic manifold of finite volume. Denote by $\text{iso}(M)$ the group of self-isometries of M and denote by $\text{out}(\pi_1(M))$*

V. Uniformization Theorem for Three-Dimensional Manifolds

the group of outer automorphisms of $\pi_1(M)$. The natural map $\mathrm{iso}(M) \to \mathrm{out}(\pi_1(M))$ is a group isomorphism.

COROLLARY 3. *Let M be as above.* $\mathrm{out}(\pi_1(M))$ *is a finite group.*

Proof. Since M is complete with finite volume, $\mathrm{iso}(M)$ is a compact Lie group. Since $\mathrm{iso}(M) \cong \mathrm{out}(\pi_1(M))$, it is discrete. Thus $\mathrm{iso}(M)$ is a finite group. ∎

CONJECTURE. *Let* $g: M^3 \to M^3$ *be a periodic, orientation-preserving diffeomorphism of a complete hyperbolic manifold of finite volume to itself. Then the fixed locus of* g, F_g, *is isotopic to a union of geodesics.*

In the case of S^3, it is easy to see that any two semifree actions of a finite cyclic group are equivariantly homotopy equivalent. In the hyperbolic case, Mostow's theorem implies that any $\mathbf{Z}/n\mathbf{Z}$-action is equivalently homotopy equivalent[1] to an action by a group of isometries. This latter action, of course, has a union of geodesics as fixed points for the generator. Thus the conjecture can be reformulated.

CONJECTURE. *There is only one differentiable action (up to isotopy) contained in a class of equivariantly homotopy equivalent actions on a complete hyperbolic manifold of finite volume.*

Unfortunately, the geometric methods used to prove the Smith conjecture do not prove these conjectures. There is hope, however, that eventually other geometric methods will prove them.[2]

4. Hierarchies and Pared Manifolds

Theorems A and B are proved by induction by more or less analogous arguments. The idea is to cut the manifold in question open repeatedly along certain nice surfaces getting at each stage simpler and simpler manifolds. By a theorem of Waldhausen, this process eventually cuts the manifold into a disjoint union of balls. Such a collection of cuts is called a *hierarchy* for the manifold. In this section we shall study this process of cutting a

---

[1] Both maps and both homotopies to the identity are equivalent.

[2] Thurston now claims to have proved these conjectures and much more. He claims that any action of a finite group on a compact, irreducible manifold, with some element having one-dimensional fixed point set, can be decomposed into isometric actions on manifolds with homogeneous geometries. In particular, this completes the classification of finite groups acting on $\mathbf{R}^3$.

manifold up in some detail with particular attention to the image of the boundary.

Let M be a compact, connected, irreducible 3-manifold with nonempty boundary. Suppose that M is not homeomorphic to the 3-ball. By irreducibility, no boundary component of M can be a 2-sphere. Thus $H_1(\partial M) \neq 0$. By Poincaré duality and Lefschetz duality, it follows that $H_2(M, \partial M) \neq 0$. As a result there are properly embedded, connected surfaces $(X, \partial X) \subset (M, \partial M)$ representing nontrivial homology classes.

DEFINITION 4.1. Let $(X, \partial X) \subset (M, \partial M)$ be a properly embedded surface. A *compressing disk* for X is an embedding $i: (D^2, \partial D^2) \hookrightarrow (M, X)$ such that

(a) $X \cap i(D^2) = i(\partial D^2)$ and
(b) $i(\partial D^2)$ is a nontrivial loop on X.

A *boundary compressing disk* for X is an embedding

$$i: (D^2, A, B) \hookrightarrow (M, \partial M, X),$$

where A and B are arcs on ∂D^2 with disjoint interiors such that

(a) $A \cup B = \partial D^2$,
(b) $i(D^2) \cap X = B$, and
(c) i is not isotopic, relative to ∂D^2, to an embedding that lies in $\partial M \cup X$ and that meets ∂M and X each in a disk (see Fig. 4.1).

A *compressing annulus* for X is an embedding

$$i: (S^1 \times I, S^1 \times \{0\}, S^1 \times \{1\}) \hookrightarrow (M, X, \partial M)$$

such that

(a) i_* is an injection on π_1,
(b) $i(S^1 \times I) \cap X = i(S^1 \times \{0\})$, and
(c) $i(S^1 \times \{0\})$ is not isotopic in X to ∂X.

LEMMA 4.2. *Let M be a compact, irreducible 3-manifold with $\partial M \neq \emptyset$. Suppose that M is not homeomorphic to B^3 and that $\partial M \subset M$ is incompressible.*

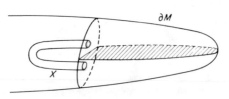

Figure 4.1

V. Uniformization Theorem for Three-Dimensional Manifolds

Let $(X, \partial X) \subset (M, \partial M)$ be a connected surface representing a nontrivial homology class. Suppose that among all such surfaces X has maximal Euler characteristic and minimal genus given its Euler characteristic. Then X has no compressing disks, no boundary compressing disks, and no compressing annuli.

The proof of this lemma is a straightforward cut and paste argument (see, for example, [6]).

DEFINITION 4.3. Let $(X, \partial X) \subset (M, \partial M)$. We say that X is a *superincompressible surface* if

(a) $X \cong D^2$ and ∂X is homotopically nontrivial in ∂M, or
(b) X is a connected, nonsimply connected surface without compressing disks, boundary compressing disks, or compressing annuli.

THEOREM 4.4 Let M be a compact, irreducible 3-manifold that is not homeomorphic to the 3-ball. If $\partial M \neq \emptyset$, then M has a superincompressible surface.

Proof. If $\partial M \neq \emptyset$ and if M is irreducible and not homeomorphic to the 3-ball, then there are surfaces $(X, \partial X) \subset (M, \partial M)$ representing nontrivial homology classes. If ∂M is incompressible, then by Lemma 4.2 a homologically nontrivial, connected surface with maximal Euler characteristic and minimal genus is superincompressible.

If ∂M is compressible, then by Dehn's lemma there is a nontrivial disk $(D^2, \partial D^2) \subset (M, \partial M)$. This is a superincompressible surface. ∎

There are homotopy conditions that are equivalent to the existence of compressing disks and annuli.

PROPOSITION 4.5. Let $(X, \partial X) \subset (M, \partial M)$ have no compressing disks or compressing annuli. Then $\pi_1(X) \to \pi_1(M)$ is an injection. If a loop $\alpha: S^1 \to X$ is freely homotopic in M to ∂M, then α is freely homotopic in X to ∂X.

These statements are proved for their geometric versions by standard techniques in 3-manifold topology [6], viz., Dehn's lemma, the loop theorem, and the annulus theorem.

DEFINITION. A compact, orientable, irreducible 3-manifold M is *Haken* if there is an orientable embedded surface $(X, \partial X) \subset (M, \partial M)$ that is superincompressible. Notice that if $\partial M \neq \emptyset$ and M is not a 3-ball, then M is Haken.

Let M be a Haken manifold. Cut it open along a superincompressible surface. The result is a union of 3-balls and Haken manifolds. We can continue the process either indefinitely or until we obtain a disjoint union of 3-balls.

THEOREM 4.6 (Waldhausen [21]). *Let M be a compact, irreducible, Haken 3-manifold. Any process of cutting M open along superincompressible surfaces terminates after a finite number of steps*

$$M = M_0, M_1, \ldots, M_n \cong \amalg \text{ 3-balls}.$$

Such a sequence of 3-manifolds, each obtained from its predecessor by cutting along superincompressible surfaces, is called a *hierarchy* for M.

PROPOSITION 4.7. *If M is an atoroidal Haken manifold and if*

$$M = M_0, M_1, \ldots, M_n$$

is a hierarchy for M, then each M_i is a compact atoroidal manifold.

Proof. Suppose that $S^2 \subset M_{i+1}$ and that in M_i it bounds a 3-ball. The cutting surface $X_i \subset M_i$ cannot meet the S^2 since the S^2 continues to exist in M_{i+1}. Thus it must either be contained in the interior of the 3-ball bounded by the S^2 or must miss that 3-ball entirely. Since X_i is superincompressible, it is not contained in the interior of a 3-ball in M_i. This implies that the $S^2 \subset M_i$ bounds the same 3-ball. From this, one proves directly by induction that the M_i are all irreducible.

The inclusion $M_i \subset M_{i-1}$ induces an injection on the fundamental group of each component of M_i. Thus, if there is a component of M_i which has a noncyclic abelian subgroup of π_1, the same is true for M_{i-1}. Again, one sees from this that when M is atoroidal all the M_i are. ∎

This result is used to put hyperbolic structures on the M_i, starting at M_n and working backwards to $M_0 = M$. In order to study the case when M has boundary tori, we must keep track of ∂M as we move through the hierarchy. To this end we provide the following definition.

DEFINITION 4.8. A *pared manifold* is a pair (M, P), where

(a) M is a compact, irreducible 3-manifold and
(b) $P \subset \partial M$ is a union of incompressible annuli and tori in M,

such that

(i) every abelian, noncyclic subgroup of $\pi_1(M)$ is peripheral with respect of P (i.e., conjugate to a subgroup of the fundamental group of a component of P) and

V. Uniformization Theorem for Three-Dimensional Manifolds

(ii) every map $\varphi: (S^1 \times I, S^1 \times \partial I) \to (M, P)$ that is injective on the fundamental groups deforms, as a map of pairs, into P.

P is called the *parabolic locus* of the pared manifold (M, P). We denote by $\partial_0 M$ the surface $\partial M - P$.

Notice that a compact manifold M^3 is irreducible and atoroidal if and only if $(M, \emptyset)$ is a pared manifold.

Let (M', P') be a Haken pared manifold.[3] Choose a superincompressible surface Z to begin the cutting process on M'. Deform this surface by isotopy until it satisfies the following condition.

CONDITION 4.9. (a) $\partial Z \cap \partial P'$ is transverse,
(b) $\#(\partial Z \cap \partial P')$ is minimal among those of surfaces isotopic to Z, and
(c) if a component of ∂Z is isotopic to a loop in P', then it already lies in P'.

Let M be M' cut open along Z. The image of P' in ∂M is a disjoint union of annuli, tori, and disks. Let $P \subset \partial M$ be the union of the annuli and tori. This produces a new pair (M, P). It is a straightforward argument using 3-manifold techniques to show that the new pair (M, P) is a pared manifold. In light of this we define a *hierarchy* for a Haken, pared manifold to be

$$(M, P) = (M_0, P_0), (M_1, P_1), \ldots, (M_k, P_k) \cong \sqcup \text{ (3-balls, } \emptyset\text{)},$$

where each (M_i, P_i) is obtained by cutting along a superincompressible surface that satisfies Condition 4.9. By Waldhausen's theorem, any procedure of cutting along superincompressible surfaces must terminate after a finite number of steps and yield a hierarchy for the pared manifold.

There is a basic topological lemma about pared manifolds that we shall need later. Since it is a purely topological result, we include it at this point.

LEMMA 4.10. *If (M_0, P_0) is a connected pared manifold, then*

(i) *M_0 is homeomorphic to the 3-ball and $P_0 = \emptyset$,*
(ii) *M_0 is homeomorphic to a solid torus and $P_0 = \emptyset$ or P_0 is an annulus,*
(iii) *M_0 is homeomorphic to $T^2 \times I$ with P_0 being $T^2 \times \{1\}$, or*
(iv) *every component of $\partial_0 M_0$ has negative Euler characteristic.*

Proof. If a component of $\partial_0 M_0$ has nonnegative Euler characteristic, then the component must be homeomorphic to S^2, D^2, $S^1 \times I$, or T^2. No component of $\partial_0 M_0$ can be a 2-disk, since P_0 is incompressible in M. If a component of $\partial_0 M_0$ is S^2, then, since M_0 is irreducible, M_0 is a 3-ball and we

---
[3] This means that M is a Haken manifold.

are in case (i). If a component of $\partial_0 M_0$ is an annulus, then that annulus must be homotopic into P_0 relative to its boundary. This implies that M_0 is a solid torus and P_0 is an annulus. If a component of $\partial_0 M_0$ is a torus, then either that torus is compressible, in which case M_0 is a solid torus and $P_0 = \emptyset$, or that torus is homotopic into P_0, in which case M_0 is homeomorphic to $T^2 \times I$ with P_0 being one end. ∎

5. Statement of the Main Theorem—The General Case

In this section we define what it means for a pared manifold (M, P) to have a hyperbolic structure. Before introducing the complexity of noncompact manifolds let us consider the case of a compact manifold M, possibly with boundary.

By a *convex hyperbolic structure on M* we mean a metric of constant sectional curvature -1 in which the boundary is *locally convex*. (Recall that the boundary is locally convex if any two points on ∂M that are sufficiently close together can be joined by a short geodesic arc in M.) If M happens to be closed, then such a structure is just an ordinary hyperbolic structure.

The theorem that generalizes Theorem A and is proved by induction on a hierarchy is Theorem A'.

THEOREM A'. *Let M be a compact, irreducible, atoroidal, Haken manifold. Then M admits a convex hyperbolic structure.*

(Notice that if $\partial M \neq \emptyset$, then M is automatically Haken.)

Let (M, P) be a pared manifold. By a *convex hyperbolic structure on (M, P)* we mean a metric of finite volume with all sectional curvatures equal to minus one on $M - P$ so that

(i) $\partial_0 M$ is locally convex and
(ii) each geodesic ray can be extended indefinitely or until it meets $\partial_0 M$.

Condition (ii) is a form of completeness. In fact, if $P = \partial M$, then (ii) says exactly that the metric on $\text{int}(M)$ is complete.

The theorem that generalizes Theorem B and is proved by induction on a hierarchy for pared manifolds is Theorem B'.

THEOREM B'. *Let (M, P) be a Haken pared manifold. Then (M, P) admits a convex hyperbolic metric of finite volume.*

V. Uniformization Theorem for Three-Dimensional Manifolds

Of course, Theorem B' generalizes Theorems A, A', and B. Throughout the rest of this chapter we shall be concerned only with Theorem B'. Briefly, one takes a hierarchy

$$(M, P) = (M_0, P_0), (M_1, P_1), \ldots, (M_n, P_n) \cong (\amalg \text{ 3-balls}, \emptyset).$$

Clearly, (M_n, P_n) has a convex hyperbolic structure. We prove by downward induction on i that the (M_i, P_i) also have such structures. The proof of the inductive step occupies the rest of this chapter.

6. Convex Hyperbolic Structures of Finite Volume

In this section we analyze further the convex hyperbolic structures introduced in the last section. Let (M, P) be a pared manifold with a convex hyperbolic structure. Let $M_0 = M - P$. As in Section 2, we define the developing map on the universal cover $\tilde{M}_0$ on M_0 $d: \tilde{M}_0 \to \mathbb{H}^3$. This map is well defined up to composition with elements in $\mathrm{PGL}_2(\mathbb{C})$. It is an isometry onto its image which is a closed convex subset of $\mathbb{H}^3$. (It is *convex* in the sense that any geodesic arc in $\mathbb{H}^3$ joining two points of $d(\tilde{M}_0)$ lies entirely in $d(\tilde{M}_0)$.) The fundamental group $\pi_1(M_0)$ is transferred by d to a subgroup $\Gamma \subset \mathrm{PGL}_2(\mathbb{C})$. The subspace $d(\tilde{M}_0)$ is invariant under Γ and M_0 is isometric with $d(\tilde{M}_0)/\Gamma$.

As we shall see presently, Γ is actually a torsion-free kleinian group. The following lemma is needed to prove that.

LEMMA 6.1. *Let $X \subset \mathbb{H}^3$ be a closed convex set invariant under a subgroup $\Gamma \subset \mathrm{PGL}_2(\mathbb{C})$. Let $\bar{X} \subset B^3$ be the closure of X, and let Ω be $S^2_\infty - S^2_\infty \cap \bar{X}$. Then there is a retraction $r: \mathbb{H}^3 \cup \Omega \to X$ that is Γ-invariant.*

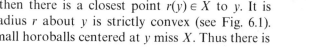

Proof. If $y \in \mathbb{H}^3 - X$, then there is a closest point $r(y) \in X$ to y. It is unique since the ball of radius r about y is strictly convex (see Fig. 6.1). If $y \in \Omega$, then sufficiently small horoballs centered at y miss X. Thus there is

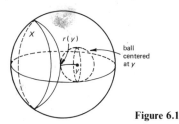

Figure 6.1

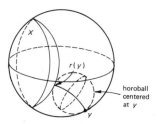

Figure 6.2

a smallest horoball centered at y meeting X. Since the horoball is strictly convex, this intersection is a point, as shown in Fig. 6.2. The definition of the retraction $r: \mathbb{H}^3 \cup \Omega \to X$ is then immediate. If $y \in \mathbb{H}^3$, then $r(y)$ is the closest point of X to y. If $y \in \Omega$, then $r(y)$ is the point of X meeting the smallest horoball centered at y.

It is easy to see that r is a continuous, Γ-invariant retraction. ∎

COROLLARY 6.2. *If Γ acts properly discontinuously on X, then it acts properly discontinuously on $\mathbb{H}^3 \cup \Omega$. In particular, Γ is a kleinian group.*

The proof is immediate.

This proves that we can embed M_0 as a locally convex subset of a complete hyperbolic manifold $N = \mathbb{H}^3/\Gamma$. The inclusion $M_0 \subset N$ induces an isomorphism on fundamental groups. Furthermore, r induces a retraction $N \to M_0$.

If $d(\partial \tilde{M}_0)$ is smooth, then for any $x \in d(\partial \tilde{M}_0)$, $r^{-1}(x)$ is a (completed) geodesic half-ray emanating from x perpendicular to $d(\partial \tilde{M}_0)$. Thus, in this case, $r: N \to M_0$ determines a decomposition of N as

$$M_0 \cup (\partial M_0 \times [0, \infty)) \cong \operatorname{int}(M_0).$$

Furthermore, $(\mathbb{H}^3 \cup \Omega)/\Gamma$ is homeomorphic to $M_0 \cup (\partial M_0 \times [0,1]) \cong M_0$. If $d(\tilde{M}_0)$ is not a smooth submanifold, then for any $\varepsilon > 0$ the ε-neighborhood of $d(\tilde{M}_0)$ will be smooth. It will be equivariantly homeomorphic to $d(\tilde{M}_0)$ for ε sufficiently small. By applying the above discussion to the ε-neighborhood, we see that N is homeomorphic to $\operatorname{int}(M_0)$ in this more general case as well.

This proves that given any convex hyperbolic structure on (M, P) there is an associated complete hyperbolic manifold N so that $(M - P)$ is isometrically embedded in N as a convex subset carrying the fundamental group. It is unique in the sense that, given any other such $(M - P) \subset N'$, there is an isometry from N to N' that is the identity of $(M - P)$. The closure of the complement $N - M_0$ is homeomorphic to $\partial M_0 \times [0, \infty)$.

V. Uniformization Theorem for Three-Dimensional Manifolds

Let us now consider the converse problem—that of constructing convex submanifolds of a given complete manifold. Let Γ be a torsion-free, kleinian group (nonelementary). We define the *Nielsen convex region* $H_\Gamma \subset \mathbb{H}^3$ to be the convex hull of the limit set L_Γ. That is to say, H_Γ is the smallest convex set in $\mathbb{H}^3$ that contains all geodesics with end points in L_Γ. It is also described as the intersection of all half-spaces in $\mathbb{H}^3$ whose closures contain L_Γ. Since L_Γ is closed and Γ-invariant, so is $H_\Gamma \subset \mathbb{H}^3$. The quotient $H_\Gamma/\Gamma \subset N = \mathbb{H}^3/\Gamma$ is called the *convex core* of N. It is denoted $C(N)$. It is a locally convex set. If L_Γ is contained in a geometric circle, e.g., if Γ is a fuchsian group, then $C(N)$ is a totally geodesic two-dimensional submanifold of N. Otherwise, $C(N)$ is a locally convex 3-manifold inside N. Clearly, $\pi_1(C(N)) \xrightarrow{\cong} \pi_1(N)$.

LEMMA 6.3. *Let N be a complete hyperbolic 3-manifold. The convex core $C(N)$ is the intersection of all connected locally convex submanifolds $M \subset N$ with $\pi_1(M) \to \pi_1(N)$ being onto.*

Proof. Since $C(N)$ is such a manifold, the lemma reduces to the assertion that any such manifold M contains $C(N)$. Let M be such a manifold and choose an isometry $N \cong \mathbb{H}^3/\Gamma$ for some kleinian group Γ. The covering $\tilde{M} \subset \mathbb{H}^3$ is convex and Γ-invariant. Let $X \subset B^3$ be its closure. Let $\Omega_M = S^2_\infty - S^2_\infty \cap X$. By Lemma 6.1, $\Omega_M \subset \Omega_\Gamma$. Thus X contains L_Γ. Since $\tilde{M}$ is convex, it contains H_Γ. Thus M contains $C(N) = H_\Gamma/\Gamma$. ∎

Let N be a complete hyperbolic manifold with holonomy representation $\Gamma \subset \text{PGL}_2(\mathbb{C})$. Let $\overline{N} = (\mathbb{H}^3 \cup \Omega_\Gamma)/\Gamma$. It is a manifold with boundary whose interior is N. By Lemma 6.1, we have a retraction $r: \overline{N} \to C(N)$. If L_Γ is not contained in a geometric circle, then the discussion follows Corollary 6.2 implies that $\overline{N}$ is homeomorphic to $C(N) \cup (\partial C(N) \times [0, 1])$. In particular, when Γ is not conjugate to a subgroup of $\text{PGL}_2(\mathbb{R})$, $\partial C(N)$ is identified with Ω_Γ/Γ by a homeomorphism well defined up to isotopy. In fact, the same arguments show that if $M \subset N$ is any locally convex submanifold that carries the fundamental group, then M and $\overline{N}$ are homeomorphic.

DEFINITION. A torsion-free kleinian group Γ is *geometrically finite* if there is an $\varepsilon > 0$ so that the ε-neighborhood of H_Γ/Γ in $\mathbb{H}^3/\Gamma$ has finite volume. A complete hyperbolic manifold N is *geometrically finite* if the image of its holonomy representation is a geometrically finite kleinian group.

THEOREM 6.4. (Ahlfors Finiteness Theorem [1]). *Let N be a complete hyperbolic 3-manifold with finitely generated fundamental group. Then the area of $\partial C(N)$ is finite.*

Remark. Topologically, this implies that $\partial C(N)$ has finitely many components, each with a finitely generated fundamental group. The deeper part of the theorem is that each component of $\partial C(N)$ has finite area.

It is a consequence of the Ahlfors finiteness theorem that in the case when L_Γ is not contained in a geometric circle, Γ is geometrically finite if and only if H_Γ/Γ has finite volume.

Geometrically finite, complete hyperbolic manifolds are exactly those that contain a convex three-dimensional submanifold of finite volume which carries the fundamental group. The object of interest for us is not so much any particular convex manifold M inside N, but rather the kleinian group $\pi_1(N) = \Gamma$. Thus we shall feel free to replace any convex submanifold of finite volume in N by any other carrying the fundamental group. In practice we usually work with the smallest one — $C(N)$.

We need to understand the ends of the convex core $C(N)$ of a geometrically finite manifold N. Suppose that $\rho: \pi_1(N) \xrightarrow{\sim} \Gamma \subset \mathrm{PGL}_2(\mathbf{C})$ is the holonomy representation. There is a continuous function $\mu: N \to \mathbf{R}^+$ that assigns to x the length of the shortest nonnull-homotopic loop in N through x. The value $\mu(x)$ is $\geq \varepsilon$ precisely if in N there is an embedded copy of the open $(\varepsilon/2)$-ball centered at x. We denote by $N_{[\varepsilon, \infty)}$ the subspace $\mu^{-1}([\varepsilon, \infty))$ and by $N_{(0, \varepsilon]}$ the subspace $\mu^{-1}((0, \varepsilon])$. Let $\mathcal{N}_\delta(C(N))$ denote the δ-neighborhood of $C(N)$ inside N.

There are results of Margulis that give us complete information about $N_{(0, \varepsilon]}$ for ε sufficiently small. Let $\Gamma_\varepsilon(z)$ be the subgroup of Γ generated by the elements that move $z \in \mathbb{H}^3$ a distance at most ε. Clearly, $\Gamma_\varepsilon(z) \neq \{e\}$ if and only if z covers a point $x \in N_{(0, \varepsilon]}$. Margulis's result [11] implies that there is $\varepsilon_0 > 0$, independent of z and N, so that $\Gamma_{\varepsilon_0}(z)$ is always an abelian subgroup. Thus $\Gamma_{\varepsilon_0}(z)$ is either a group of parabolic translations with a common fixed point, or a cyclic group generated by a hyperbolic element. Translating this into geometry, we find that $N_{(0, \varepsilon_0]}$ is a disjoint union of pieces isometric with one of the following:

(i) a torus cusp or equivalently a $(\mathbf{Z} \times \mathbf{Z})$-cusp (a horoball modulo a rank-two subgroup of parabolic motions),

(ii) a $\mathbf{Z}$-cusp (a horoball modulo a rank-one subgroup of parabolic motions),

(iii) $\mathcal{N}_\delta(g)/\langle\alpha\rangle$, where g is a geodesic in $\mathbb{H}^3$, $\mathcal{N}_\delta(g)$ is its δ-neighborhood, and α is a hyperbolic element with axis g.

Of course, in (iii) α must move points along the axis g a distance less than or equal to ε_0. The constant δ depends on ε_0 and α. For example, if α moves points on g exactly a distance ε_0, then $\delta = 0$.

V. Uniformization Theorem for Three-Dimensional Manifolds

We wish to study the intersection of the convex core $C(N)$ with $N_{(0,\varepsilon]}$ in the case in which N is geometrically finite.

LEMMA 6.5. *Let N be a geometrically finite complete hyperbolic 3-manifold. Let $\varepsilon > 0$, and let $C(N)_{[\varepsilon,\infty)} = C(N) \cap N_{[\varepsilon,\infty)}$. Then $C(N)_{[\varepsilon,\infty)}$ is compact.*

Proof. For any $\varepsilon > 0$, the ε-neighborhood of $C(N)$ has finite volume. Consider finite sets $\{x_1, \ldots, x_k\}$ of $C(N)_{[\varepsilon,\infty)}$ with the property that

$$d(x_i, x_j) > \varepsilon$$

for all $i \neq j$. Associated to any such set is a collection of k disjoint balls of radius $\varepsilon/2$ in the $(\varepsilon/2)$-neighborhood of $C(N)$. This neighborhood has finite volume. Hence, there is an upper bound for the cardinality of such a set. If $\{x_1, \ldots, x_k\}$ is a maximal such set, then the k closed balls of radius ε centered at the $\{x_i\}$ cover $C(N)_{[\varepsilon,\infty)}$. Thus $C(N)_{[\varepsilon,\infty)}$ is compact for all $\varepsilon > 0$. ∎

THEOREM 6.6. *Let N be a geometrically finite manifold. Let $0 < \varepsilon \leq \varepsilon_0$. Then $N_{(0,\varepsilon]}$ has only finitely many components. If ε is sufficiently small, then these components are all cusps. The convex core $C(N)$ contains a neighborhood of infinity of each $\mathbf{Z} \times \mathbf{Z}$-cusp and meets a neighborhood of infinity of each $\mathbf{Z}$-cusp in a set isometric to*

$$\{(x_1, x_2, x_3) \in \mathbb{H}^3 \mid x_3 \geq K \text{ and } A_1 \leq x_2 \leq A_2\}/(\gamma),$$

where γ is the parabolic motion that is translation by 1 in the x_1-direction and $K, A_1,$ and A_2 are suitable real constants depending on N.

We break the proof of Theorem 6.6 into several steps.

LEMMA 6.7. *Let N be a complete hyperbolic manifold and let $E \subset N_{(0,\varepsilon_0]}$ be a torus cusp. The convex core $C(N)$ contains a neighborhood of infinity of E.*

Proof. Arrange the upper half-space model of $\mathbb{H}^3$ so that $\pi_1(E)$ fixes ∞. Since $\pi_1(N)$ is not elementary, there is at least one other point in $L_{\pi_1(N)}$. We choose coordinates so that this additional point is $\mathbf{0} = (0,0,0)$. The translates of $\mathbf{0}$ under $\pi_1(E)$ form a lattice in the plane $\{x_3 = 0\}$, which is contained in $L_{\pi_1(N)}$. There is an upper bound K to the radius of any circle in the plane $\{x_3 = 0\}$, which contains no point of the lattice in its interior. It follows immediately that the convex hull of $L_{\pi_1(N)}$ contains the horoball $\{(x_1, x_2, x_3) \in \mathbb{H}^3 \mid x_3 \geq K\}$. ∎

LEMMA 6.8. *Let N be a complete hyperbolic manifold and let $E \subset N_{(0,\varepsilon_0]}$ be a $\mathbf{Z}$-cusp. There are constants depending on E, $0 < \varepsilon \leq \varepsilon_0$, $L > 0$, and $-\infty \leq A_1 \leq A_2 \leq +\infty$ so that $C(N) \cap E \cap N_{(0,\varepsilon]}$ is isometric to*

$$\{(x_1, x_2, x_3) \in \mathbb{H}^3 \mid x_3 \geq L \text{ and } A_1 \leq x_2 \leq A_2\}/\langle \gamma \rangle,$$

where γ is translation by 1 in the x_1-direction.

Proof. Arrange the upper half-space model for $\mathbb{H}^3$ so that $\pi_1(E) \subset \pi_1(N) \subset \mathrm{PGL}_2(\mathbf{C})$ fixes $+\infty$, and so that its generator is γ. Suppose that $E \subset N$ is the image of the horoball $B = \{(x_1, x_2, x_3) \mid x_3 \geq K\}$. (Of course, E is isometric to $B/\langle \gamma \rangle$.) The limit set $L_{\pi_1(N)}$ contains points in $\mathbf{C}$. The subset $L_{\pi_1(N)} \cap \mathbf{C} \subset \mathbf{C}$ is invariant under γ. Choose a minimal strip $S \subset \mathbf{C}$, $S = \{(x_1, x_2) \mid A_1 \leq x_2 \leq A_2\}$ such that $L_{\pi_1(N)} \cap \mathbf{C}$ is contained in S. (We allow the possibility that either or both A_1 and A_2 are nonfinite.) Clearly, $H_{\pi_1(N)}$ is contained in the part of $\mathbb{H}^3$ above S. If A_1 (say) is finite, then the line $\{x_2 = A_1\}$ contains points of $L_{\pi_1(N)}$. The largest possible gap between these points is of length 1, since $L_{\pi_1(N)}$ is periodic under translation by 1 in the x_1-direction. Thus any circle in the plane $\{x_3 = 0\}$ that contains no points of $L_{\pi_1(N)}$ in its interior can meet the line $\{x_2 = A_1\}$ in an arc of length at most 1. This proves that the part of the vertical plane $\{(x_1, x_2, x_3) \mid x_2 = A_1\}$ at height above $1/2$ is contained in $\partial H_{\pi_1(N)}$.

Next we wish to show that for some $L > 0$ the set

$$\{(x_1, x_2, x_3) \mid x_3 \geq L, A_1 \leq x_2 \leq A_2\}$$

is contained in $H_{\pi_1(N)}$.

If a point (x_1, x_2, x_3) with $A_1 \leq x_2 \leq A_2$ is not in $H_{\pi_1(N)}$, then there is a circle $c \subset \mathbf{C}$ that contains no points of $L_{\pi_1(N)}$, that contains (x_1, x_2) in its interior, and so that $d((x_1, x_2), C) > x_3$. Since C intersects the lines $x_2 = A_1$ and $x_2 = A_2$ in arcs of at most length 1, if $x_3 > 1$, then the center of C is contained in the strip $A_1 \leq x_2 \leq A_2$ (see Fig. 6.3). The translates of C by powers of γ are also circles that miss $L_{\pi_1(N)}$. If the radius of C (which is at least x_3) is sufficiently large, then the union of the translates of C under powers of γ contains a substrip of $A_1 \leq x_2 \leq A_2$ of width at least $\frac{1}{2}$(radius of C) $\geq \frac{1}{2} x_3$ containing no points of $L_{\pi_1(N)}$, as shown in Fig. 6.4.

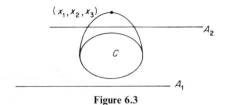

Figure 6.3

V. Uniformization Theorem for Three-Dimensional Manifolds

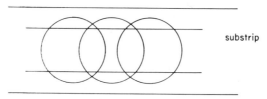

Figure 6.4

This shows that if Lemma 6.8 is false then one of A_1 and A_2 must be non-finite and inside $\{(x_1, x_2) | A_1 \leq x_2 \leq A_2\}$ there are arbitrarily wide substrips $\{\alpha_1 \leq x_2 \leq \alpha_2\}$ containing no points of $L_{\pi_1(N)}$.

Let $S' \subset S$ be a maximal substrip of S containing no points of $L_{\pi_1(N)}$ in its interior: $S' = \{(x_1, x_2) | A'_1 \leq x_2 \leq A'_2\}$. Suppose that the width of S' is greater than $2K$. Let P be the vertical plane in $\mathbb{H}^3$ above the center line of S'. Then $P \cap H_{\pi_1(N)} \subset B$. The space $P \cap H_{\pi_1(N)}/(\gamma) \subset E \subset N$ is topologically a half infinite, totally geodesic annulus that goes off to infinity in E. Its boundary is a simple closed curve on $\partial C(N)$.

Now consider the annuli produced from all maximal substrips $S' \subset \mathbf{C}$ of width $\geq 2K$ whose interiors miss $L_{\pi_1(N)}$. This is a family of disjoint annuli in $C(N) \cap E \subset N$ with boundaries that are circles in $\partial C(N)$. We claim that the boundaries of no two of these annuli can be parallel on ∂C. To see this consider the vertical projection from $H_{\pi_1(N)}$ to the plane $\{x_3 = 0\}$. The preimage of any point (x_1, x_2) is either empty, a half-open interval going off to infinity or the full geodesic $\{(x_1, x_2, t) | 0 < t < \infty\}$. The only points whose preimages are the third type are points of $L_{\pi_1(N)}$. For such a geodesic to meet $\partial H_{\pi_1(N)}$ it must be contained in $\partial H_{\pi_1(N)}$ and the corresponding point of $L_{\pi_1(N)}$ must be an extremal point of the set $L_{\pi_1(N)} - \{\infty\}$ in the plane $\{x_3 = 0\}$. In particular, the corresponding point of $L_{\pi_1(N)}$ must be contained in the boundary of S. This proves that if we restrict the vertical projection to $\partial H_{\pi_1(N)}$, then the image is contained in $S - (\text{int } S) \cap L_{\pi_1(N)}$.

Let S'_1 and S'_2 be substrips of S that have width at least $2K$ and that contain no points of $L_{\pi_1(N)}$. Let C_1 and C_2 be their center lines. Let γ_1 and γ_2 be the arcs in $\partial H_{\pi_1(N)}$ lying above C_1 and C_2. The quotients of γ_1 and γ_2 by $\pi_1(E)$ form simple closed curves embedded in $\partial C(N)$. Suppose that these simple closed curves are freely homotopic in $\partial C(N)$. If we lift this homotopy to $\partial H_{\pi_1(N)}$, we get a proper homotopy from γ_1 to a translate of γ_2 by some element in $\pi_1(N)$ fixing infinity. But the only elements in $\pi_1(N)$ that fix infinity are $\pi_1(E)$. Thus we get a proper homotopy from γ_1 to γ_2. If we project this homotopy into the plane $\{x_3 = 0\}$, we obtain a proper homotopy from C_1 to C_2 which misses all the points of $L_{\pi_1(N)}$ contained in the interior of S. This implies that C_1 and C_2 bound a substrip of S containing no points of $L_{\pi_1(N)}$. Hence S'_1 and S'_2 belong to the same maximal substrip of S containing no points of

$L_{\pi_1(N)}$. This proves that if γ_1 and γ_2 come from different maximal strips, then they are not freely homotopic in $\partial C(N)$.

By the Ahlfors finiteness theorem, $\partial C(N)$ has finite topological type. Hence, it can have only finitely many nonparallel simple closed curves. This implies that there are only finitely many maximal substrips of $\mathbf{C} - L_{\pi_1(N)}$ of width $\geq 2K$. In particular (since each substrip must be of finite width), there is an upper bound to the width of any substrip of $\mathbf{C} - L_{\pi_1(N)}$. Hence

$$\{(x_1, x_2, x_3) | x_3 \geq L\} \cap H_{\pi_1(N)} = \{(x_1, x_2, x_3) | x_3 \geq L \text{ and } A_1 \leq x_2 \leq A_2\}$$

for some $L \geq K$. Let ε be the distance from $(0, 0, L)$ to $(1, 0, L)$. Then

$$C(N) \cap E \cap N_{(0,\varepsilon]}$$

is isometric to

$$\{(x_1, x_2, x_3) \in \mathbb{H}^3 | x_3 \geq L \text{ and } A_1 \leq x_2 \leq A_2\}/(\gamma). \blacksquare$$

DEFINITION. Let $E \subset N_{(0,\varepsilon_0]}$ be a **Z**-cusp. Choose coordinates so that $\pi_1(E)$ fixes ∞. We say that E is a *finite* **Z***-cusp* if $L_{\pi_1(N)} - \{\infty\}$ is contained in a strip of finite width $S \subset \mathbf{C}$.

LEMMA 6.9. *Let ε_1 be the minimum of ε_0 and the hyperbolic distance from $(0, 0, \frac{1}{2})$ to $(1, 0, \frac{1}{2})$. If E is a finite **Z**-cusp in N, then inside $\partial C(N)$ there is a totally geodesic, properly embedded, half-infinite annulus that goes to infinity in E. Its boundary is a circle in $\partial C(N) \cap C(N)_{[\varepsilon_1, \infty)}$.*

Proof. E is isometric to $\{(x_1, x_2, x_3) | x_3 \geq K\}/(\gamma)$, where γ is translation by 1 in the x_1-direction and where the hyperbolic distance from $(0, 0, K)$ to $(1, 0, K)$ is ε_0.

Since E is a finite **Z**-cusp, $L_{\pi_1(N)} - \{\infty\}$ is contained in some minimal strip $\{(x_1, x_2) | A_1 \leq x_2 \leq A_2\}$ with A_1 and A_2 finite. The geodesic plane

$$T = \{(x_1, x_2, x_3) | x_2 = A_1 \text{ and } x_3 \geq \max(K, \tfrac{1}{2})\}$$

is contained in $\partial H_{\pi_1(N)}$. Since E embeds in N, we see that $T/(\gamma)$ embeds in $E \cap \partial C(N)$. It is a properly embedded, totally geodesic, half-open annulus whose boundary is contained in $C(N)_{[\varepsilon_1, \infty)}$. (In fact, its boundary is contained in $C(N)_{[\varepsilon_1, \infty)} \cap C(N)_{(0,\varepsilon_1]}$.) $\blacksquare$

Proof of Theorem 6.6. By hypothesis, $C(N)_{[\varepsilon_0, \infty)}$ is compact. Since $C(N)$ contains a neighborhood of infinity of each torus cusp of $N_{(0,\varepsilon_0]}$ and contains all closed geodesics, there is a boundary component (a torus) in $C(N)_{[\varepsilon_0, \infty)}$ for each torus cusp and each closed geodesic of length $\leq \varepsilon_0$. Hence, there are only finitely many components of $N_{(0,\varepsilon_0]}$ which are torus cusps or neighborhoods of geodesics of length $< \varepsilon_0$.

V. Uniformization Theorem for Three-Dimensional Manifolds

Next we show that all the Z-cusps of $N_{(0,\varepsilon_0]}$ are finite Z-cusps. Let $E \subset N_{(0,\varepsilon_0]}$ be a Z-cusp. Then for ε sufficiently small $C(N) \cap E \cap N_{(0,\varepsilon]}$ is isometric to

$$\{(x_1, x_2, x_3) | x_3 \geq K \text{ and } A_1 \leq x_2 \leq A_2\}/(\gamma).$$

In particular,

$$C(N) \cap E \cap N_{(0,\varepsilon]} \cap N_{[\varepsilon,\infty)}$$
$$= \{(x_1, x_2, x_3) | x_3 = K \text{ and } A_1 \leq x_2 \leq A_2)\}/(\gamma).$$

Since E and $N_{(0,\varepsilon]}$ are closed, the above intersection is a closed subset of $C(N) \cap N_{[\varepsilon,\infty)} = C(N)_{[\varepsilon,\infty)}$. By hypothesis, $C(N)_{[\varepsilon,\infty)}$ is compact. Thus, so is this intersection. That implies that A_1 and A_2 are finite, and hence that E is a finite Z-cusp.

By Lemma 6.9, there is $\varepsilon_1 > 0$ so that $\partial C(N) \cap C(N)_{[\varepsilon_1,\infty)}$ contains an embedded loop carrying the class of each Z-cusp. This loop is contained in the Z-cusp. Thus, the loops associated to the various Z-cusps are disjoint. They are not freely homotopic in N since the different Z-cusps of N correspond to distinct conjugacy classes of $\pi_1(N)$.

Since $C(N)_{[\varepsilon_1,\infty)}$ is compact, there can be at most finitely many disjoint, nonparallel, nontrivial, simple closed curves in its boundary. It follows immediately that $N_{(0,\varepsilon_0]}$ has only finitely many Z-cusps.

This proves that $N_{(0,\varepsilon_0]}$ has finitely many components and that all its Z-cusps are finite Z-cusps. Once we know this, Theorem 6.6 follows easily from Lemmas 6.7 and 6.8. ∎

Let N be a geometrically finite manifold. For any $0 < \varepsilon \leq \varepsilon_0$, define $P_N(\varepsilon)$ to be $C(N)_{(0,\varepsilon]} \cap C(N)_{[\varepsilon,\infty)}$. If we choose ε sufficiently small (depending on N), then $P_N(\varepsilon)$ is a disjoint union of tori and compact annuli and $C(N)_{(0,\varepsilon]}$ is homeomorphic to $P_N(\varepsilon) \times [0,\infty)$. Furthermore, for $\varepsilon' < \varepsilon$, $C(N)_{(0,\varepsilon']} \subset C(N)_{(0,\varepsilon]}$ is a smaller product. Thus for all ε sufficiently small,

$$(C(N)_{[\varepsilon,\infty)}, P_N(\varepsilon))$$

are naturally homeomorphic. We denote this topological pair by $(C(N)_0, P_N)$.

COROLLARY 6.10. *$(C(N)_0, P_N)$ is a pared manifold. N is homeomorphic to $C(N)_0 - P_N$.*

Proof. We know that $C(N)_0$ is compact and irreducible and that $P_N \subset \partial C(N_0)$ is a disjoint union of incompressible tori and annuli. We also know that $(C(N)'_0, P_N) \subset (N, N_{(0,\varepsilon]})$ is a homotopy equivalence of pairs. Thus any abelian, noncyclic subgroup of $\pi_1(C(N)_0) = \pi_1(N)$ is a group of parabolic elements. As such, it is conjugate to a subgroup of the fundamental group of a torus component of P_N.

Let $\pi\colon \tilde{N} \to N$ be the covering of N corresponding to the fundamental group of a cusp component E of $N_{(0,\varepsilon_0]}$. Then $\tilde{N}_{(0,\varepsilon_0]}$ is connected and maps isometrically onto E. Consider any cusp of $N_{(0,\varepsilon_0]}$ and a component of its preimage $\tilde{E}$. It is the quotient of a horoball by some (possibly trivial) parabolic group. If this group is nontrivial, then $\tilde{E}$ must meet $\tilde{N}_{(0,\varepsilon_0]}$; that is $\tilde{E} = \tilde{N}_{(0,\varepsilon_0]}$. This shows that π^{-1} (cusps of $N_{(0,\varepsilon_0]}$) is a cusp $\tilde{N}_{(0,\varepsilon_0]}$ together with a disjoint union of horoballs embedded in $\tilde{N}$. In particular, only one component is nonsimply connected. This implies that any annulus $\varphi\colon (S^1 \times I, S^1 \times \partial I) \to (C(N)_0, P_N)$ for which φ_* is an injection is homotopic relative to its boundary into P_N. These are exactly the properties required of a pared manifold. ∎

As a result of this we can reformulate Theorems A′ and B′ as follows:

THEOREM A′. *Let M be a compact, irreducible, atoroidal Haken manifold. There is a geometrically finite, complete hyperbolic manifold N so that $C(N)$ is homeomorphic to M.*

THEOREM B′. *Let (M, P) be a Haken pared manifold. There is a geometrically finite, complete hyperbolic manifold N so that $(C(N)_0, P_N)$ is homeomorphic to (M, P).*

These are the theorems that are proved inductively over a hierarchy.

7. The Gluing Theorem—Statement and First Reduction

For this and the next six sections (Sections 8–13) we adopt the following notation: $(M', \partial M')$ is a connected pared manifold; $(Z, \partial Z) \subset (M', \partial M')$ is a connected superincompressible surface that satisfies Condition 4.9; (M, P) is the pared manifold that results from cutting $(M', \partial M')$ open along Z. The two copies of Z in $\partial_0 M$ are denoted by Z_0 and Z_1. The map

$$\varphi\colon Z_0 \to Z_1$$

is the gluing homeomorphism. The involution $\tau\colon \partial_0 M \to \partial_0 M$ is defined to be φ on Z_0 and φ^{-1} on Z_1. However, note that since M' is oriented, $\varphi\colon Z_0 \to Z_1$ is orientation-reversing when Z_0 and Z_1 are given the induced orientation from M.

Here we shall study the problem of finding a hyperbolic structure for (M, P) that can be "glued together" to form one for $(M', \partial M')$. Of course, the general inductive step involves a more complicated picture in that we must glue along compact surfaces which are not boundary components. Luckily, there is a clever trick that shows that it is unnecessary to analyze in detail the general case, but rather that it suffices to prove the gluing theorem

V. Uniformization Theorem for Three-Dimensional Manifolds

in the case which we are considering here. The next six sections (Sections 8–13) concern the following *gluing theorem*.

THEOREM (Gluing Theorem). *If the pared manifold (M, P) has a convex hyperbolic structure, then $M' = M/\tau$ has a complete hyperbolic structure of finite volume.*

The general case of the gluing theorem and the reduction to the case above will be dealt with in the discussion of orbifolds (Sections 14–16). The theorem involves a kind of doubling process; many facets are chiseled out of the boundary of the 3-manifold, which is viewed as a kaleidoscope.

In order to construct a hyperbolic structure on $(M', \partial M')$, we shall deform the hyperbolic structure on (M, P) until the involution τ becomes an isometry (in an appropriate sense) and then apply the Maskit combination theorem to glue via τ. First, though, we need to analyze the structure of the boundary of the convex core. The following result is contained in [20].

PROPOSITION 7.1. *Let N be a geometrically finite hyperbolic manifold such that $\partial C(N)$ is nonempty. Then every covering space N' of N with finitely generated, nonelementary, fundamental group is also geometrically finite.*

Proof. Let us suppose first that N has no cusps. Then $C(N)$ is compact. Thus there is a number $R > 0$ so that every point $x \in C(N)$ has distance at most R from $\partial C(N)$. Let $\widetilde{C(N)}$ be the universal cover of $C(N)$. It also is true that every point in $\widetilde{C(N)}$ has distance at most R from $\partial \widetilde{C(N)}$. The manifold $\widetilde{C(N)}/\pi_1(N')$ is a convex submanifold of N' that carries the fundamental group. Hence, it contains $C(N')$. It follows that every point of $C(N')$ is at most R from $\partial C(N')$. By the Ahlfors finiteness theorem, $\partial C(N')$ has finite area. Since $\pi_1(N') \subset \pi_1(N)$ has no parabolic elements, $\partial C(N')$ must be compact. Thus $C(N')$ is within distance R of a compact set and is hence compact. This proves that in this case N' is geometrically finite.

Now we turn to the general case. Let ε, $0 < \varepsilon \leq \varepsilon_0$, be given so that all components of $N_{(0,\varepsilon]}$ are cusps. The space $C(N)_{[\varepsilon, \infty)}$ is compact and connected, and $\partial C(N) \cap C(N)_{[\varepsilon, \infty)} \neq \emptyset$. Thus there is $R > 0$ so that every point of $C(N)_{[\varepsilon, \infty)}$ can be joined to a point of $\partial C(N) \cap C(N)_{[\varepsilon, \infty)}$ by a path of length $\leq R$ in $C(N)_{[\varepsilon, \infty)}$. Let $Y \subset N'$ be the preimage of $C(N)_{[\varepsilon, \infty)}$ under the covering $\pi: N' \to N$. Clearly, $Y \subset N'_{[\varepsilon, \infty)}$ and any point of Y is joined to $\pi^{-1}(\partial C(N) \cap C(N)_{[\varepsilon, \infty)})$ by a path of length $\leq R$ in Y. Since $C(N') \subset \pi^{-1}(C(N))$, this means that any point of $C(N') \cap Y$ can be joined to a point of $\partial C(N') \cap Y$ by a path of length $\leq R$. By the Ahlfors finiteness theorem, $\partial C(N')$ has finite area. Since $Y \subset N'_{[\varepsilon, \infty)}$, we have $\partial C(N') \cap Y \subset \partial C(N')_{[\varepsilon, \infty)}$. This means that $Y \cap \partial C(N')$ is compact. Consequently $Y \cap C(N')$ is compact.

All of $C(N')$ consists of $Y \cap C(N')$ union $C(N') \cap \pi^{-1}(N_{(0,\varepsilon]})$. These two pieces meet along $C(N') \cap \pi^{-1}(N_{(0,\varepsilon]} \cap N_{[\varepsilon,\infty)})$. We know that

$$C(N') \cap \pi^{-1}(N_{(0,\varepsilon]} \cap N_{[\varepsilon,\infty)})$$

(which is a closed subset of $Y \cap C(N')$) is compact. We also know that in N' the various components of $\pi^{-1}(N_{(0,\varepsilon]})$ are separated by at least a fixed positive distance. Thus there are only finitely many components of $\pi^{-1}(N_{(0,\varepsilon]})$ which $Y \cap C(N')$ meets. The convex hull $H_{N'} \subset \mathbb{H}^3$ is connected and has a closure which meets more than one point in $S^2_\infty = \partial \mathbb{H}^3$. Thus a horoball $B \subset \mathbb{H}^3$ meets $H_{N'}$ if and only if its boundary does. Translating this into N', we see that $C(N')$ meets a component of $\pi^{-1}(N_{(0,\varepsilon]})$ if and only if it meets the boundary of its component, i.e., if and only if $Y \cap C(N')$ meets the boundary of this component.

This proves that $C(N')$ meets only finitely many of the components of $\pi^{-1}(N_{(0,\varepsilon]})$. Any component of $\pi^{-1}(N_{(0,\varepsilon]})$ that has fundamental group $\mathbf{Z} \times \mathbf{Z}$ has finite volume. Let E be a component of $\pi^{-1}(N_{(0,\varepsilon]})$ that has group $\mathbf{Z}$ and that contains a finite $\mathbf{Z}$-cusp of N'. Then E is isometric to $\{(x_1, x_2, x_3) | x_3 \geq L\}/(\gamma)$. The intersection $C(N') \cap E$ is contained in $\{(x_1, x_2, x_3) | x_3 \geq L \text{ and } A_1 \leq x_2 \leq A_2\}/(\gamma)$, where A_1 and A_2 are finite. This intersection is also of finite volume.

Suppose that $E \subset N'$ is a component of $\pi^{-1}(N_{(0,\varepsilon]})$ that contains a nonfinite $\mathbf{Z}$-cusp. The subspace E is isometric to $\{(x_1, x_2, x_3) | x_3 \geq L\}/(\gamma)$, where γ is translation by 1 in the x_1-direction and L is some appropriate positive constant. Furthermore, $L_{\pi_1(N)} \cap \mathbf{C} \subset \mathbf{C}$ is unbounded in the x_2-direction. Thus $C(N') \cap E$ contains annuli

$$A_t = \{(x_1, x_2, x_3) | x_2 = t, x_3 \geq \max(L, \tfrac{1}{2})\}/(\gamma)$$

for an unbounded set of $t \in \mathbf{R}$. This means that $C(N')_{[\varepsilon,\infty)} \cap E$ is not compact. But $C(N')_{[\varepsilon,\infty)} \cap E$ is a closed subset of $\pi^{-1}(C(N)_{[\varepsilon',\infty)})$ for some appropriate ε'. As we have seen, $\pi^{-1}(C(N)_{[\varepsilon',\infty)}) \cap C(N')$ is compact. This contradiction implies that N' has no nonfinite $\mathbf{Z}$-cusps.

Finally, we must consider horoball components of $\pi^{-1}(N_{(0,\varepsilon]})$. Let $B \subset N'$ be such a component. We know that $C(N') \cap \partial B$ is compact. Lift to the universal cover, where we have $H_{\pi_1(N')} \cap \partial B$ is compact. If the point $x \in S^2_\infty$, which is the center of B, is in $L_{\pi_1(N')}$, then it must be isolated in the limit set. Since the horoball embeds in N', no nontrivial element of $\pi_1(N')$ can have x as a fixed point. Thus x cannot be an isolated point of $L_{\pi_1(N')}$. This implies that $H_{\pi_1(N')} \cap B$ is compact.

This completes the analysis of all possible components of

$$C(N') \cap \pi^{-1}(N_{(0,\varepsilon]}).$$

There are only finitely many of them and each is of finite volume. Thus $C(N')$ is a union of a compact piece and finitely many pieces of finite volume. It is then of finite volume. This implies that N' is geometrically finite. ∎

V. Uniformization Theorem for Three-Dimensional Manifolds

PROPOSITION 7.2 (Maskit). *Let N be a geometrically finite manifold and Z a component of $\partial C(N)$ that is incompressible in $C(N)$. Then either $\pi_1(Z) \subset \pi_1(N) \subset \mathrm{PGL}_2(\mathbf{C})$ is quasi-fuchsian or $\pi_1(Z) \subset \mathrm{PGL}_2(\mathbf{C})$ has an accidental parabolic, i.e., there is a nonperipheral loop in Z whose trace under the given representation is two.*

Proof. There are several definitions of quasi-fuchsian groups. We shall see in Section 9 that they are all equivalent. For the moment we take the following definition: A discrete and faithful representation $\rho \colon \pi_1(Z) \to \mathrm{PGL}_2(\mathbf{C})$ is *quasi-fuchsian* if the convex hull of the associated hyperbolic manifold is homeomorphic to $Z \times [0, 1]$.

Under our hypothesis the representation of $\pi_1(Z)$ is discrete and faithful. Furthermore, if $l \subset Z$ is a peripheral loop, then l is represented arbitrarily far out in some end of $Z \subset \partial C(N)$. These ends are totally geodesic and lie in cusps of N. Hence l represents a parabolic conjugacy class in $\pi_1(N)$. It remains to show that if $N' \to N$ is the covering corresponding to $\pi_1(Z)$, then $C(N')$ is homeomorphic to $Z \times I$. We know that $C(N')$ is of finite volume. Hence, if we consider $C(N')_{[\varepsilon, \infty)}$, it is a compact manifold with fundamental group equal to $\pi_1(Z)$. It follows that $C(N')_{[\varepsilon, \infty)}$ is homotopy equivalent to Z.

Under this homotopy equivalence there is a component of $P_N(\varepsilon)$ corresponding to each cusp of Z. Either there are additional components of $P_N(\varepsilon)$, in which case Z has accidental parabolics, or we can construct a homotopy equivalence of pairs $(C(N')_{[\varepsilon, \infty)}, P_{N'}(\varepsilon)) \to (\bar{Z}, \partial\bar{Z})$, where $\bar{Z}$ is Z minus its cusps. In this latter case, since $C(N')_{[\varepsilon, \infty)}$ is compact, orientable, and irreducible, the homotopy equivalence is realized by a homeomorphism $(C(N')_{[\varepsilon, \infty)}, P_{N'}(\varepsilon)) \xrightarrow{\cong} (\bar{Z} \times I, \partial\bar{Z} \times I)$. This implies that $C(N')$ is homeomorphic to $Z \times I$. ∎

Recall that $(M', \partial M')$ is a pared manifold and that $Z \subset M'$ is a superincompressible surface. The pared manifold (M, P) results from cutting $(M', \partial M')$ open along Z. The surfaces Z_0 and Z_1 in $\partial_0 M$ are the two resulting copies of Z.

COROLLARY 7.3. *Let N be a complete hyperbolic manifold so that $(C(N)_0, P_N)$ is homeomorphic to (M, P). The representations*

$$\pi_1(Z_i) \to \pi_1(M) = \pi_1(N) \subset \mathrm{PGL}_2(\mathbf{C})$$

are faithful and have images that are quasi-fuchsian groups.

Proof. Since Z is superincompressible in M', $\pi_1(Z) \to \pi_1(M')$ is an injection. It follows that the inclusions of Z_i into M must be injective on the fundamental group. The only elements of $\pi_1(M)$ that have parabolic image under $\pi_1(M) \xrightarrow{\cong} \pi_1(N) \subset \mathrm{PGL}_2(\mathbf{C})$ are the conjugates to elements in $\pi_1(P)$.

Thus, if a loop $\alpha\colon S^1 \to Z_i \subset M$ is parabolic, then it must be freely homotopic in M to P and hence freely homotopic in M' to $\partial M'$. Since Z is superincompressible, this implies that α is freely homotopic in Z to ∂Z. Consequently, the only parabolic elements in $\pi_1(Z_i) \to \pi_1(M) \cong \pi_1(N) \subset \mathrm{PGL}_2(\mathbf{C})$ are the peripheral elements of Z_i. Invoking Proposition 7.2, we see that $\pi_1(Z_i)$ becomes a quasi-fuchsian group under the representation described in the statement of Corollary 7.3. ∎

8. Combination Theorems

Let (M, P) be an oriented pared manifold. Consider pairs $\{N, \psi\}$, where N is a complete hyperbolic manifold and $\psi\colon C(N) \to M - P$ is an orientation-preserving homeomorphism. We say that two such structures $\{N, \psi\}$ and $\{N', \psi'\}$ are equivalent if there is an orientation-preserving isometry

$$I\colon N \to N'$$

so that

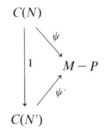

commutes up to isotopy.

We define $Q(M, P)$ as the set of equivalence classes. In the case for which (M, P) is $(S \times I, \partial S \times I)$ for S a compact surface, we also denote $Q(M, P)$ by $\mathrm{QF}(S)$. The reason for the other notation is that the hyperbolic manifolds in question are exactly the quasi-fuchsian ones associated to S. Notice that $Q(M, P)$ is nonempty if and only if (M, P) has a convex hyperbolic structure.

There is another description of $Q(M, P)$ in the case for which $\partial M \neq \varnothing$, which we find useful. The objects are pairs $\{N, \psi\}$, where N is a complete hyperbolic manifold and

$$\psi\colon (C(N), \partial C(N)) \to (M - P, \partial_0 M)$$

is an orientation-preserving proper map that is a homotopy equivalence of pairs. Two such objects $\{N, \psi\}$ and $\{N', \psi'\}$ are equivalent if there is an orientation-preserving isometry $I\colon N \to N'$ so that $\psi' \circ I$ and ψ are properly homotopic as maps of pairs. Clearly, the first set of equivalence classes maps

V. Uniformization Theorem for Three-Dimensional Manifolds

naturally to the second. To show this function is onto we show that any proper map that is a homotopy equivalence of pairs $\psi: (C(N), \partial C(N)) \to (M - P, \partial_0 M)$ is properly homotopic (as maps of pairs) to a homeomorphism.

First, $\psi: \partial C(N) \to \partial_0 M$ is a proper map that is a homotopy equivalence of surfaces. As such, it is properly homotopic to a homeomorphism. The homotopy extension lemma lets us then deform ψ by a proper homotopy of pairs until $\psi | \partial C(N)$ is a homeomorphism. Next we consider $\psi | C(N)_{(0,\varepsilon]}$. From the standard nature of these pieces it is easy to deform ψ, relative to $\partial C(N)$, by a proper homotopy until $\psi | C(N)_{(0,\varepsilon]} \to M - P$ is a homeomorphism onto the ends of $M - P$. Then, using the product structure for these ends of $M - P$, we push $\psi | C(N)_{[\varepsilon,\infty)}$ off of them. We are left with a homotopy equivalence

$$\psi: C(N)_{[\varepsilon,\infty)} \to M - (\text{neighborhood of } P),$$

which is a homeomorphism on the boundary. By Waldhausen's theorem [21], this map is homotopic relative to $\partial(C(N)_{[\varepsilon,\infty)})$ to a homeomorphism. This proves that the map between the two structure sets is onto. A similar argument shows that it is an injection.

Now let us return to the situation of the previous section: $(M', \partial M')$ is an oriented, pared manifold; $Z \subset M'$ is a superincompressible surface; (M, P) is $(M', \partial M')$ cut open along Z; Z_0 and Z_1 are the oriented surfaces which are the copies of Z in $\partial_0 M$; and $\varphi: Z_0 \to Z_1$ is the orientation reversing gluing homeomorphism that induces a map $\bar{\varphi}_*: \mathrm{QF}(Z_0) \to \mathrm{QF}(Z_1)$, which is defined as follows. If $\psi: C(N_\alpha) \to Z_0 \times I$ represents $\alpha \in \mathrm{QF}(Z_0)$, then

$$C(N_\alpha) \xrightarrow{\psi} Z_0 \times I \xrightarrow{\varphi \times \text{flip}} Z_1 \times I$$

represents $\bar{\varphi}_*(\alpha)$. The reason for the flip in the second factor is that the map $C(N_\alpha) \to Z_1 \times I$ must induce $\varphi_* \circ \psi_*$ on fundamental groups and must be orientation-preserving. Thus the map $Z_0 \times (I, \partial I) \to Z_1 \times (I, \partial I)$ must be orientation-preserving and induce φ_* in π_1. The only map, up to homotopy of maps of pairs, which does this is $\varphi \times \text{flip}$.

More generally, if $\tau: \partial_0 M \to \partial_0 M$ is the gluing involution, then we have the involution $\bar{\tau}_*: \mathrm{QF}(\partial_0 M) \to \mathrm{QF}(\partial_0 M)$. It is induced by the topological involution

$$(\tau \times \text{flip}): \partial_0 M \times I \to \partial_0 M \times I.$$

We denote by $Q(M, P)$ the product of the $Q(M_i, P \cap M_i)$ as M_i runs over the components of M.

If N is a hyperbolic structure for (M, P), then, by Corollary 7.3, lifting to the covering space corresponding to $\pi_1(Z_i) \subset \pi_1(M)$ produces a quasi-fuchsian representation of $\pi_1(Z_i)$. Such a representation corresponds to a complete hyperbolic manifold $\tilde{N}$ so that

(a) $\pi_1(\tilde{N})$ is identified with $\pi_1(Z_i)$ and
(b) $(C(\tilde{N})_0, P_{\tilde{N}})$ is homeomorphic to $(Z_i \times I, \partial Z_i \times I)$ by an orientation-preserving homeomorphism inducing the identification in (a).

This process defines maps $c_i : Q(M, P) \to QF(Z_i)$.

THEOREM 8.1. *$Q(M', \partial M')$ is nonempty; that is,* int M' *has a complete hyperbolic structure if and only if there is an* $\alpha \in Q(M, P)$ *that satisfies*

$$\bar{\varphi}_*(c_0(\alpha)) = c_1(\alpha).$$

Since we shall only need the "if" direction in what follows, we shall discuss only that implication. It is a consequence of two combination theorems of Maskit's and a theorem of Waldhausen's on Haken 3-manifolds. There are two cases to consider, depending on whether M is connected or disconnected (we assume that M' and Z are connected). Let us consider the case in which M is disconnected first.

THEOREM 8.2 (Maskit [12]). *Let G_0 and G_1 be kleinian groups with $G_0 \cap G_1 = H$. Suppose that H is quasi-fuchsian with limit set Λ_H and with $S^2 - \Lambda_H = B_0 \sqcup B_1$. Suppose in addition that for $i = 0$ and 1 the only elements $g_i \in G_i$ with the property that $g_i B_i \cap B_i \neq \emptyset$ are those in H. Then the group*

$$G \subset \mathrm{PGL}_2(\mathbf{C})$$

generated by G_0 and G_1 is a kleinian group, which as an abstract group is isomorphic to $G_0 *_H G_1$.

There is a direct geometric description of the complete hyperbolic manifold associated to $G_0 *_H G_1$ and also of its convex core. Let N_0 and N_1 be the complete hyperbolic manifolds associated to G_0 and G_1. Let N be the covering space of N_0 and N_1 associated to H. The domain of discontinuity $\Omega_{\pi_1(N)}$ is $B \sqcup B_1$. There is a unique harmonic function $\tilde{f} : \mathbb{H}^3 \to \mathbf{R}^1$ that takes boundary values 0 on B_0 and 1 on B_1. It is invariant under $\pi_1(N)$ and defines a harmonic function $f : N \to \mathbf{R}^1$. Let $N = N_- \cup N_+$, where

$$N_- = f^{-1}([0, \tfrac{1}{2}])$$

and $N_+ = f^{-1}([\tfrac{1}{2}, 1])$. Let $X = f^{-1}(\tfrac{1}{2})$. It turns out that if $p_i : N \to N_i$ is the covering projection, then $p_0 | N_-$ and $p_1 | N_+$ are embeddings. We form $N = (\overline{N_0 - p_0(N_-)}) \cup_X (\overline{N_1 - p_1(N_+)})$. This is a complete hyperbolic manifold whose fundamental group is $G_0 *_H G_1$. The image $p_i(X)$ is embedded in $C(N_i)$. The convex core of N is

$$\overline{[C(N_0) - (p_0(N_-) \cap C(N_0))]} \cup_X \overline{[C(N_1) - (p_1(N_+) \cap C(N_1))]}.$$

V. Uniformization Theorem for Three-Dimensional Manifolds

Let us deduce Theorem 8.1 in the case for which M has two components from Theorem 8.2. Let $(M, P) = (M_0, P_0) \sqcup (M_1, P_1)$ with $Z_i = \partial M_i - P_i$. We have hyperbolic structures α_i on (M_i, P_i) so that $\bar{\varphi}_*(c_0(\alpha_0)) = c_1(\alpha_1)$. Choose representative $\rho_i : \pi_1(M_i) \xrightarrow{\cong} \Gamma_i \subset \mathrm{PGL}_2(\mathbb{C})$ for the α_i. Let $N_i = \mathbb{H}^3/\Gamma_i$. Inside N_i we have a convex manifold homeomorphic to $M_i - P_i$. In particular, its boundary is Z_i. Thus associated to this boundary is a simply connected component $D_i \subset \Omega_{\Gamma_i}$ that is invariant under $\pi_1(Z_i)$. Furthermore, $D_i/\pi_1(Z_i)$ embeds in $\Omega_{\Gamma_i}/\Gamma_i$, i.e., the stabilizer of D_i in Γ_i is exactly $\pi_1(Z_i)$.

The subgroup $\pi_1(Z_i)$ is quasi-fuchsian and represents $c_i(\alpha_i)$. The component of discontinuity associated to the 0 end of $Z_i \times I$ is D_i. We denote the other by D_i'. Since $\bar{\varphi}_*(c_0(\alpha_0)) = c_1(\alpha_1)$, we can conjugate $\pi_1(Z_1) \subset \Gamma_1$ by an element $\beta \in \mathrm{PGL}_2(\mathbb{C})$ until, for each $g \in \pi_1(Z_0) \subset \Gamma_0, g$ and

$$\varphi_*(g) \in \pi_1(Z_1) \subset \Gamma_1$$

are the same element in $\mathrm{PGL}_2(\mathbb{C})$. Since φ flips the ends, β throws D_1 onto D_0' and D_1' onto D_0. Conjugate all of Γ_1 by β. This allows us to assume that

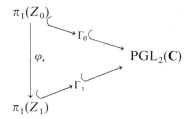

commutes and that $D_0 = D_1'$.

This establishes all the hypotheses of Theorem 8.2 for the groups Γ_0 and Γ_1. Hence, there is a kleinian group

$$\Gamma_0 *_{\varphi_*} \Gamma_1 \cong \pi_1(M_0) *_{\varphi_*} \pi_1(M_1).$$

Let N be the associated complete hyperbolic manifold. By van Kampen's theorem, $\pi_1(M')$ is isomorphic to $\pi_1(N)$. Since $(M', \partial M')$ is a pared manifold and M' is not homeomorphic to $S^1 \times D^2$ or $T^2 \times I$, it must be the case that N has finite volume. We can realize the identification of fundamental groups by a homotopy equivalence $\psi : M' \to N$. Since the tori in M have fundamental groups that correspond to $\mathbb{Z} \times \mathbb{Z}$-cusps in N, we can deform ψ until it maps $(M', \partial M') \to (N_{[\varepsilon, \infty)}, \partial N_{[\varepsilon, \infty)})$ and is a homotopy equivalence of pairs. (Here we choose ε sufficiently small so that all components of $N_{(0, \varepsilon]}$ are $\mathbb{Z} \times \mathbb{Z}$-cusps.) By Waldhausen's theorem [21], ψ is homotopic to a homeomorphism. Thus int M' is homeomorphic to N. Actually, one can produce the homeomorphism int $M' \to N$ directly from the geometric description of N given above without reference to Waldhausen's theorem.

To study the case for which M is connected we need a second combination theorem.

THEOREM 8.3 (Maskit [12]). *Let G be a kleinian group. Let H_0 and H_1 be quasi-fuchsian subgroups of G that are stabilizers of distinct components B_0 and B_1 of Ω_G. Suppose that there is an element $A \in \mathrm{PGL}_2(\mathbf{C})$ such that $AH_0 A^{-1} = H_1$ with $A(B_0) = S^2 - \bar{B}_1$. Let G' be the group generated by G and $\{A\}$. Then G' is a kleinian group isomorphic to $G *_A$.*

There is an analogous description of the complete hyperbolic manifold associated to $G *_A$. We have a complete hyperbolic manifold N_G and two covering projections $p_0 : N_0 \to N_G$ and $p_1 : N_1 \to N_G$ corresponding to the subgroups H_0 and H_1 in G. There is an identification $I_A : N_0 \cong N_1$ induced by conjugation by A. As before, there are harmonic functions $f_i : N_i \to [0, 1]$. We let $X_i \subset N_i$ be $f_i^{-1}(\frac{1}{2})$ and let $(N_i)_\pm$ be $f_i^{-1}([\frac{1}{2}, 1])$ and $f_i^{-1}([0, \frac{1}{2}])$ respectively. The map I_A induces an isomorphism $X_0 \cong X_1$. It turns out that $p_0|(N_0)_-$ and $p_0|(N_1)_+$ are embeddings. We form

$$N - \overline{[p_0((N_0)_-) \amalg p_1((N_1)_+)]}$$

and identify $p_0(X_0)$ with $p_1(X_1)$ via I_A. The argument that from this result establishes Theorems 8.1 in the case for which M' is connected is similar to the argument in the other case.

In light of Theorem 8.1, to prove the gluing theorem we need only find an element $\alpha \in Q(M, P)$ with the property that $\bar{\varphi}_* c_0(\alpha) = c_1(\alpha)$. In more invariant form, we are searching for an $\alpha \in Q(M, P)$, so that $C(\alpha) \in \mathrm{QF}(\partial_0 M)$ satisfies $\bar{\tau}_* c(\alpha) = c(\alpha)$. Here, we adopt the notation

$$c(\alpha) = (c_0(\alpha), c_1(\alpha)) \in \mathrm{QF}(Z_0) \times \mathrm{QF}(\partial_0 M).$$

In order to find such an α, we must be able to deform the hyperbolic structure on (M, P). The general deformation theory is the subject of the next section.

9. Deformation Theory

We shall work with manifolds with a given fundamental group and given homeomorphism type of boundary. The correct setting in which to work is the following. Let N and M be irreducible 3-manifolds with boundary. Any identification $\pi_1(N) \xrightarrow{\cong} \pi_1(M)$ determines a homotopy equivalence (up to homotopy). Conversely, a homotopy equivalence $N \to M$ determines an identification $\pi_1(N) \xrightarrow{\cong} \pi_1(M)$ up to conjugation. This sets up a bijection between conjugacy classes of isomorphisms from $\pi_1(N)$ to $\pi_1(M)$ and homotopy classes of homotopy equivalences from N to M.

Let $f : (N, \partial N) \to (M, \partial M)$ be a proper map that is a homotopy equivalence of pairs. When restricted to ∂N it induces a proper map $f : \partial N \to \partial M$ which is a homotopy equivalence. Any such map is properly homotopic to a

V. Uniformization Theorem for Three-Dimensional Manifolds

homeomorphism. This allows us to deform $f: (N, \partial N) \to (M, \partial M)$ through a proper homotopy of pairs until it induces a homeomorphism $\partial N \to \partial M$. An identification $\pi_1(N) \xrightarrow{\cong} \pi_1(M)$ (up to conjugacy) and a homeomorphism $\partial N \to \partial M$ (up to isotopy) are said to be *compatible* if they have a common source that is a proper map $f: (N, \partial N) \to (M, \partial M)$, which is a homotopy equivalence of pairs.

Let N be a geometrically finite kleinian manifold. We choose an isometry $N \cong \mathbb{H}^3/\Gamma$. We denote by Ω_N the domain of discontinuity of Γ. By a theorem of Ahlfors [2], Ω_N is either empty or of full measure. If $\Omega_N = \emptyset$, then $\partial C(N) = \emptyset$, and hence $N = C(N)$. Thus N is of finite volume. According to Mostow's rigidity theorem, any other structure on the same underlying manifold is equivalent to N. In particular, $Q(N)$ is a point.

We wish to study the case in which Ω_N has full measure in S^2_∞. Let N be a geometrically finite manifold with $\Omega_N \neq \emptyset$. Recall that

$$\bar{N} = (\mathbb{H}^3 \cup \Omega_N)/\pi_1(N)$$

is a topological 3-manifold with boundary. The interior of the manifold is N. Recall from Lemma 6.1 that if $X_N \subset N$ is a convex submanifold of finite volume, then there is a retraction $r: (\bar{N}, \partial \bar{N}) \to (X_N, \partial X_N)$, which is a proper map and a homotopy equivalence of pairs and determines a homeomorphism $\partial N \to \partial X$ that is compatible with the natural identification $\pi_1(N) = \pi_1(X_N)$.

A direct study of the ends of $C(N)$ allows us to construct a homeomorphism $s: C(N) \to C(N)_{[\varepsilon, \infty)} - P_N(\varepsilon)$, well defined up to isotopy. Thus $\partial \bar{N}$, ∂X_N, $\partial C(N)$, and $\partial(C(N)_{[\varepsilon, \infty)} - P_N(\varepsilon))$ are all homeomorphic in a manner compatible with the obvious identifications on fundamental groups.

DEFINITION. Let A and B be complete metric spaces. A map $f: A \to B$ is a *quasi-isometry* if f is onto and if there is $K > 1$ so that for all $x, y \in A$, $d(x, y)/K \leq d(f(x), f(y)) \leq K \cdot d(x, y)$. Notice that a quasi-isometry is a homeomorphism, and its inverse is also a quasi-isometry.

PROPOSITION 9.1. *Let N be a geometrically finite hyperbolic manifold with fundamental group $\Gamma \subset \mathrm{PGL}_2(\mathbb{C})$. Let N' be a complete hyperbolic manifold with group Γ'. Let $I: \Gamma \to \Gamma'$ be an isomorphism and $\beta: \partial \bar{N} \to \partial \bar{N}'$ a compatible orientation-preserving homeomorphism. Then:*

(1) *N' is geometrically finite.*
(2)[4] *There is an orientation-preserving homeomorphism*

$$f: (C(N)_{[\varepsilon, \infty)}, P_N(\varepsilon)) \to (C(N')_{[\varepsilon, \infty)}, P_{N'}(\varepsilon))$$

that induces I and is isotopic to β on $\partial(C(N)_{[\varepsilon, \infty)} - P_N(\varepsilon))$.

---

[4] (2) holds only provided that neither L_Γ nor $L_{\Gamma'}$ is contained in a geometric circle. If either is so contained, then we must replace the convex hull by a δ-neighborhood.

(3) *There is an orientation-preserving quasi-isometry $f: N \to N'$ that induces I and that extends to a homeomorphism $\bar{f}: \bar{N} \to \bar{N}'$ with $\bar{f} | \partial \bar{N}$ isotopic to β.*

(4) *There is an orientation-preserving quasi-conformal homeomorphism $\tilde{f}: S^2 \to S^2$, so that $\tilde{f}(\gamma(x)) = I(\gamma) \tilde{f}(x)$ for all $x \in S^2$ and all $\gamma \in \Gamma$ and so that the induced map $[f]: \Omega_N/\Gamma \to \Omega_{N'}/\Gamma'$ is isotopic to β.*

Proof. To show that N' is geometrically finite, we study its cusps. For simplicity we assume that neither L_Γ nor $L_{\Gamma'}$ is contained in a geometric circle. We choose a map $g: (C(N), \partial C(N)) \to (C(N'), \partial C(N'))$, so that $g|\partial C(N) = \beta$ and $g_*: \pi_1(C(N)) \to \pi_1(C(N'))$ agrees with $I: \Gamma \to \Gamma'$.

The maximal noncyclic abelian subgroups of Γ or Γ' up to conjugacy are in natural one-to-one correspondence with the $\mathbf{Z} \times \mathbf{Z}$-cusps of N or N'. Any such cusp is automatically contained in $C(N)$ or $C(N')$. These groups must correspond under $I: \Gamma \to \Gamma'$. This allows us to deform $g: C(N) \to C(N')$ relative to $g|\partial C(N)$ until g maps the $\mathbf{Z} \times \mathbf{Z}$-cusps of $C(N)$ into the $\mathbf{Z} \times \mathbf{Z}$-cusps of $C(N')$. By working from the standard model for $\mathbf{Z} \times \mathbf{Z}$-cusps, it is easy to see that we can make g on these cusps a quasi-isometric diffeomorphism

$$g: (\mathbf{Z} \times \mathbf{Z}\text{-cusps of } N_{(0,\varepsilon]}) \xrightarrow{\cong} (\mathbf{Z} \times \mathbf{Z}\text{-cusps of } N'_{(0,\varepsilon]}).$$

Next, consider a $\mathbf{Z}$-cusp of N. Since N is geometrically finite, this is a finite $\mathbf{Z}$-cusp and meets $C(N)$ in an end homeomorphic to $S^1 \times I \times [0, \infty)$. In particular, there are two ends of $\partial C(N)$ that are contained in this cusp. In this way, the ends of $\partial C(N)$ are divided into disjoint pairs—one pair for each $\mathbf{Z}$-cusp of N. By the Ahlfors finiteness theorem, all the ends of $C(N')$ correspond to parabolic elements in $\pi_1(N')$. Furthermore, these ends are properly embedded as geodesic annuli in the corresponding $\mathbf{Z}$-cusp of N'. Thus, if E is a $\mathbf{Z}$-cusp of N, then there are two cusps of $\partial C(N')$ that represent $g_\#(\pi_1(E))$. In particular, $g_\# \pi_1(E)$ is parabolic in N'. Since the $\mathbf{Z}$-cusp in N' carrying this cyclic group meets $\partial C(N')$ in two cusps, it must be a finite $\mathbf{Z}$-cusp of N'. In this way, we see that g sets up a one-to-one correspondence between the $\mathbf{Z}$-cusps of N and those of N', and all the $\mathbf{Z}$-cusps of N' are finite $\mathbf{Z}$-cusps.

We can choose $g: C(N) \to C(N')$ to be a quasi-isometry on all the cusps as well as on $\partial C(N)$. Thus, after this deformation, g restricts to a map

$$g': (C(N)_{[\varepsilon, \infty)}, \partial(C(N)_{[\varepsilon, \infty)})) \to (C(N')_{[\varepsilon, \infty)}, \partial(C(N')_{[\varepsilon, \infty)}))$$

that is a homotopy equivalence and a homeomorphism on the boundary. Since the domain is compact and irreducible and the range is irreducible, g' deforms relative to its boundary to a homeomorphism [21]. This shows that there is a homeomorphism $f: C(N) \to C(N')$ which is a quasi-isometry.

V. Uniformization Theorem for Three-Dimensional Manifolds

Since $C(N)$ has finite volume, so does $C(N')$. Hence, N' is geometrically finite. Notice that the homeomorphism f constructed above is homotopic to g and that $f|\partial C(N)$ is isotopic to β. This proves (1) and (2).

Next we consider δ-neighborhoods X_N of $C(N)$ and $X_{N'}$ of $C(N')$ in N and N', respectively. The constant δ is chosen so that X_N and $X_{N'}$ are C^∞-submanifolds of N and N' and contain $C(N)$ and $C(N')$ as deformation retracts.

Now apply the above constructions to X_N and $X_{N'}$. This allows us to construct a diffeomorphism $f: X_N \to X_{N'}$ which is a quasi-isometry.

Since X_N and $X_{N'}$ are smooth, the retraction maps $r: \overline{N - X_N} \to \partial X_N$ and $r': \overline{N' - X_{N'}} \to \partial X_{N'}$ are C^∞ and are transverse to every point of ∂X_N and $\partial X_{N'}$ with preimage a geodesic ray. We parametrize these rays by distance from ∂X_N or $\partial X_{N'}$. This gives us canonical product structures $\overline{N - X_N} \cong \partial X_N \times [0, \infty)$ and $\overline{N' - X_{N'}} \cong \partial X_{N'} \times [0, \infty)$. We extend the quasi-isometric diffeomorphism $f: X_N \to X_{N'}$ to $\bar{f}: N \to N'$ by using these product structures. A simple computation shows that $\bar{f}$ is still a quasi-isometry.

By Mostow's theorem, any lift $\tilde{f}$ of $\bar{f}$ to $\mathbb{H}^3$ extends to a quasi-conformal homeomorphism $\tilde{f}: S^2_\infty \to S^2_\infty$. This extension conjugates Γ to Γ' and realizes the isomorphism I (up to inner automorphism). Thus it maps $\Omega_\Gamma \to \Omega_{\Gamma'}$ and induces a homeomorphism $[\tilde{f}]: \Omega_\Gamma/\Gamma \to \Omega_{\Gamma'}/\Gamma'$. This mapping, together with $\bar{f}$, determines a homeomorphism $\overline{N} \to \overline{N}'$ that agrees with f on X_N. Hence $[\tilde{f}]$ is isotopic to the original map β. This proves both (3) and (4). ∎

Now we introduce the classical definition of a quasi-fuchsian group. Let S be a complete hyperbolic surface of finite volume. A map

$$\rho: \pi_1(S) \to \text{PGL}_2(\mathbb{C})$$

is *quasi-fuchsian* if there is a quasi-conformal homeomorphism

$$f: S^2_\infty \to S^2_\infty$$

which conjugates ρ to a fuchsian representation.

COROLLARY 9.2. *Let S be as above and let $\rho: \pi_1(S) \to \text{PGL}_2(\mathbb{C})$ be a discrete and faithful representation. The following are equivalent:*

(1) *The action $\rho(\pi_1(S))$ on S^2_∞ is quasi-conformally conjugate to a fuchsian representation.*

(2) *ρ is geometrically finite, and the conjugacy classes in $\pi_1(S)$ represented by the cusps in S are exactly the conjugacy classes that are parabolic under ρ.*

(3) *$C(N_\rho)$ is homeomorphic to $S \times I$.*

(4) *$L_{\rho(\pi_1(S))}$ is a Jordan curve in S^2_∞, and no element of $\rho(\pi_1(S))$ interchanges its complementary components.*

Proof. Clearly, (1) $\Rightarrow$ (4). Let N_f be a fuchsian manifold with $\pi_1(N_f) \cong \pi_1(S)$. Any of the four conditions implies the existence of a homeomorphism $\partial \overline{N}_\rho \xrightarrow{\cong} \partial \overline{N}_f$ compatible with the identification of groups. Thus, by Proposition 9.1, any of the four conditions implies (1), (2), and (3). Thus all the conditions are equivalent. ∎

Let N be a geometrically finite hyperbolic manifold. According to Proposition 9.1, $Q(C(N)_0, P_N)$ is the set of equivalence classes of pairs (N', ψ), where N' is a complete hyperbolic manifold and $\psi: N' \to N$ is a quasi-isometry. (N', ψ) is equivalent to (N'_1, ψ_1) if there is an isometry from N' to $N'_1, I: N' \to N'_1$, so that $\psi_1 \circ I$ is homotopic to ψ by a proper homotopy which extends to one of pairs $(\overline{N}', \partial \overline{N}') \times I \to (\overline{N}'_1, \partial \overline{N}'_1)$.

We equip $Q(C(N)_0, P_N)$ with the Teichmüller metric. If $\varphi: S^2_\infty \to S^2_\infty$ is a quasi-conformal homeomorphism, then we let $D(\varphi)$ be $\frac{1}{2}$ the log of the maximal dilation constant for φ (maximum meaning the essential maximum). Let N' and N'' be elements of $Q(C(N)_0, P_N)$. The *Teichmüller distance* from N' to N'' is the infimum of $D(\psi)$ as ψ ranges over all orientation-preserving quasi-conformal homeomorphisms $\psi: S^2_\infty \to S^2_\infty$ that conjugate Γ' to Γ'', realizing the given isomorphism, and which realize the correct homotopy class of mappings $\Omega_{\Gamma'/\Gamma'} \to \Omega_{\Gamma''/\Gamma''}$. It is a theorem of Teichmüller that this infimum is realized by a unique quasi-conformal mapping $T(\Gamma', \Gamma'')$. The map $T(\Gamma', \Gamma'')$ induces $T: \Omega_{\Gamma'}/\Gamma' \to \Omega_{\Gamma''}/\Gamma''$, which is real analytic except at a finite set of points. The singularities of T will be nonempty unless Γ' and Γ'' are conjugate in $\mathrm{PGL}_2(\mathbf{C})$. All these facts are contained in [4].

The Teichmüller distance defines a complete metric, and hence a topology, on $Q(C(N)_0, P_N)$. Clearly, any quasi-conformal deformation of N gives rise to one of $\Omega_N/\pi_1(N)$. Thus we have a function $\partial: Q(C(N)_0, P_N) \to \mathcal{T}(\partial C(N))$. The following is an immediate corollary of Proposition 9.1, together with the existence theorem for solutions to the Beltrami equation.

THEOREM 9.2 (Bers [4]). *Let (M, P) be a pared manifold. The map $\partial: Q(M, P) \to \mathcal{T}(\partial M - P)$ is a homeomorphism provided that $Q(M, P)$ is nonempty.*

Notice that the result includes Mostow's theorem. The reason is that if N is of finite volume, then any other hyperbolic structure on the underlying manifold of N automatically has finite volume. Thus, in this case,

$$Q(C(N)_0, P_N)$$

contains all the hyperbolic structures on the underlying manifold of N. On the other hand, if N has finite volume, then $C(N) = N$ and $\partial C(N) = \emptyset$. Hence, $\mathcal{T}(\partial C(N))$ is a single point.

V. Uniformization Theorem for Three-Dimensional Manifolds

Suppose that (M, P) is a pared manifold so that every component of $\partial_0 M$ is incompressible in M and no nonperipheral loop in $\partial_0 M$ is freely homotopic in M to P. According to Proposition 7.2, if $[N, \psi] \in Q(M, P)$, then the covering space of N associated to the fundamental group of a component of $\partial_0 M$ is quasi-fuchsian. This defines for each component Z of $\partial_0 M$ a mapping.

$$c_Z : Q(M, P) \to QF(Z).$$

Let $c: Q(M, P) \to QF(\partial_0 M)$ be the product of all the c_Z as Z ranges over the components of $\partial_0 M$.

Using Bers's theorem, we can form the composition

$$\mathscr{T}(\partial_0 M) \xrightarrow{\partial^{-1}} Q(M, P) \xrightarrow{c} QF(\partial_0 M) \xrightarrow{\partial} \mathscr{T}(\partial_0 M) \times \mathscr{T}(\partial_0 M).$$

This composition sends α to $(\alpha, \sigma_{(M,P)}(\alpha))$. The induced map

$$\sigma_{(M,P)} : \mathscr{T}(\partial_0 M) \to \mathscr{T}(\partial_0 M)$$

is called the *skinning map*. The picture we have is the following. Specifying a hyperbolic structure α at infinity (i.e., on $\Omega_{\pi_1(M)}/\pi_1(M)$) determines a unique geometrically finite hyperbolic 3-manifold N_α. There is a covering of N_α corresponding to $\pi_1(\partial_0 M) \subset \pi_1(M)$ that is a quasi-fuchsian group. The hyperbolic structure on one end of the quasi-fuchsian group agrees with the structure at infinity for N_α. The other end of the quasi-fuchsian group is "buried" in the 3-manifold N_α. It is the structure on this buried end that is $\sigma_{(M,P)}(\alpha)$.

We now reformulate Theorem 8.1 in terms of the skinning map $\sigma_{(M,P)}$. Recall that $\tau : \partial_0 M \to \partial_0 M$ is the gluing involution on $\partial_0 M = Z_0 \amalg Z_1$. It induces an involution $\tau : \mathscr{T}(\partial_0 M) \to \mathscr{T}(\partial_0 M)$.

THEOREM 9.4. *If there is a structure $\alpha \in \mathscr{T}(\partial_0 M)$, so that $\tau \circ \sigma_{(M,P)}(\alpha) = \alpha$, then $(M', \partial M') = (M/\tau, P/\tau)$ has a complete hyperbolic structure of finite volume.*

This is simply a reformulation, in new coordinates, of Theorem 8.1. To see this, let $\alpha = (\alpha_0, \alpha_1) \in \mathscr{T}(Z_0) \times \mathscr{T}(Z_1) = \mathscr{T}(\partial_0 M)$. Then $\sigma_{(M,P)}(\alpha)$ becomes $(\sigma_0(\alpha_0, \alpha_1), \sigma_1(\alpha_0, \alpha_1)) \in \mathscr{T}(Z_0) \times \mathscr{T}(Z_1)$. We also have $\tau(\alpha_0, \alpha_1) = (\varphi_*^{-1}(\alpha_1), \varphi_*(\alpha_0))$, where $\varphi : Z_0 \to Z_1$ is the gluing homeomorphism and $\varphi_* : \mathscr{T}(Z_0) \to \mathscr{T}(Z_1)$ is the map induced by φ. Under the identifications

$$\partial : Q(M, P) \cong \mathscr{T}(\partial_0 M) = \mathscr{T}(Z_0) \times \mathscr{T}(Z_1)$$

and

$$\partial : QF(Z_i) \cong \mathscr{T}(Z_i) \times \mathscr{T}(Z_i),$$

the maps $c_i : Q(M, P) \to QF(Z_i)$ send (α_0, α_1) to $(\alpha_i, \sigma_i(\alpha_0, \alpha_1))$ for $i = 0, 1$. Under these same identifications the map $\bar{\varphi}_* : QF(Z_0) \to QF(Z_1)$ of Theorem

8.1 sends $(\beta_0, \beta_0') \in \mathcal{T}(Z_0) \times \mathcal{T}(Z_0)$ to $(\varphi_*(\beta_0'), \varphi_*(\beta_0)) \in \mathcal{T}(Z_1) \times \mathcal{T}(Z_1)$. The equation $\tau \circ \sigma_{(M,P)}(\alpha) = \alpha$ translates to

$$\tau(\sigma_0(\alpha_0, \alpha_1)), \sigma_1(\alpha_0, \alpha_1)) = (\alpha_0, \alpha_1)$$

or

$$(\varphi_*^{-1}(\sigma_1(\alpha_0, \alpha_1)), \varphi_*(\sigma_0(\alpha_0, \alpha_1))) = (\alpha_0, \alpha_1)$$

or

(9.5) $\quad \varphi_* \alpha_0 = \sigma_1(\alpha_0, \alpha_1) \quad$ and $\quad \alpha_1 = \varphi_* \sigma_0(\alpha_0, \alpha_1).$

On the other hand,

$$\bar{\varphi}_* c_0(\alpha_0, \alpha_1) = \bar{\varphi}_*(\alpha_0, \sigma_0(\alpha_0, \alpha_1)) = (\varphi_*(\sigma_0(\alpha_0, \alpha_1)), \varphi_*(\alpha_0))$$

and

$$c_1(\alpha_0, \alpha_1) = (\alpha_1, \sigma_1(\alpha_0, \alpha_1)).$$

Hence, $\bar{\varphi}_* c_0(\alpha_0, \alpha_1) = c_1(\alpha_0, \alpha_1)$ if and only if equations (9.5) hold.

The sketch of the proof of the existence of a fixed point α in $\mathcal{T}(\partial_0 M)$ for the mapping $\tau \circ \sigma_{(M,P)}$ forms Sections 10–12.

10. The Fixed Point Theorem

THEOREM 10.1. *Let (M', P') and (M, P) be as in Section 7. Suppose that no component Z of $\partial_0 M$ has the property that the inclusion of it into the component of M containing it induces a map of fundamental groups with image of finite index. Then $\tau \circ \sigma_{(M,P)} \colon \mathcal{T}(\partial_0 M) \to \mathcal{T}(\partial_0 M)$ has the property that for any $g \in \mathcal{T}(\partial_0 M)$ the sequence $\{(\tau \circ \sigma_{(M,P)})^k(g)\}_{k \geq 0}$ converges to a limit point in $\mathcal{T}(\partial_0 M)$. This limit point is independent of g and is the unique fixed point of $\tau \circ \sigma_{(M,P)}$.*

As a consequence of this theorem and Theorem 8.1, we can solve the gluing problem for (M, P) and τ as long as no component of $\partial_0 M$ has a fundamental group of finite index in that of the component of M containing it. By a theorem of Waldhausen's [21], any such component of (M, P) will be homeomorphic to an I-bundle over $(S, \partial S)$, where S is a compact surface. It follows that the inclusion induces a map on fundamental groups that is of index one or two. The exceptional case of the gluing theorem is dealt with in Section 13.

V. Uniformization Theorem for Three-Dimensional Manifolds

We define a distance function on $\mathscr{T}(\partial_0 M) = \mathscr{T}(Z_0) \times \mathscr{T}(Z_1)$ to be the maximum of the Teichmüller distance functions on $\mathscr{T}(Z_0)$ and $\mathscr{T}(Z_1)$. This defines a complete metric on $\mathscr{T}(\partial_0 M)$.

The proof of Theorem 10.1 is divided into two parts:

(I) *Show that* $\tau \circ \sigma_{(M,P)} \colon \mathscr{T}(\partial_0 M) \to \mathscr{T}(\partial_0 M)$ *is strictly distance decreasing.*

(II) *Show that for any* $g \in \mathscr{T}(\partial_0 M)$ *the sequence* $\{(\tau \circ \sigma_{(M,P)})^k(g)\}$ *is bounded.*

As we shall show presently, the fixed point theorem is an immediate consequence of these two results. Later in this section we shall prove part (I). It is a fairly direct corollary of the classical work of Teichmüller. Part (II), which we call the *bounded image theorem*, is of a different order of difficulty. In Sections 11 and 12 we present an outline, and an outline only, of the proof of this result. It is an extremely difficult result to prove (at least as of now). Its proof uses an extraordinary range of novel geometric notions, all introduced by Thurston.

Let us deduce Theorem 10.1 from statements (I) and (II). The argument presented here was shown to us by both John Franks and Peter Shalen. Consider in $\mathscr{T}(\partial_0 M)$ the subset L of all limit points of the sequence

$$\{(\tau \circ \sigma_{(M,P)})^k(g)\}_{k \geq 0}.$$

A standard diagonalization argument shows that L is closed. By (II), L is also bounded. Hence L is compact. Clearly, $\tau \circ \sigma_{(M,P)} \colon L \to L$ is a homeomorphism. Since L is compact, there is a pair of points l_1 and l_2 in L of maximal distance apart. Let m_1 and m_2 be points of L so that

$$\tau \circ \sigma_{(M,P)}(m_i) = l_i.$$

If $l_1 \neq l_2$, then, since $\tau \circ \sigma_{(M,P)}$ is strictly distance decreasing, the distance between m_1 and m_2 is greater than that between l_1 and l_2. This contradicts the choice of l_1 and l_2. It follows that $l_1 = l_2$, and hence that L itself is a point. This means that the sequence $\{(\tau \circ \sigma_{(M,P)})^k(g)\}_{k \geq 0}$ converges to a point of $\mathscr{T}(\partial_0 M)$. The limit point is a fixed point for $\tau \circ \sigma_{(M,P)}$. Since $\tau \circ \sigma_{(M,P)}$ is strictly distance decreasing, it can have at most one fixed point.

Actually, (II) is a consequence of a stronger boundedness statement:

(II') *For some $K > 0$ depending on (M, P) and τ such that the image of*

$$(\tau \circ \sigma_M)^K(\mathscr{T}(\partial_0 M))$$

is bounded.

We shall sketch the proof of (II) but not of (II').

The fact that $\tau \circ \sigma_{(M, P)} \colon \mathcal{T}(\partial_0 M) \to \mathcal{T}(\partial_0 M)$ is distance decreasing is equivalent to the fact that $\sigma_{(M, P)}$ is distance decreasing (since τ is an isometry). To see that $\sigma_{(M, P)}$ is distance decreasing, one begins with two structures α and β in $\mathcal{T}(\partial_0 M)$. Let $(\partial_0 M)_\alpha$ and $(\partial_0 M)_\beta$ denote the resulting hyperbolic surfaces of finite area. By Theorem 9.3 these structures give rise to two hyperbolic structures $\partial^{-1}\alpha$ and $\partial^{-1}\beta$ in $Q(M, P)$. Let M_1 be a component of M, and let Γ_α and Γ_β in $\mathrm{PGL}_2(\mathbf{C})$ be the kleinian groups associated with $\partial^{-1}\alpha$ and $\partial^{-1}\beta$ restricted to M_1. Let $T \colon (\partial_0 M)_\alpha \to (\partial_0 M)_\beta$ be the Teichmüller mapping, Let μ be the associated Beltrami differential on $(\partial_0 M)_\alpha$. The map T will have a finite number of singularities (points where T is not real analytic) but at least one if $\alpha \neq \beta$. The essential maximum of $(1 + |\mu|)/(1 - |\mu|)$ is $d(\alpha, \beta)$. Since $\Omega_{\Gamma_\alpha}/\Gamma_\alpha$ is conformally equivalent to $(\partial_0 M)_\alpha \cap M_1$, we can lift μ to a Γ_α-invariant Beltrami differential $\tilde{\mu}$ on Ω_{Γ_α}. Its essential maximum is at most $d(\alpha, \beta)$. Since Beltrami differentials are only measurable objects and since $\Omega_{\Gamma_\alpha} \subset S^2_\infty$ is of full measure, we can view $\tilde{\mu}$ as a Beltrami differential on all of S^2_∞. Let $\tilde{\Phi} \colon S^2_\infty \to S^2_\infty$ be a quasi-conformal homeomorphism whose Beltrami differential is $\tilde{\mu}$. Since $\tilde{\mu}$ is Γ_α-invariant, $\tilde{\Phi}$ conjugates Γ_α to another kleinian group. The new kleinian group determines an element in $Q(M_1, P_1)$ whose boundary is conformally equivalent to $(\partial_0 M)_\beta \cap (M_1 - P_1)$. By Bers's theorem (Theorem 9.3), the new kleinian group is conjugate to Γ_β. By varying $\tilde{\Phi}$ by a conformal isomorphism, we can assume $\tilde{\Phi}\Gamma_\alpha\tilde{\Phi}^{-1} = \Gamma_\beta$.

Now suppose that Z is a component of $\partial M_1 - P_1$. Let $D(\alpha) \subset \Omega_{\Gamma_\alpha}$ be the component stabilized by $\pi_1(Z) \subset \Gamma_\alpha$. Let $D'(\alpha)$ be the other component of $\Gamma_{\pi_1(Z)}$. (Recall that $\pi_1(Z)$ is quasi-fuchsian.) The surface $D'(\alpha)/\pi_1(Z)$ is by definition $\sigma_{(M, P)}(\alpha)|Z$. The map $\tilde{\Phi}$ restricts a quasi-conformal homeomorphism $D'(\alpha) \to D'(\beta)$, which passes to the quotients $T' \colon \sigma_{(M, P)}(\alpha)|Z \to \sigma_{(M, P)}(\beta)|Z$. Clearly, the Beltrami differential of T' is $(\tilde{\mu}|D'(\alpha))/\pi_1(Z)$. Thus the maximal dilation of T' is at most $d(\alpha, \beta)$. This shows that

$$d(\sigma_{(M, P)}(\alpha)|Z, \sigma_{(M, P)}(\beta)|Z) \leq d(\alpha, \beta).$$

To show that the Teichmüller distance from $\sigma_{(M, P)}(\alpha)|Z$ to $\sigma_{(M, P)}(\beta)|Z$ is less than $d(\alpha, \beta)$, we need only show that T' is not the Teichmüller mapping. Consider the covering $\tilde{T}' \colon D'(\alpha) \to D'(\beta)$. Its Beltrami differential is $\tilde{\mu}|D'(\alpha)$. Let $\gamma \in \Gamma_\alpha - \pi_1(Z)$, and let $x \in D(\alpha)$. We denote by $[x]$ the image of x in $D(\alpha)/\pi_1(Z) = (\partial_0 M)_\alpha|Z$. We have $\gamma x \in (\Omega_\Gamma - D(\alpha)) \subset D'(\alpha)$. The Beltrami differentials $\tilde{\mu}$ at x and $\tilde{\mu}$ at γx are locally equivalent (multiplication by γ sends $\tilde{\mu}$ near x to $\tilde{\mu}$ near γx). Since the solution of the Beltrami equation

$$(\partial f/\partial \bar{z})(\partial f/\partial z)^{-1} = \tilde{u}$$

is unique up to postcomposition with a complex analytic mapping, T near $[x] \in (\partial_0 M)_\alpha|Z$ and $\tilde{T}'$ near γx are conjugate by complex analytic mappings. Thus if $[x]$ is a singularity for T, then γx is a singularity for $\tilde{T}'$. Since

$$\pi_1(Z) \subset \Gamma_\alpha$$

V. Uniformization Theorem for Three-Dimensional Manifolds

has infinite index, each singularity of T gives rise to infinitely many singularities for $\tilde{T}'$ that are not identified under $\pi_1(Z)$. Thus T' itself has infinitely many singularities. It cannot, as a result, be the Teichmüller mapping.

Applying this argument for both Z_0 and Z_1, we see that

$$d(\sigma_{(M,P)}(\alpha), \sigma_{(M,P)}(\beta)) < d(\alpha, \beta).$$

This proves that $\sigma_{(M,P)}$ is strictly distance decreasing under the hypotheses of Theorem 10.1.

The heart of the matter is (II): the bounded image theorem.

11. The First Step in the Proof of the Bounded Image Theorem

Let M be a compact 3-manifold. Let $\mathcal{R}(\pi_1(M))$ be the space of all representations $\rho: \pi_1(M) \to \text{PGL}_2(\mathbf{C})$ with the compact-open topology. In this topology, a sequence $\{\rho_k\}$ converges to ρ if and only if $\lim_{k \to \infty} \rho_k(\gamma) = \rho(\gamma)$ for all $\gamma \in \pi_1(M)$. There is a subspace $D(\pi_1(M)) \subset \mathcal{R}(\pi_1(M))$ consisting of all discrete and faithful representations. Unless $\pi_1(M)$ is an elementary group, this subspace is closed. The group $\text{PGL}_2(\mathbf{C})$ acts by conjugation on $\mathcal{R}(\pi_1(M))$, preserving $D(\pi_1(M))$. Let $\text{AH}(M)$ be the quotient of $D(\pi_1(M))$ by this action. It is a Hausdorff space if $\pi_1(M)$ is not elementary. $\text{AH}(M)$ sits naturally as a closed subset of the complex affine variety of $\text{PGL}_2(\mathbf{C})$-*characters* of $\pi_1(M)$.

If (M, P) is a pared manifold, then $\text{AH}(M, P)$ is the subspace of $\text{AH}(M)$ where all conjugacy classes of $\pi_1(M)$ represented by loops in P are required to map to parabolic elements of $\text{PGL}_2(\mathbf{C})$.

If (M, P) is oriented, then we have the space $Q(M, P)$ with its Teichmüller metric. Forgetting the fact of geometric finiteness leads to a function

$$Q(M, P) \to \text{AH}(M, P).$$

This map is continuous and an embedding into its image, which is an open subset of $\text{AH}(M, P)$.

Briefly, the proof of the bounded image theorem has two parts. First, letting $\sigma = \sigma_{(M,P)}$, one shows that for any $\alpha \in Q(M, P)$ the sequence

$$\{\alpha_n\}_{n \geq 0} = \{\partial^{-1}(\tau \circ \sigma)^n(\partial \alpha)\}_{n \geq 0} \in Q(M, P)$$

forms a bounded subset of $\text{AH}(M, P)$. This means that any subsequence of this sequence has yet a further subsequence which converges in $\text{AH}(M, P)$. Second, we take a subsequence $\{\alpha_{i_j}\}$ converging in $\text{AH}(M, P)$. By studying the sequence of quasi-fuchsian groups $c(\alpha_{i_j})$, we show that in fact the limit point lies in $Q(M, P)$. That is to say, the sequence converges in $Q(M, P)$. From these two statements it follows immediately that $\{\alpha_n\}$ is bounded in $Q(M, P)$ or, equivalently, that $\{\partial \alpha_n\} = \{(\tau \circ \sigma)^n(\partial \alpha)\}$ is bounded in $\mathcal{T}(\partial_0 M)$.

In this section we discuss the proof of the fact that every subsequence $\{\alpha_{i_j}\}$ of $\{\alpha_n\}$ has a further subsequence converging in $AH(M, P)$. This discussion makes use in an essential way of the canonical decomposition of a pared 3-manifold.

Let M be an irreducible 3-manifold, and let $Q \subset \partial M$ be an incompressible surface. There is a subpair $(\Sigma, S) \subset (M, Q)$ called the *characteristic subpair* of (M, Q) (see [8] or [9]). It is a possibly disconnected submanifold that is uniquely determined up to isotopy of pairs by the following conditions:

(a) Each component of (Σ, S) either is an (I-bundle, ∂I-subbundle)-pair or has a Seifert fibration structure in which S is a union of fibers in the boundary.

(b) The components of $\partial_1 \Sigma = \overline{\partial \Sigma - S}$ are essential[5] annuli and tori in (M, Q). Each is either a component of $\overline{\partial M - Q}$ or is not parallel into ∂M.

(c) No component of (Σ, S) is homotopic in (M, Q) into a distinct component.

(d) Any map of the torus $(T, \varnothing)$ or the annulus $(S^1 \times I, S^1 \times \partial I)$ into (M, Q), which is injective on the fundamental group and essential as a map of pairs, is homotopic as a map of pairs into (Σ, S).

In the case of $(M, \overline{\partial_0 M})$ for which (M, P) is a pared manifold as constructed in Section 7 (and as usual $\partial_0 M = \partial M - P$), we can limit the types of components which occur in (Σ, S). They are of three types, up to homeomorphism:

(a) $(T^2 \times I, \varnothing)$—a neighborhood of a torus component of P.

(b) $(S^1 \times D^2, A)$—where A is a nonempty union of essential annuli in $\partial(S^1 \times D^2)$.

(c) (I-bundles, ∂I-subbundles).

For each component of type (b), $\overline{\partial(S^1 \times D^2) - A}$ is a union of essential annuli in $(M, \overline{\partial_0 M})$. Thicken these up and consider them as I-bundles over the annulus with the ∂I-subbundle lying in $\partial_0 M$. Add to this all components of type (c). The result, denoted by $(W, \partial_0 W) \subset (M, \overline{\partial_0 M})$, is an ($I$-bundle, ∂I-subbundle)-pair (usually disconnected). It is called the *window* of $(M, \overline{\partial_0 M})$. It is unique up to isotopy of pairs. Its frontier $\partial_1 W = \overline{\partial W - (\partial_0 W \cup P)}$ is a collection of essential annuli in $(M, \overline{\partial_0 M})$.

It is an easy exercise to show that if L is a component of $\overline{M - W}$, then either

(i) L is homeomorphic to a solid torus and $\partial_1 W \cap L$ contains at least one essential annulus on ∂L or

---

[5] *Essential* means injective on the fundamental group and not null homotopic as a map of pairs.

V. Uniformization Theorem for Three-Dimensional Manifolds

(ii) all essential annuli in $(L, \overline{\partial_0 M \cap L})$ are parallel in L into
$$(\partial_1 W \cap L, \partial(\partial_1 W \cap L)).$$

DEFINITION 11.1. Let $A \subset \mathrm{AH}(M, P)$ be a subset and let $l \subset M$ be a loop We say that l is *bounded under A* if there is a constant K_l so that for every $\alpha \in A$ with associated complete hyperbolic manifold N_α either l is represented by a parabolic conjugacy class in $\pi_1(N_\alpha)$ or the geodesic in N_α freely homotopic to l has length $\leq K_l$.

The first step toward showing that the sequence $\{\alpha_n\} \in \mathrm{QM}(M, P)$ is bounded is a general result on the boundedness of loops in $\partial_1 W$ under all discrete and faithful representations.

THEOREM 11.2. *Let l be the core of a component of $\partial_1 W$ in M. Then l is bounded under all of $\mathrm{AH}(M, P)$.*

We say nothing about the proof of this theorem except that it uses the ergodicity of the geodesic flow on a surface and certain elementary growth estimates for the area of certain "branched surfaces" in $\mathbb{H}^3$.

Once we have boundedness for the loops in $\partial_1 W$, we can generalize to a much stronger boundedness theorem. The method of doing this involves a setup that generalizes the notion of an acylindrical pared manifold. Let (M, P) be a pared manifold. It is *acylindrical* if any map
$$\varphi: (S^1 \times I, S^1 \times \partial I) \to (M, \overline{\partial_0 M}),$$
which is injective on fundamental groups and nonnull homotopic as a map of pairs, deforms (as a map of pairs) into $(P, \partial P)$. This is equivalent to requiring that the characteristic submanifold of $(M, \partial_0 M)$ be a regular neighborhood of P. More generally, let (M, P) be a pared manifold and let $X \subset (\overline{\partial_0 M})$ be a subsurface. We say that (M, P, X) is *acylindrical* if any map φ
$$\varphi: (S^1 \times I, S^1 \times \partial I) \to (M, \overline{\partial_0 M - X}),$$
which is injective on fundamental groups and is nontrivial as a map of pairs, deforms (as a map of pairs) into $(X \cup P, \partial(X \cup P))$.

THEOREM 11.3. *Let (M, P) be a pared manifold, and let $X \subset \partial_0 M$ be a subsurface. Suppose that*

(1) *(M, P, X) is acylindrical,*
(2) *each component of $\partial_0 M - X$ is incompressible in M, and*
(3) *any map $\varphi: (S^1 \times I, S^1 \times \{0\}, S^1 \times \{1\}) \to (M, \partial_0 M - X, P)$, which is injective on the fundamental group, deforms as a map of triples into $(P, \partial P, P)$.*

Then any subset of AH(M, P) *that is bounded in all loops of X is bounded on all loops of M.*

This result is a relative version of a theorem that is proved by Thurston [18].

The statement that a subset S of AH(M, P) is bounded on all loops of M means that for each $g \in \pi_1(M)$, $\{\operatorname{tr} \rho(g)\}_{\rho \in S}$ is bounded. This is equivalent to saying that in the complex affine variety $X(\pi_1(M))$ of $\operatorname{PGL}_2(\mathbf{C})$-characters on $\pi_1(M)$ the subset S is bounded. Thus any sequence in S will have a subsequence converging to a point of $X(\pi_1(M))$. Since AH(M, P) is a closed subset of $X(\pi_1(M))$, this implies that the subsequence in question converges to a point of AH(M, P).

This proves the following simple fact.

FACT 11.4. *If a subset* $S \subset$ AH(M, P) *is bounded on all loops of M, then S has compact closure in* AH(M, P).

COROLLARY 11.5. *Let* (M, P) *be a pared manifold constructed as in Section 7. Let* $(W, \partial_0 W) \subset (M, \overline{\partial_0 M})$ *be its window. Then the image* AH(M, P) → AH($\overline{M - W}$) *is bounded.*

Proof. Let L be a component of $\overline{M - W}$. We know that $\pi_1(L) \to \pi_1(M)$ injects so that there is induced a map AH(M, P) → AH(L). The statement of Corollary 11.5 means that for each component L this map has bounded image.

Consider first components of L that are solid tori and meet $\partial_1 W$ in a non-empty collection of essential annuli in ∂L. Since we know that the cores of these annuli are bounded in length under all of AH(M, P), it follows immediately that the same is true for the core of L. This is exactly the result of Corollary 11.5 for this component.

Now consider a component L so that $(L, \overline{\partial_0 M} \cap L)$ has no essential annuli except those parallel into $(P \cap L, \partial P \cap L)$. Let P_L be the union of the torus components in P that lie in L. It is easy to see that (L, P_L) is a pared manifold, and that $(L, P_L, \partial_1 W \cap L)$ satisfies the hypotheses of Theorem 11.3. Using Theorem 11.3, we conclude that any subset of AH(L, P_L) that is bounded on all loops in $\partial_1 W \cap L$ is bounded in AH(L, P_L). According to Theorem 11.2, the image of the restriction map r_L: AM(M, P) → AH(L, P_L) is such a map. Hence Im(r_L) ⊂ AH(L, P_L) is bounded. This is equivalent to saying that its image in AH(L) is bounded. ∎

DEFINITION. Let Z be a compact surface, and let $\{r_i\}$ be a sequence in AH($Z \times I, \partial Z \times I$). Suppose that $B \subset Z$ is a compact subsurface. We say that B is a *bounded surface for the sequence* if each loop l in B is bounded

V. Uniformization Theorem for Three-Dimensional Manifolds

under the sequence. In addition, if any loop bounded under the sequence is homotopic to a loop on B, then we say that B is a *maximal bounded surface*.

THEOREM 11.6. *For any sequence* $\{r_i\} \in \mathrm{AH}(Z \times I, \partial Z \times I)$ *there is compact subsurface* $B \subset S$ *unique up to isotopy such that*

(1) B *is a maximal bounded surface for the sequence* $\{r_i\}$;
(2) *no boundary component of* B *is homotopically trivial in* Z;
(3) *if two boundary components of* B *are parallel in* Z, *then they form the boundary of an annular component of* B;
(4) $\partial Z \subset B$.

(N.B.: B *can be empty*).

For a proof of this result of Thurston's see [20]. Using it and the previous results in this section behind us, we are ready to establish the main result of this section. Let (M, P) be a pared manifold as constructed in Section 7.

THEOREM 11.7. *Suppose that* (M, P) *is not homeomorphic to* $(E, E|\partial S)$, *where* E *is an I-bundle over a surface* S. *Then all the loops in* $\partial_0 M$ *remain bounded under* $\{\partial^{-1}(\tau \circ \sigma)^n(\partial \alpha)\}_{n \geq 0}$ *for any* $\alpha \in \mathcal{T}(\partial_0 M)$.

Proof. Consider the maximal bounded surface $B \subset \overline{\partial_0 M}$. By Corollary 11.5, B contains $(\overline{M - W}) \cap (\overline{\partial_0 M})$. Thus $\partial B \subset \partial_0 W$.

On $\partial_0 W \subset \overline{\partial_0 M}$, there is the involution determined by switching the endpoints of the intervals in the I-bundle structure. We call this involution $\iota: \partial_0 W \to \partial_0 W$. For any loop l in $\partial_0 W$, $\iota(l)$ and l are freely homotopic in M.

Thus if $l \subset \partial_0 W$ is bounded under the sequence, then $\iota(l)$ is also bounded. Hence, $\iota(B) \cap (\partial_0 W)$ is isotopic to $B \cap \partial_0 W$. We can arrange that $B \cap \partial_0 W$ is ι-invariant.

Finally, there is the gluing involution $\tau: \overline{\partial_0 M} \to \overline{\partial_0 M}$. We claim that $\tau(B) = B$ up to isotopy.

In order to establish this, we need the following lemma.

LEMMA 11.8. *Let* (L, P) *be a pared manifold. Let* α, β *be elements of* $Q(L, P)$ *so that the quasi-conformal distance from* $\partial \alpha$ *to* $\partial \beta$ *in* $\mathcal{T}(\partial_0 L)$ *is* $\leq d$. *Then for any loop* m *in* L, *the closed geodesics* γ_α *and* γ_β *in* N_α *and* N_β *that are freely homotopic to* m *satisfy*

$$d^{-1} \cdot l_{N_\beta}(\gamma_\beta) \leq l_{N_\alpha}(\gamma_\alpha) \leq d \cdot l_{N_\beta}(\gamma_\beta),$$

where $l_{N_\alpha}(\gamma_\alpha)$ *and* $l_{N_\beta}(\gamma_\beta)$ *denote the lengths of* γ_α *and* γ_β.

Proof. Let N_α and N_β be the hyperbolic 3-manifolds associated to α and β. We claim that there is a d-quasi-conformal homeomorphism $S_\infty^2 \to S_\infty^2$ conjugating the action of $\pi_1(N_\alpha)$ to that of $\pi_1(N_\beta)$ and inducing the natural identification of the groups. Since α and β are in $Q(L, P)$, Proposition 9.1 implies that they are quasi-conformally equivalent.

Now let $\psi: \partial \overline{N}_\alpha \to \partial \overline{N}_\beta$ be a d-quasi-conformal homeomorphism whose existence is guaranteed by the hypotheses. Lift this to $\tilde{\psi}: \Omega_\alpha \to \Omega_\beta$. Let μ be its Beltrami differential. Since $\tilde{\psi}$ conjugates $\pi_1(N_\alpha)$ to $\pi_1(N_\beta)$, μ is invariant under $\pi_1(N_\alpha)$. Since $S_\infty^2 - \Omega_\alpha$ has 0 measure, we can view μ as a Beltrami differential on all of S_∞^2 invariant under $\pi_1(N_\alpha)$. Let $\psi: S_\infty^2 \to S_\infty^2$ be a quasi-conformal map with $(d\psi/d\bar{z})(d\psi/dz)^{-1} = \mu$. It is unique up to postcomposition with an element in $\mathrm{PGL}_2(\mathbf{C})$. Thus for any $g_\alpha \in \pi_1(N_\alpha)$, $\psi \circ g_\alpha$ and ψ differ by $h_\alpha \in \mathrm{PGL}_2(\mathbf{C})$. This shows that ψ conjugates $\pi_1(N_\alpha) \subset \mathrm{PGL}_2(\mathbf{C})$ to another geometrically finite kleinian group. Let N_β' be the associated manifold. From the construction of ψ, it is clear that $\partial \overline{N}_\beta' = \partial \overline{N}_\beta$ in $\mathcal{T}(\partial_0 L)$. By Bers's theorem, this implies that N_β' and N_β are conjugate, i.e., after composing ψ with an element of $\mathrm{PGL}_2(\mathbf{C})$, it conjugates $\pi_1(N_\alpha)$ to $\pi_1(N_\beta)$. Clearly, it is a d-guasi-conformal mapping.

For any $g \in \pi_1(L)$, let g_α and g_β denote the images of g in $\pi_1(N_\alpha)$ and $\pi_1(N_\beta)$. We have just constructed a d-quasi-conformal homeomorphism on the 2-sphere at infinity Ψ so that $\Psi \circ g_\alpha \circ \Psi^{-1} = g_\beta$. Clearly, this implies that g_α is parabolic if and only if g_β is. We wish to compare the lengths of the closed geodesics determined by g_α and g_β in the case when they are both loxodromic transformations. To do this, we study how their lengths are determined by the actions on S_∞^2. Choose the upper-half space model so that the axis of g_α is $\{0, \infty\}$ and so that g_α is represented by the matrix

$$\begin{pmatrix} x_\alpha & 0 \\ 0 & x_\alpha^{-1} \end{pmatrix}.$$

Then $l_{N_\alpha}(\gamma_\alpha) = |\log \|x_\alpha\|^2|$.

If we restrict to the boundary, then the cyclic group generated by g_α acts freely and properly discontinuously on $\mathbf{C}^* = \mathbf{C} - \{0\}$. A fundamental domain for the action is the annulus $\{z \mid 1 \leq |z| \leq \|x_\alpha\|^2\}$, shown in Fig. 11.1. The quotient is the torus obtained by gluing the inner circle to the outer by multiplication by x_α^2.

Thus, from the action of g_α on S_∞^2, we produce a conformally flat, oriented torus together with a homomorphism $\pi_1(T^2) \to \pi_1(N_\alpha)$ with image, the infinite cyclic group generated by g_α. Such conformally flat, oriented tori together with a fixed homomorphism $\pi_1(T^2) \to$ (infinite cyclic group) $\to 0$ have exactly one conformal invariant. This invariant is the gluing map when the torus is described, as above, as an annulus with its boundary components

V. Uniformization Theorem for Three-Dimensional Manifolds

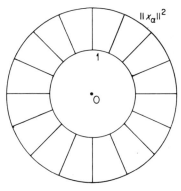

Figure 11.1

glued together. Thus, for the tori associated to g_α and g_β, the invariants are x_α^2 and x_β^2. If we deform such a torus with invariant ω by a d-quasi-conformal homeomorphism to another torus with invariant ω', then

$$d^{-1} \leq |\log\|\omega\|/\log\|\omega'\|| \leq d.$$

In our case, we conclude that

$$d^{-1} \leq |\log\|x_\alpha^2\||/|\log\|x_\beta^2\|| \leq d.$$

This proves Lemma 11.8. ∎

That a loop m in $\partial M - P$ is bounded for $\{\alpha_n\}$ if and only if $\tau(m)$ is bounded follows from Lemma 11.8 and the next lemma.

LEMMA 11.9. *There is $d > 0$ independent of n so that*

$$d(\partial_i C(\alpha_n), \partial_i \bar{\tau}_*(C(\alpha_n))) \leq d$$

for all n and $i = 0, 1$.

Proof

$$\partial C(\alpha_n) = ((\tau \circ \sigma)^n(\partial\alpha), \sigma(\tau \circ \sigma)^n(\partial\alpha))$$

and

$$\begin{aligned}\partial \bar{\tau}_* C(\alpha_n) &= ((\tau \circ \sigma)^{n+1}(\partial\alpha), \tau \circ (\tau \circ \sigma)^n(\partial\alpha)) \\ &= ((\tau \circ \sigma)^{n+1}(\partial\alpha), \sigma(\tau \circ \sigma)^{n-1}(\partial\alpha)).\end{aligned}$$

Thus the two distances which we must bound independent of n are

$$d((\tau \circ \sigma)^n(\partial \alpha), (\tau \circ \sigma)^{n+1}(\partial \alpha)) \quad \text{and} \quad d(\sigma(\tau \circ \sigma)^n(\partial \alpha), \sigma(\tau \circ \sigma)^{n-1}(\partial \alpha)).$$

Since τ is an isometry and σ is distance decreasing, $(\tau \circ \sigma)$ is distance decreasing. Thus

$$d((\tau \circ \sigma)^n(\partial \alpha), (\tau \circ \sigma)^{n+1}(\partial \alpha)) \leq d(\partial \alpha, (\tau \circ \sigma)(\partial \alpha)).$$

Also,

$$d(\sigma(\tau \circ \sigma)^n(\partial \alpha), \sigma(\tau \circ \sigma)^{n-1}(\partial \alpha)) = d((\tau \circ \sigma)^{n+1}(\partial \alpha), (\tau \circ \sigma)^n(\partial \alpha))$$
$$\leq d((\tau \circ \sigma)(\partial \alpha), \partial \alpha).$$

Thus we can take d equal to $d(\partial \alpha, (\tau \circ \sigma)(\partial \alpha))$. ∎

Lemma 11.9 implies that there is a d-quasi-conformal homeomorphism from $\partial c(\alpha_n)$ to $\partial c(\alpha_n)$ that flips the ends and induces $\tau_\#$ on the fundamental groups. (The fact that the map flips the ends is irrelevant here.) It follows from Lemma 11.8 that for any loop m in $\partial_0 M$, the geodesics γ_n in $c(\alpha_n)$ representing m and γ'_n in $c(\alpha_n)$ representing $\tau(m)$ satisfy

$$d^{-1} \|\gamma'_n\| \leq \|\gamma_n\| \leq d \cdot \|\gamma'_n\|.$$

But the length of m (or γ) in $c(\alpha_n)$ is the same as the length of $\tau(m)$ (or $\tau(\gamma))$ in M under α_n.

It follows immediately that $m \subset \partial_0 M$ is bounded under the $\{\alpha_n\}$ if and only if $\tau(m) \subset \partial_0 M$ is bounded under the same sequence.

This proves that $\tau B \equiv B$ up to isotopy. It is an easy matter to deform τ until $\tau B = B$.

Now, at last, we are ready to show that $B \subset \overline{\partial_0 M}$ is in fact all of $\overline{\partial_0 M}$.

We are supposing that (M, P) has no component fibering over a surface. Thus $\overline{M - W}$ meets every component of $\partial_0 M$. Hence B contains a nonperipheral loop in every component of $\partial_0 M$. Thus if we restrict B to a component Z of $\partial_0 M$, there are two possibilities:

(1) $B \cap Z = Z$ or
(2) ∂B contains a loop in $Z \cap W$ nonperipheral in Z.

We denote the set of all nonperipheral components of ∂B by $\partial_{np}(B)$. Under both the gluing involution τ and the window involution ι, the space $\partial_{np} B$ is invariant. Since $\partial_{np}(B)$ is invariant under ι, the components of $\partial_{np}(B)$ come in pairs. Each pair is the boundary of a nontrivial annulus in W. These annuli are not homotopic as maps of pairs into $(P, \partial P)$. Since $\partial_{np}(B)$ is a finite set of curves, they form the boundary of a finite set of nontrivial, nonperipheral annuli in $W \subset M$. Since $\partial_{np}(B)$ is invariant under τ, these annuli

V. Uniformization Theorem for Three-Dimensional Manifolds

are glued together under τ to form a family of disjoint embedded tori in (M', P').

One sees easily that these tori are nontrivial and nonperipheral from the fact that the annuli were nontrivial and nonperipheral. By assumption, there are no such tori in (M', P'). This means that $\partial_{np}(B)$ is empty. Thus it must be the case that $B = \partial_0 M$, i.e., every loop in $\partial_0 M$ remains bounded under $\{c(\alpha_n)\}$.

Equivalently, for any loop l in $\partial_0 M$, there is a constant K_l, independent of n, bounding the length of the closed geodesic homotopic to l in

$$\partial^{-1}((\tau \circ \sigma)^n(\partial \alpha)) \in Q(M, P).$$

This completes the proof of Theorem 11.7. ∎

COROLLARY 11.10. *Let (M, P) be the pared manifold obtained by cutting $(M', \partial M')$ open. Let $\alpha \in Q(M, P)$. The sequence $\{\partial^{-1}(\tau \circ \sigma)^n(\partial \alpha)\}_{n \geq 0} \in Q(M, P)$ has compact closure in $AH(M, P)$ provided that (M, P) is not homeomorphic to an I-bundle over a surface.*

Proof. Under these hypotheses, we have seen that

$$\{\partial^{-1}(\tau \circ \sigma)^n(\partial \alpha)\}_{n \geq 0} \in AH(M, P)$$

is bounded on all of $\partial_0 M$. We let $X = \partial_0 M$. Clearly, $(M, P, \partial_0 M)$ is acylindrical. (There are no annuli to consider since the boundary of the annuli must map to the empty set.) Likewise, conditions (2) and (3) of Theorem 11.3 are vacuously satisfied. It follows from Theorem 11.3 and Fact 11.4 that the sequence $\{\partial^{-1}(\tau \circ \sigma)^n(\partial \alpha)\}_{n \geq 0} \in AH(M, P)$ has compact closure in $AH(M, P)$. ∎

12 Completion of the Proof of the Bounded Image Theorem

Let us recap our progress. We began with a pared manifold (M', P') and cut it open along a connected incompressible surface Z that satisfies Condition 4.9. The result is a pared manifold (M, P) with either one or two components. $\partial_0 M$ has the gluing involution τ. We suppose that no component of (M, P) is finitely covered by $(S \times I, \partial S \times I)$ for S a surface. We have shown that for any $\alpha \in Q(M, P)$ the sequence $\{\alpha_n\} = \{\partial^{-1}(\tau \circ \sigma)^n(\partial \alpha)\}$ in $Q(M, P)$ is bounded in $AH(M, P)$.

To study the limit point of a subsequence, we need to understand the quasi-fuchsian covering spaces $c_0(\alpha_n) \in QF(Z_0)$ and $c_1(\alpha_n) \in QF(Z_1)$. We begin this study by exposing some basic facts about the topology of the hyperbolic 3-manifolds N_α associated to $\alpha \in AH(S \times I, \partial S \times I)$, where S

is a compact, connected, oriented surface of negative Euler characteristic. Such a 3-manifold comes equipped with an isomorphism $\pi_1(S) \xrightarrow{\cong} \pi_1(N_\alpha)$ that sends the conjugacy classes in $\pi_1(S)$ that are represented by peripheral loops to parabolic elements in $\pi_1(N_\alpha) \subset \mathrm{PGL}_2(\mathbf{C})$.

The isomorphism $\pi_1(S) \xrightarrow{\cong} \pi_1(N_\alpha)$ is realized by a homotopy equivalence $f: S \to N_\alpha$. Since the boundary components of S represent parabolic elements in $\pi_1(N_\alpha)$, we can deform f until $f|\partial S: \partial S \to \partial(N_\alpha)_{(0,\varepsilon]}$. Here we take $f|\partial S$ to be a standard embedding. Standard techniques in 3-manifold topology allow us to deform f until it is an embedding

$$f: (S, \partial S) \hookrightarrow ((N_\alpha)_{[\varepsilon, \infty)}, \partial(N_\alpha)_{[\varepsilon, \infty)}).$$

We can complete S to $\hat{S}$ by adding cusps. Then we complete f to $\hat{f}: \hat{S} \hookrightarrow N_\alpha$, an embedding that is totally geodesic in the cusps of $\hat{S}$. Such an embedding is called an *incompressible splitting surface* for N_α.

Such a surface $\hat{S}$ divides N_α into two pieces. It is the boundary of each, and the inclusion of $\hat{S}$ into each side is a homotopy equivalence. Since $\hat{S}$ and N_α are oriented, we can label the sides of $\hat{S}$ in N_α by $(N_\alpha)_\pm$. If $\hat{S}' \hookrightarrow N_\alpha$ is another splitting surface, then the decompositions $(N_\alpha)'_\pm$ and $(N_\alpha)_\pm$ differ by the union of a compact set and a cylinder $S^1 \times I \times [0, \infty)$ in each cusp of N_α that corresponds to a boundary component of $\hat{S}$.

Let (M, P) be a pared manifold and Z a component of $\partial_0 M$. We have $c_Z: \mathrm{AH}(M, P) \to \mathrm{AH}(Z \times I, \partial Z \times I)$. As we saw in Section 7, in the case under consideration there is an induced map $c_Z: Q(M, P) \to \mathrm{QF}(Z)$. We have chosen our conventions so that the following diagram commutes:

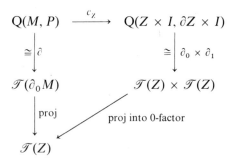

(Here ∂_0 and ∂_1 refer to the 0- and 1-ends of the unit interval.)

Also recall that if $\alpha \in Q(M, P)$, then $\partial c_Z(\alpha) = (\partial\alpha|Z, \sigma_{(M,P)}(\partial\alpha)|Z)$, where $\sigma_{(M,P)}$ is the skinning map of Section 9.

THEOREM 12.1. *Let $\{\beta_i\} \in \mathrm{QF}(S)$ be a sequence of quasi-fuchsian representations converging in $\mathrm{AH}(S \times I, \partial S \times I)$ to a representation β_∞. Suppose the limit hyperbolic manifold N is a covering of another hyperbolic manifold*

V. Uniformization Theorem for Three-Dimensional Manifolds

N' via $\pi: N \to N'$. Let $\hat{S} \hookrightarrow N$ be an incompressible splitting surface with sides $N_{\pm}$. Then one of the following four conditions must hold:

(1) $\pi|N_+$ is finite-to-one.
(2) N' is of finite volume and has a finite cover fibering over the circle.
(3) There is an accidental parabolic in N.
(4) The end N_+ of N determines a simply connected region of discontinuity in S_∞^2 invariant under β_∞. The resulting hyperbolic structure on S produces a point $\partial_+ \beta_\infty \in \mathcal{T}(S)$ which is the limit of the $\partial_+(\beta_i)$.

The analogous result holds for N_-.

Explanations. Condition (3). We say that β_∞ (or N) has *accidental parabolics* if there is a nonperipheral loop C in S whose image in N represents a parabolic conjugacy class in $\pi_1(N)$.

Condition (4). We say that there is a simply connected region of discontinuity associated to N_+ if $\tilde{N}_+ \subset \mathbb{H}^3$ has closure that contains a simply connected region of discontinuity for $\beta_\infty(\pi_1(S))$. Associated to such a region is a point $\partial_+\beta_\infty \in \mathcal{T}(S)$.

Like those of Theorems 11.2, 11.3, and 11.6, the proof of Theorem 12.1 involves much geometric analysis. Some of this analysis can be found in Sections 8 and 9 of [19]. We do not discuss the proof here; using the result, we shall complete the proof of the bounded image theorem.

For the rest of this section we fix a subsequence $\{\alpha_{i_j}\}$ of $\{\alpha_n\}$ that converges there to a point $\alpha_\infty \in \mathrm{AH}(M, P)$. We form $\beta_{i_j} = c(\alpha_{i_j})$ and $\beta_\infty = c(\alpha_\infty)$ in

$$\mathrm{AH}(\overline{\partial_0 M}, \partial(\overline{\partial_0 M})).$$

The β_{i_j} are quasi-fuchsian and converge to β_∞. Let N be the hyperbolic 3-manifold associated to β_∞. Let N' be the hyperbolic 3-manifold associated to α_∞. Clearly, each component of N covers a component of N'. This then is exactly the setup of Theorem 12.1. Suppose for the moment that we have eliminated conditions (1)–(3) of Theorem 12.1 for all ends of all components of N. Thus $\{\partial \alpha_{i_j}\} \in \mathcal{T}(Z_0) \times \mathcal{T}(Z_1)$ converges. This means that

$$\{\alpha_{i_j}\} \in Q(M, P)$$

converges to a point of $Q(M, P)$. It follows from Corollary 11.10 that the entire sequence $\{\alpha_n\}_{n \geq 0}$ is bounded in $Q(M, P)$.

Of course, it remains to rule out Conditions (1)–(3) of Theorem 12.1 for the ends $(c_0(\alpha_\infty))_\pm$ and $(c_1(\alpha_\infty))_\pm$. The first step in that direction is the following lemma.

LEMMA 12.2. *Let $\beta \in \mathrm{AH}(S \times I, \partial S \times I)$ be given and let N be the associated complete hyperbolic 3-manifold. Let $\hat{S} \subset N$ be an incompressible* splitting surface with $N_\pm$ as the two sides. There are two families of disjoint, nonparallel, nonspheripheral, simple closed curves $\Pi_+(N)$ and $\Pi_-(N)$ in $\hat{S}$ such that

(1) a nonperipheral loop $\gamma \subset \hat{S}$ is parabolic in N if and only if γ is freely homotopic to a power of some loop in $\Pi_+(N)$ or $\Pi_-(N)$;

(2) the loops in $\Pi_\pm(N)$ bound disjointly embedded proper open annuli ($S^1 \times [0, \infty)$) in $N_\pm$; the infinite end of any one of these annuli is totally geodesic in a cusp component of $N_{(0,\varepsilon]} \subset N_\pm$.

Proof. Since $\hat{S}$ is the union of a compact piece and ends contained in the cusp components of $N_{(0,\varepsilon]}$ that correspond to the boundary loops of S, we can choose $\varepsilon > 0$ sufficiently small so that any cusp component of $N_{(0,\varepsilon]}$ representing an accidental parabolic misses $\hat{S}$. Then each such component of $N_{(0,\varepsilon]}$ lies in N_+ or N_-. We focus attention on all those in (say) N_+. Since $\hat{S} = \partial N_+ \hookrightarrow N_+$ is a homotopy equivalence, for each accidental cusp l in N_+ there is a map $\varphi_l: (S^1 \times [0, \infty), S^1 \times \{0\}) \to (N_+, \hat{S})$ that is a totally geodesic embedding near infinity into the component l of $N_{(0,\varepsilon]}$ in question. Standard 3-manifold techniques allow us to deform φ_l relative to a neighborhood of infinity until φ_l is an embedding. Furthermore, any finite set of such φ_l can be made disjoint. If we have a disjoint family of these annuli, coming from distinct cusps, then the resulting curves on $\hat{S}$ are disjointly embedded and nonparallel. As a result, there is an upper bound to the number of cusps in N_+. Applying the same argument to N_- completes the proof. ∎

COROLLARY 12.3. *There are at most $3g - 3 + p$ curves in $\Pi_+(N)$ or in $\Pi_-(N)$, where g is the sum of the genera of $\partial_0 M$ and p is the number of cusps in $\partial_0 M$.*

Proof. Any family of nontrivial, nonparallel, nonperipheral disjoint simple closed curves in $\partial_0 M$ has at most $3g - 3 + p$ members. ∎

Let N'_{i_j} be the hyperbolic manifold associated to α_{i_j}, and let N_{i_j} be the covering corresponding to $\beta_{i_j} = c(\alpha_{i_j})$. We have the covering projection $\pi_{i_j}: N_{i_j} \to N'_{i_j}$. Recall that N_{i_j} has two components that are interchanged by τ. Thus the universal cover of each N_{i_j} and of N is $\mathbb{H}^3 \cup \mathbb{H}^3$.

LEMMA 12.4. *There is $d > 0$ so that for each j there is an orientation preserving d-quasi-isometry $\bar{G}_{i_j}: N_{i_j} \to N_{i_j}$, which induces $\tau_\#$ on the fundamental groups and which flips the ends. There are lifts $G_{i_j}: (\mathbb{H}^3 \cup \mathbb{H}^3) \to (\mathbb{H}^3 \cup \mathbb{H}^3)$ of the $\bar{G}_{i_j}$ so that a subsequence converges to $G_\infty: (\mathbb{H}^3 \cup \mathbb{H}^3) \to (\mathbb{H}^3 \cup \mathbb{H}^3)$. It is a d-quasi-isometry that induces a map $\bar{G}_\infty: N \to N$.*

V. Uniformization Theorem for Three-Dimensional Manifolds

Proof. We saw in the proof of Lemma 11.9 that there is a d, independent of n, and a d-quasi-conformal homeomorphism $f_n: S_\infty^2 \to S_\infty^2$ conjugating $c_0(\alpha_n)$ to $c_1(\alpha_n)$ and inducing $\varphi_\#$ on the fundamental groups. It flips the regions of discontinuity. By the extension result [20, Section 11], the f_n extend to d-quasi-isometries $F_n: \mathbb{H}^3 \to \mathbb{H}^3$ that conjugate $c_0(\alpha_n)$ to $c_1(\alpha_n)$ and induce $\varphi_\#$. By postcomposing the F_n by elements of $PGL_2(\mathbb{C})$, we can assume each F_n fixes the origin in $\mathbb{H}^3$. This has the effect of conjugating the representative for $c_1(\alpha_n)$.

The sequence $\{F_{i_j}\}$ forms a normal family that fixes the origin in $\mathbb{H}^3$. Thus there is a further subsequence converging to $F_\infty: \mathbb{H}^3 \to \mathbb{H}^3$. Clearly, F_∞ is a d-quasi-isometry conjugating $c_0(\alpha_\infty)$ to $c_1(\alpha_\infty)$ and inducing $\varphi_\#$ on fundamental groups. The d-quasi-conformal maps $\bar{G}_{i_j}$ and $\bar{G}_\infty$ are simply $F_{i_j} \sqcup F_{i_j}^{-1}$ and $F_\infty \sqcup F_\infty^{-1}$. ∎

COROLLARY 12.5. *For any $\varepsilon > 0$ and for all j sufficiently large, every element of $\Pi_\pm(N) \subset \partial_0 M$ is represented in N_{i_j} by a closed curve of length $\leq \varepsilon/d$. The map $G_{i_j}: N_{i_j} \to N_{i_j}$ sends the components of $N_{(0, \varepsilon/d]}$ representing elements in $\Pi_\pm(N)$ into components of $N_{(0, \varepsilon]}$ representing elements in $\Pi_\mp(N)$.*

PROPOSITION 12.6. *For all j sufficiently large there are a maps*

$$f_{i_j}: (M - P) \to N'_{i_j} \quad \text{and} \quad \tilde{f}_{i_j}: \partial_0 M \to N_{i_j}$$

with $\pi_{i_j} \circ \tilde{f}_{i_j} = f_{i_j}|\partial_0 M$ and the following properties:

(a) *f_{i_j} restricted to a neighborhood of P in $M - P$ is a homeomorphism onto a totally geodesic subspace of the cusps of N'_{i_j}.*

(b) *f_{i_j} induces the natural identification of fundamental groups.*

(c) *The components of $(N'_{i_j})_{(0, \varepsilon]}$ corresponding to the loops of $\Pi_\pm(N)$ are outside the image $f_{i_j}(M - P)$.*

(d) *There is a homotopy $H_{i_j}: (\partial_0 M) \to N_{i_j}$ from $\tilde{f}_{i_j} \circ \tau$ to $\bar{G}_{i_j} \circ \tilde{f}_{i_j}$ that misses all components of $(N_{i_j})_{(0, \varepsilon]}$ corresponding to loops of $\Pi_\pm(N)$.*

Proof. Since we have an identification of $\pi_1(M)$ with $\pi_1(N')$, there is a map $f_\infty: (M - P) \to N'$ satisfying properties (a) and (b) above. The lift of f_∞ to $\tilde{f}_\infty: \partial_0 M \to N$ is homotopic to the incompressible splitting surface. Thus there is a homotopy $H_\infty: (\partial_0 M) \times I \to N$ from $\tilde{f}_\infty \circ \tau$ to $\bar{G}_\infty \circ \tilde{f}_\infty$. We can take this homotopy to be proper and to move the cusps of $\partial_0 M$ through totally geodesic annuli in $N_{(0, \varepsilon]}$.

Decompose $M - P$ as a compact space M_0 and a regular neighborhood of P. We do this so that the annuli separating the pieces are mapped by f_∞ to annuli in $\partial N'_{(0, \varepsilon]}$. Let F be a fundamental domain for M_0 in its universal cover. We have a lifting $l_\infty: F \to \mathbb{H}^3$ (or $\mathbb{H}^3 \sqcup \mathbb{H}^3$ if N' is disconnected) covering f_∞. There are finitely many elements $g_1,\ldots,g_T$ in $\pi_1(M)$ so that $\alpha_\infty(g_i)l_\infty(F) \cap l_\infty(F) \neq \emptyset$. It is precisely these elements that glue faces of F together to form M_0.

Now let $\alpha': \pi_1(M) \to \mathrm{PGL}_2\mathbb{C}$ be a representation near α_∞. There is a mapping $l': F \to \mathbb{H}^3$ (or $\mathbb{H}^3 \amalg \mathbb{H}^3$) near to l_∞ with the property that the $\alpha'(g_i)$ make the same identifications on $l'(F)$ as the $\alpha_\infty(g_i)$ do on $l_\infty(F)$. This map l' induces a map $f': M_0 \to N_\alpha$, which induces the natural identification on fundamental groups. Furthermore, if α' sends the loops of P to parabolic elements, we can choose l' so that the induced map f' sends the annuli of ∂M_0 near P to annuli on $\partial(N_\alpha)_{(0,\varepsilon]}$.

In this way we construct maps $f_{i_j}: M - P \to N_{i_j}$ satisfying (a) and (b) of Proposition 12.6 so that there are lifts of $f_{i_j}|M_0$ to $l_{i_j}: F \to \mathbb{H}^3$ (or $\mathbb{H}^3 \cup \mathbb{H}^3$) with all the $l_{i_j}(F)$ contained in a fixed compact set in $\mathbb{H}^3$ (or $\mathbb{H}^3 \cup \mathbb{H}^3$).

Similarly, we have the homotopy $H_\infty: (\partial M - P) \times I \to N$ from $\tilde{f}_\infty \circ \tau$ to $\overline{G}_\infty \circ \tilde{f}_\infty$. We truncate $\partial_0 M$ near the cusps to form ∂M_0 and choose a fundamental domain T for $\partial M_0 \times I$. The homotopy lifts to $m_\infty: T \to (\mathbb{H}^3 \cup \mathbb{H}^3)$. There are approximating maps $m_{i_j}: T \to \mathbb{H}^3 \cup \mathbb{H}^3$ that induce the homotopies H_{i_j} in N_{i_j} from $\tilde{f}_{i_j} \circ \tau$ to $G_{i_j} \circ \tilde{f}_{i_j}$. Once again, the $m_{i_j}(T)$ are all contained in a fixe compact set.

For any compact set $K \subset (\mathbb{H}^3 \cup \mathbb{H}^3)$, there is $\varepsilon > 0$ so that the image of K in each N_{i_j} is contained in $(N_{i_j})_{(0,\varepsilon]}$. Thus; in light of the above, there is $\varepsilon > 0$ so that $f_{i_j}(M - P) \subset N'_{i_j}$ meets $(N^i_{i_j})_{(0,\varepsilon]}$ only in the cusp components corresponding to P, and $H_{i_j}((\partial M - P) \times I) \subset N_{i_j}$ meets $(N_{i_j})_{(0,\varepsilon]}$ only in the cusp components corresponding to P. Hence, for j sufficiently large, these images do not meet the components of $(N'_{i_j})_{(0,\varepsilon]}$ or $(N_{i_j})_{(0,\varepsilon]}$ corresponding to accidental parabolics. This proves Proposition 12.6. ∎

Choose $\varepsilon > 0$ sufficiently small and j sufficiently large so that Corollary 12.5 holds for N_{i_j} and ε/d, and also so that Proposition 12.6 holds for ε and j. Let $n = i_j$ for this j. We shall study the structure of $f_n(M - P)$ in N'_n and $\tilde{f}_n(\partial_0 M)$ in N_n. We know that $\tilde{f}_n: (\partial_0 M) \to N_n$ is properly homotopic to a splitting surface. In particular, it represents in each component of N_n the generating class for $H_2(N_n, \text{cusps}) \cong \mathbb{Z}$. We say that a point $x \notin \tilde{f}_n(\partial_0)$ is *on the plus side* of $\tilde{f}_n(\partial_0 M)$ if the ray joining x to $+\infty$ has even intersection number with $\tilde{f}_n(\partial_0 M)$. Otherwise x is *on the minus side* of $\tilde{f}_n(\partial_0 M)$. Let E_n^- be all those points on the minus side of $\tilde{f}_n(\partial_0 M)$.

LEMMA 12.7. *The projection map $\pi_n: E_n^- \to N'_n$ is finite-to-one. Furthermore, $\pi_n(E_n^-) \cup f_n(M - P) = N'_n$.*

Proof. First consider the case of a proper orientation-preserving embedding $\varphi_n: (M - P) \hookrightarrow N'_n$. The complement of this embedding is homeo-

V. Uniformization Theorem for Three-Dimensional Manifolds

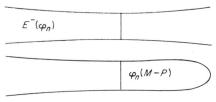

Figure 12.1

morphic to $(\partial_0 M) \times [0, \infty)$. In this case, $E^-(\varphi_n)$ maps homeomorphically onto $N'_n - \varphi_n(M - P)$ as in Fig. 12.1. Thus in this case the result is clear.

Unfortunately, since f_n had to be constructed by deforming a map in the limit and since the limit group can be associated to a compact 3-manifold of a different homeomorphism type, there is no hope of making the f_n embeddings while keeping the conclusions of Proposition 12.6.

On the other hand, f_n is homotopic to the embedding φ_n. Since f_n is already a proper embedding near P, we can take the homotopy to be compactly supported. Thus E_n^- differs from $E^-(\varphi_n)$ by a compact set. Hence,

$$\pi : E_n^- \to N'_n$$

is finite-to-one.

Let C be a locally finite (lf) chain that is a cycle generating $H_3^{1f}(M - P, \partial_0 M)$ (for example, triangulate $M - P$ and let C be the sum of the 3-simplices). Likewise, choose a locally finite chain C^- in $E^-(\varphi_n)$ that generates

$$H_3^{1f}(E^-(\varphi_n), \partial_0 M)$$

and whose boundary agrees with that of C. The chain $\varphi_n(C) - \pi_n(C^-)$ generates $H_3^{1f}(N'_n)$.

Let $D: (M - P) \times I \to N'_n$ be the deformation from φ_n to f_n. This induces a compactly supported deformation $\tilde{D}: E^-(\varphi_n) \times I \to N_n$ that begins at the identity, and so that $\tilde{D}|(\partial E^-(\varphi_n) \times I)$ covers $D|(\partial_0 M) \times I$. The image of $\tilde{D}_t(E^-(\varphi_n) \times I)$ is exactly the minus side of the lifting $\tilde{D}_t(\partial_0 M)$. Thus the chain $\pi_n(\tilde{D}_t(C^-)) - D_t(C)$ is always a cycle generating $H_3^{1f}(N'_n)$. Thus it must cover every point of M. In particular $\pi_n(\tilde{D}_1(E^-(\varphi_n))) \cup D_1(M - P) = N'_n$. Since $\tilde{D}_1(E^-(\varphi_n)) = E^-(f_n)$, it follows that $\pi_n(E^-(f)) \cup f_n(M - P) = N'_n$. ∎

Now at last we are ready to rule out Conditions (1)–(3) of Theorem 12.1 for $\pi : N \to N'$.

Condition (3) *cannot occur for a component of* $\pi : N \to N'$.

Proof. Suppose $\Pi_+(N) \cup \Pi_-(N)$ were nonempty. Pick $n = i_j$ so that Corollary 12.5 and Proposition 12.6 hold for $\pi_n : N_n \to N'_n$. Each accidental parabolic is represented in N_n by a component of $(N_n)_{(0,\varepsilon]}$. This component is disjoint from $\tilde{f}_n(\partial_0 M)$ and hence is either in the plus side or the minus side of $\tilde{f}_n(\partial_0 M)$. This defines a decomposition $\Pi_+(N) \cup \Pi_-(N)$ as $\Pi_+^n(N) \cup \Pi_-^n(N)$. (In fact, for n sufficiently large, these decompositions agree. We have no need for this fact so we do not prove it.) Since $\bar{G}_n : N_n \to N_n$ flips ends and $\bar{G}_n \circ \tilde{f}_n$ is homotopic to $\tilde{f}_n \circ \tau$ by a homotopy missing the components of $(N_n)_{(0,\varepsilon]}$ corresponding to accidental parabolics, we see that $\tau_\#(\Pi_\pm^n(N)) = \Pi_\mp^n(N)$.

Let $l_+ \subset N$ represent an element in $\Pi_+^n(N)$. Its image in N_n' lies in a component of $(N_n')_{(0,\varepsilon]}$ and is some power γ^t of the generator γ of this component. By Proposition 12.6, this component of $(N_n')_{(0,\varepsilon]}$ is outside of $f_n(M - P)$. Hence by Lemma 12.7 it is in the image of $\pi_n | E_n^-$. Since $\pi_n | E_n^-$ is finite-to-one, some power of γ, γ^s, lifts to a loop l_- in E_n^-. We choose l_- so that $s > 0$ and so that s is minimal.

We have loops l_+ and l_- in N so that $(l_+)^s$ is freely homotopic to $(l_-)^t$ in M. In $\partial_0 M$, l_+ is represented by a simple closed curve. By the annulus theorem, l_- is homotopic in $\partial_0 M$ to the (s/t)-power of a simple closed curve. Choosing base points on γ and l_-, we have $[\gamma] \in \pi_1(M)$ with $[\gamma^s] \in \pi_1(\partial_0 M)$. Since we have arranged that s is minimal, no smaller positive power of $[\gamma]$ is contained in $\pi_1(\partial_0 M)$.

If $s = 1$, then $t = 1$ and l_+ and l_- both cover γ and are freely homotopic. If $s > 1$, then we have an element $[\gamma] \in \pi_1(M)$ that is not in $\pi_1(\partial_0 M)$, but so that its sth power is in $\pi_1(\partial_0 M)$. Using the characteristic submanifold of M, one can show that there are exactly two ways in which this can happen:

(1) γ is freely homotopic to the core of a solid torus in Σ, which meets $\partial_0 M$ in a union of nontrivial annuli or

(2) γ is a loop in a component of Σ that is a twisted I-bundle over a nonorientable surface.

In the first case the minimal power of γ, which is contained in $\pi_1(\partial_0 M)$, is represented there by a simple closed curve. In the second case, the square of $[\gamma]$ is in $\pi_1(\partial_0 M)$ and if this class is represented by a power of a simple closed curve in $\partial_0 M$, then it is represented by a simple closed curve.

Applying this result to l_-, we see that l_- itself is represented by a simple closed curve in $\partial_0 M$ (see Fig. 12.2). Thus $s = t$ and l_+ and l_- cover the same loop in N'. Thus the lengths of l_+ and l_- are the same. Hence l_- is contained in $N_{(0,\varepsilon]}$ and represents an element of $\Pi_-^n(N)$. Let $x_\pm$ be the simple closed curves on $\partial M - P$ representing $l_\pm$. There is an annulus in M connecting them. We claim that the annulus $(A, \partial A) \to (M, \partial_0 M)$ connecting x_+ to x_- is essential. Certainly, it is nontrivial on fundamental groups. If it could be deformed, relative to ∂A, into $\partial M - P$, then x_+ and x_- would be freely homotopic in the covering space N corresponding to $\pi_1(\partial_0 M)$. This would

V. Uniformization Theorem for Three-Dimensional Manifolds

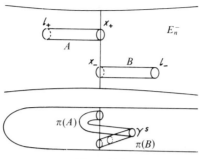

Figure 12.2

mean that l_+ and l_- were freely homotopic in N. But if N is any hyperbolic 3-manifold and l_+ and l_- are nontrivial loops in distinct components of $N_{(0,\varepsilon]}$ (ε less than the Margulis constant), then l_+ and l_- are not freely homotopic. This contradiction establishes that the annulus in M connecting x_+ to x_- is essential.

Now consider $m_+ = \tau(l_-) \in \Pi^n_+(N)$. Performing the same construction, we find $m_- \in \Pi^n_-(N)$ and a nontrivial annulus on $(M, \partial_0 M)$ connecting them.

Continuing in this way, we can find an arbitrarily long string of annuli, $A_1, A_2, A_3, \ldots, A_R$ in $(M, \partial_0 M)$, so that

(a) each A_i is nontrivial,
(b) $\partial_0(A_i) \in \Pi^n_+(N)$ and $\partial_1(A_i) \in \Pi^n_-(N)$, and
(c) $\tau\partial_1(A_i) = \partial_0(A_{i+1})$.

If R is greater than the cardinality of $\Pi^n_+(N)$, then among the $\{A_i\}$ there must be two A_r and A_s, $r < s$, so that $\tau\partial_1(A_s) = \partial_0 A_r$ up to isotopy. The chain of annuli $A_r, A_{r+1}, \ldots, A_s$ glue together in $M' \doteq M/\tau$ to form a torus T. Since each A_i is nontrivial, it follows easily that $\pi_1(T) \to \pi_1(M')$ is an injection. Since none of the boundary curves of the A_i are peripheral in $\partial_0 M$, it follows that $T \to M'$ is not homotopic to a map $T \to P' = P/\tau$. This contradicts the fact that (M', P') is a pared manifold. This contradiction shows that $\Pi_{\pm}(N)$ must be empty.

Condition (2) cannot occur for N'.

Proof. Since $\pi_1(M) = \pi_1(N')$, there is induced a natural one-to-one correspondence between the torus components of P and the $(\mathbf{Z} \times \mathbf{Z})$-cusps of N'. Hence there is a homotopy equivalence of pairs

$$(M, T(P)) \to (N', (\mathbf{Z} \times \mathbf{Z})\text{-cusps of } N_{(0,\varepsilon]}),$$

where $T(P)$ is the union of the torus components of P. Consequently, $H_3(N', (\mathbf{Z} \times \mathbf{Z})$-cusps of $N_{(0,\varepsilon]}) \cong H_3(M, T(P))$. The latter group is zero since each component of M has boundary points outside of P. If N' had a component of finite volume, then $H_3(N', (\mathbf{Z} \times \mathbf{Z})$-cusps) would be nonzero.

Condition (1) *cannot occur for any end of* N.

Since we have shown that Conditions (2) and (3) do not occur, for each end of N there are two remaining possibilities: (1) there is a simply connected component in the sphere at infinity associated to that end or (2) the end maps finite-to-one into N'. It can not happen that both ends of a component of N map finite-to-one into N', for if that happened, then N' would have a component finitely covered by a component of N. This would imply that the corresponding component of (M, P) was an I-bundle over a surface. Thus for each component N_Z of N there are three possibilities:

12.8.

(I) Associated to each end of N_Z there is a simply connected, invariant region of discontinuity.

(II) Associated to $(N_Z)_-$ there is a simply connected, invariant region of discontinuity and $(N_Z)_+$ maps finite-to-one into N'.

(III) Associated to $(N_Z)_+$ there is a simply connected, invariant region of discontinuity and $(N_Z)_-$ maps finite-to-one into N'.

CLAIM 12.9. *If associated to* $(N_Z)_-$ *there is a simply connected, invariant region of discontinuity, then* $(N_Z)_-$ *maps finite-to-one into* N'.

Proof. The quasi-fuchsian groups $\beta_{i_j}|\pi_1(Z)$ are converging to $\pi_1(N_Z)$. If the negative end of N_Z corresponds to a simply connected region of discontinuity Ω_- for $\pi_1(N_Z)$, then the negative domain of discontinuities for $\beta_{i_j}|\pi_1(Z)$ are converging to Ω_-. Since Ω_-^n is a component of the region of discontinuity for all of $\rho_n(\pi_1(M))$ (here is where $+$ and $-$ are distinguished), it follows immediately that Ω_- is a component of the region of discontinuity for the limit representation ρ_∞ of $\pi_1(M)$. The region in N_Z on the negative side of $C(N_Z)$ then embeds isometrically in N' (see Fig. 12.3). Clearly, then, the negative end of N_Z maps finite-to-one into N'. ∎

From Claim 12.9 it follows directly that (II) of 12.8 cannot occur. To rule out (III) of 12.8 we need one last lemma.

LEMMA 12.10. *Let* Z_0 *be a component of* $\partial_0 M$, *and let* $Z_1 = \tau(Z_0)$. *Associated to* $(N_{Z_0})_+$ *is a simply connected region of discontinuity for* $\pi_1(N_{Z_0})$ *if and only if associated to* $(N_{Z_1})_-$ *there is such a region for* $\pi_1(N_{Z_1})$.

Proof. An end $(N_{Z_0})_\pm$ has an associated simply connected region of discontinuity if and only if the ends $\partial_\pm(\beta_{i_j}|Z) \in \mathcal{T}(Z_0)$ converge to a point in

V. Uniformization Theorem for Three-Dimensional Manifolds

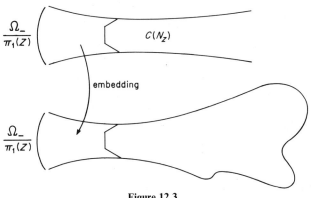

Figure 12.3

$\mathcal{T}(Z_0)$. On the other hand, we know from Lemma 11.9 that there is a fixed constant d so that the quasi-conformal distance from $\partial_{\pm}(\beta_{i_j}|Z_0)$ to $\partial_{\mp}(\beta_{i_j}|Z_1)$ is at most d. Thus if the sequence $\partial_{\pm}(\beta_{i_j}|Z_0)$ converges, then the sequence $\partial_{\mp}(\beta_{i_j}|Z_1)$ is bounded. Since $\beta_{i_j}|Z_1$ converges to a discrete and faithful representation of $\pi_1(Z_1)$, the sequence $\partial_{\mp}(\beta_{i_j}|Z_1)$ is bounded if and only if it converges. This shows that a simply connected region of discontinuity associated to $(N_{Z_0})_{\pm}$ implies the existence of such a region associated to $(N_{Z_1})_{\mp}$. Replacing Z_0 by Z_1 gives the symmetric result. Together these implications establish Lemma 12.10. ∎

Now suppose there were a component Z_0 for which N_{Z_0} were not of type (I). Then by Claim 12.9, it would have to be the case that $(N_{Z_0})_-$ had no associated region of discontinuity, whereas $(N_{Z_0})_+$ had such a region. Considering N_{Z_1}, Lemma 12.10 tells us that $(N_{Z_1})_-$ has an associated region of discontinuity while $(N_{Z_1})_+$ does not. This means that N_{Z_1} is of type (II). Coverings of this type were ruled out by Claim 12.9. It follows that all the components N_{Z_0} are of type (I).

Components of type (I) have two simply connected, invariant regions of discontinuity, i.e., they are quasi-fuchsian. We have just shown that the hyperbolic 3-manifold corresponding to β_∞ is a quasi-fuchsian manifold, i.e., that $\beta_\infty \in \mathrm{QF}(\partial_0 M)$. Thus (again invoking Claim 12.9) the manifold N' corresponding to $\alpha_\infty \in \mathrm{AH}(M, P)$ is geometrically finite, i.e., $\alpha_\infty \in Q(M, P) \subset \mathrm{AH}(M, P)$.

This completes the proof that α_∞ and hence all limit points of
$$\{\partial^{-1}(\tau \circ \sigma)^n(\partial \alpha)\}_{n \geq 0}$$
lie in $Q(M, P)$. Thus $\{(\tau \circ \sigma)^n(\partial \alpha)\}_{n > 0} \in \mathcal{T}(\partial_0 M)$ is a bounded set. This, of course, completes the proof of the bounded image theorem—at last.

13. Special Cases

In this section we discuss the exceptional cases in which a component of (M, P) is homeomorphic to $(E, E|\partial S)$, where E is an I-bundle over a surface. If E is nontrivial, then there is a two-sheeted covering of E which is a product $(S \times I, \partial S \times I)$. Thus we can replace (M, P) by a two-sheeted cover $(\tilde{M}, \tilde{P})$ (which is a trivial cover of all components that are not twisted I-bundles), so that all the exceptional components are products $(S \times I, \partial S \times I)$. Of course, the gluing involution τ lifts to a gluing involution $\tilde{\tau}$ on $\partial_0 \tilde{M}$. If we wish in $(\tilde{M}, \tilde{P})$ to glue a product component to a different component, this can be achieved by simply eliminating the product component from the decomposition. Performing this simplification on $(\tilde{M}, \tilde{P})$, we are left with one of two possibilities:[6]

(a) $(\tilde{M}, \tilde{P})$ has no exceptional components,
(b) $(\tilde{M}, \tilde{P}) \cong (S \times I, \partial S \times I)$.

If we can solve the gluing problem for $\tilde{\tau}$ on $(\tilde{M}, \tilde{P})$, then we shall construct a complete hyperbolic structure of finite volume on a two-sheeted covering $\tilde{M}'$ of $M' = M/\tau$. According to Mostow's rigidity theorem, the involution of $\tilde{M}'$ that switches the sheets of $\tilde{M}'$ is homotopic to an isometry T of order 2. The quotient $\tilde{M}'/T$ will be a complete hyperbolic manifold of finite volume whose fundamental group is isomorphic to $\pi_1(M')$. Since M' is Haken, this implies that $\tilde{M}'/T$ is homeomorphic to $\text{int}(M')$.

As a result, if, in addition to Theorem 10.1, we can solve the gluing problem for a product $(S \times I, \partial S \times I)$, then we can solve the general gluing problem. The case in which $(M, P) \cong (S \times I, \partial S \times I)$ is the case in which $M' = M/\tau$ fibers over the circle. Thus the last remaining case of the gluing problem is dealt with by the following theorem.

THEOREM 13.1. *Let M' be a compact, irreducible 3-manifold that fibers over S^1, with fiber of negative Euler characteristic. Suppose that every map $T^2 \to M'$ that is injective on the fundamental group is homotopic to a map to $\partial M'$. Then M' has a complete hyperbolic structure of finite volume.*

The condition that M' be atoroidal is equivalent to the conditions that (a) the fiber have negative Euler characteristic and (b) the monodromy map ψ on the fiber be pseudo-Anosov. If we cut M' open along a fiber, we obtain a pared manifold of the form $(S \times I, \partial S \times I)$, where S is the fiber. Since $\chi(S) < 0$, int S has a complete hyperbolic structure. Theorem 13.1 is a consequence of the following theorem.

---

[6] We are assuming here that the result of gluing $(\tilde{M}, \tilde{P})$ together is connected.

V. Uniformization Theorem for Three-Dimensional Manifolds

THEOREM 13.2. *Let S be a surface with $\chi(S) < 0$. Let $\varphi: S \to S$ be a pseudo-Anosov homeomorphism. Let $(g_1, g_2) \in \mathcal{T}(S) \times \mathcal{T}(S) = \mathrm{QF}(S)$ be an arbitrary element. The sequence $\{\varphi^n(g_1), \varphi^{-n}(g_2)\} \in \mathcal{T}(S) \times \mathcal{T}(S) = \mathrm{QF}(S)$ converges to a limit in $\mathrm{AH}(S \times I, \partial S \times I)$. This limit group has an empty region of discontinuity on S_∞^2. The associated 3-manifold admits a quasi-conformal homeomorphism which induces φ_* on the fundamental group.*

Given this, one applies Sullivan's rigidity theorem [17] to conclude that the quasi-conformal homeomorphism is actually conformal, i.e., an isometry. Thus we have produced a hyperbolic 3-manifold N with $\pi_1(N) \cong \pi_1(S)$ and an isometry $\varphi: N \to N$ inducing φ_* on $\pi_1(N)$. Since φ_* is of infinite order in $\mathrm{out}(\pi_1(N))$, no power of φ is close to the identity, i.e., the infinite cyclic group generated by φ acts properly discontinuously (and hence freely) on N. The quotient $N/(\varphi)$ is a hyperbolic 3-manifold homotopy equivalent to M'. Hence $N/(\varphi)$ is homeomorphic to int M'. The quotient automatically has finite volume.

This result is outlined in more detail in [16].

14. Kleinian Groups with Torsion

Let $\hat{S}$ be a closed surface and let $S \subset \hat{S}$ be the complement of a finite set of points. Let τ be a triangulation of $\hat{S}$ so that every point of $\hat{S} - S$ is a vertex of τ. There is on $\hat{S}$ the dual cell composition to τ, D_τ. Restricting D_τ to S, we find a decomposition of S into *faces*, which are cells or once punctured cells. We call $D_\tau | S$ *the decomposition dual to τ of S into faces*.

Let M be a 3-manifold and $T \subset \partial M$ be a compact subsurface. Let $Y = \overline{\partial M - T}$, and form $\hat{S}$ by collapsing the boundary circles of Y to points. This surface $\hat{S}$ contains $\partial M - T$ as the complement of a finite set of points. Let τ be a triangulation of $\hat{S}$ as before, and let D_τ be the decomposition of $\partial M - T$ into faces, which is dual to τ.

A *boundary pattern* $\mathscr{P}$ of (M, T) is (see Fig. 14.1)

(1) a decomposition D_τ of $\partial M - T$ into faces, which is dual to a triangulation,
(2) a subcomplex $\partial_0 \mathscr{P}$ which is a subsurface of $\partial M - T$, and
(3) a function $\theta_\mathscr{P}: (\text{edges of } D_\tau) \to \{\pi/2, \pi/3, \ldots, 0\}$.

These are required to satisfy the following:

(a) If three edges e_1, e_2, and e_3 are incident at a vertex v, then

$$\sum_{i=1}^{3} \theta_\mathscr{P}(e_i) > \pi \quad \text{or} \quad (\theta_\mathscr{P}(e_1), \theta_\mathscr{P}(e_2), \theta_\mathscr{P}(e_3)) = (0, \pi/2, \pi/2)$$

up to cyclic permutation.

(b) There is no edge of $\partial_0 \mathscr{P}$ in $\theta_\mathscr{P}^{-1}(0)$.

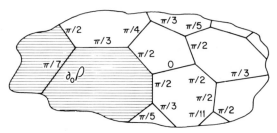

Figure 14.1

We are most interested in the special case of this construction in which (M, T) forms a pared manifold. Let (M, P) be a pared manifold with boundary pattern $\mathscr{P}$. We define its *parabolic locus* to be P union a regular neighborhood of $\theta_{\mathscr{P}}^{-1}(0)$ in $\partial_0 M = \partial M - P$. This regular neighborhood is required to meet $\partial_0 \mathscr{P}$ in a regular neighborhood of $\theta_{\mathscr{P}}^{-1}(0) \cap \partial(\partial_0 \mathscr{P})$ in $\partial(\partial_0 \mathscr{P})$, shown in Fig. 14.2.

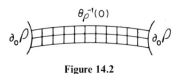

Figure 14.2

Let $\mathscr{P}$ be a boundary pattern for (M, T). There is associated a type of generalized Coxeter group denoted $\Gamma(M, \mathscr{P})$. To construct this group, we take the universal cover $\tilde{M}$ of M and induce on it the boundary pattern $\tilde{\mathscr{P}}$ covering $\mathscr{P}$. Let $R_{\tilde{\mathscr{P}}}$ be the "Coxeter group" generated by $\{r_{\tilde{f}}\}$ as $\tilde{f}$ ranges over the faces of the decomposition of $\tilde{\mathscr{P}}$ which lie outside $\partial_0 \tilde{\mathscr{P}}$. The relations that these generators are required to satisfy are (1) $(r_{\tilde{f}})^2 = 1$, and (2) $(r_{\tilde{f}} r_{\tilde{f}'})^l = 1$ if $\tilde{f}$ and $\tilde{f}'$ share an edge $\tilde{e}$ with $\tilde{\theta}_{\tilde{\mathscr{P}}}(\tilde{e}) = \pi/l$. The group $\pi_1(M)$ acts on the faces of $\tilde{\mathscr{P}}$. This induces an action of $\pi_1(M)$ on $R_{\tilde{\mathscr{P}}}$, namely, $\gamma^{-1}(r_{\tilde{f}})\gamma = r_{\gamma(\tilde{f})}$ for $\gamma \in \pi_1(M)$. The group $\Gamma(M, \mathscr{P})$ is the semidirect product $R_{\tilde{\mathscr{P}}} \rtimes \pi_1(M)$. Notice that if $\mathscr{P}'$ is another boundary pattern for (M, P) identical in all respects with $\mathscr{P}$ except that $\partial_0 \mathscr{P}' \subset \partial_0 \mathscr{P}$, then there is a natural inclusion $\Gamma(M, \mathscr{P}) \subset \Gamma(M, \mathscr{P}')$.

Since the decomposition of $\partial M - T$ determined by $\mathscr{P}$ is dual to a triangulation, any two distinct faces share at most one edge. This means that there is another Coxeter group associated with the boundary pattern $\mathscr{P}$ on $\partial M - T$. This group $R_{\mathscr{P}}$ is generated by elements r_f as f ranges are the faces of $\mathscr{P} - \partial_0 \mathscr{P}$. They are subject to relations analogous to those above: $(r_f)^2 = 1$ and $(r_f r_{f'})^l = 1$ if f and f' share an edge with $\theta_{\mathscr{P}}(e) = \pi/l$.

V. Uniformization Theorem for Three-Dimensional Manifolds

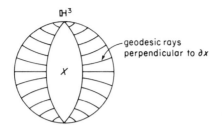

Figure 14.3

There is an obvious surjective homeomorphism

$$\Gamma(M, \mathcal{P}) = R_{\tilde{\mathcal{P}}} \rtimes \pi_1(M) \to R_{\mathcal{P}}.$$

When restricted to $R_{\tilde{\mathcal{P}}}$ this map is just induced by the covering projection thought of as a map from the faces of $\tilde{\mathcal{P}}$ to those of $\mathcal{P}$. Any splitting of $\pi_1(M)$ back into $\Gamma(M, \mathcal{P})$ is sent trivially.

In the case in which $\mathcal{P}$ is a boundary pattern for a pared manifold (M, P), we define a subgroup of $\Gamma(M, \mathcal{P})$ to be *peripheral* if it is conjugate to a subgroup of $\Gamma(N_\infty, \mathcal{P}|N_\infty)$, where N_∞ is a component of a regular neighborhood in M of $P \cup \theta_{\mathcal{P}}^{-1}(0)$.

Let (M, P) be a pared manifold with boundary pattern $\mathcal{P}$. A *geometric realization* of it is a convex hyperbolic structure of finite volume on $M_0 = M - (P \cup \theta_{\mathcal{P}}^{-1}(0))$ so that ∂M_0 is made up of totally geodesic faces that realize the decomposition D_τ of $\mathcal{P} - \partial_0 \mathcal{P}$ and that meet at the angles prescribed by $\theta_{\mathcal{P}}$. We shall call this a geometric realization of $(M, \mathcal{P})$. Given a geometric realization of $(M, \mathcal{P})$, there is a convex subspace X inside $\mathbb{H}^3$ and a representation $\rho \colon \Gamma(M, \mathcal{P}) \to \mathrm{PGL}_2(\mathbb{C})$. It is faithful onto a discrete group that leaves X invariant (Fig. 14.3). The quotient $X/\rho(\Gamma(M, \mathcal{P}))$ is identified with M_0 (see Fig. 14.4). It sits in $\mathbb{H}^3/\rho(\Gamma(M, \mathcal{P}))$ as a locally convex subset of finite volume. Its frontier is $\partial_0 \mathcal{P}$.

The image of ρ is a kleinian group with torsion. Its elements of finite order are exactly the ones that act with fixed points on $\mathbb{H}^3$. Since ρ is faithful, it follows that the only elements of finite order in $\Gamma(M, \mathcal{P})$ are conjugates of (a) the reflections $r_{\tilde{f}}$ and (b) powers of products of reflections $(r_{\tilde{f}_1} r_{\tilde{f}_2})^h$ in adjacent faces whose common edge has nonzero angle.

If we take all parabolic elements in $\rho(\Gamma(M, \mathcal{P}))$ and take an invariant family of sufficiently small horoballs centered at the fixed points of these elements, then the image of this set of horoballs in $\mathbb{H}^3/\mathrm{Im}\,\rho$ meets M_0 in the intersection of M_0 with a regular neighborhood of $P \cup \theta_{\mathcal{P}}^{-1}(0)$ in M. We call this image $(M, \mathcal{P})_{\mathrm{thin}}$ and the closure of its complement $(M, \mathcal{P})_{\mathrm{thick}}$. The underlying space of $(M, \mathcal{P})_{\mathrm{thick}}$ is compact and homeomorphic to M. The space underlying the intersection of the thick and thin parts is isotopic to $P \cup$ (regular neighborhood of $\theta_{\mathcal{P}}^{-1}(0)$) in $\partial_0 M$. According to Selberg's

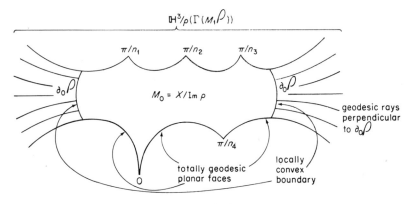

Figure 14.4

theorem [5], any kleinian group Γ has a torsion-free, normal subgroup of finite index Γ_0. We say that Γ is *geometrically finite* if Γ_0 is. (This is independent of the choice of Γ_0.) If L_Γ is not contained in a geometric circle, then Γ is geometrically finite if and only if H_Γ/Γ is of finite volume.

If we have a geometric realization of $(M, \mathscr{P})$, then the kleinian group $\rho(\Gamma(M, \mathscr{P}))$ acts on $\mathbb{H}^3$. In the quotient there is the convex manifold M_0, which is of finite volume. Since its preimage in $\mathbb{H}^3$ is convex, M_0 contains H_Γ/Γ. Thus $\rho(\Gamma(M, \mathscr{P}))$ is a geometrically finite kleinian group. If we pass to a torsion-free, normal subgroup of finite index $\Gamma_0 \subset \rho(\Gamma(M, \mathscr{P}))$, then $\mathbb{H}^3/\Gamma_0$ is a complete hyperbolic 3-manifold. The preimage of M_0 in $\mathbb{H}^3/\Gamma_0$ is a convex manifold $\tilde{M}_0$ of finite volume carrying the fundamental group. Its thin part is the preimage of the space underlying $(M, \mathscr{P})_{\text{thin}}$. The finite group $\rho(\Gamma(M, \mathscr{P}))/\Gamma_0$ acts (nonfreely) on $\tilde{M}_0$ as a group of isometries. The quotient space inherits a metric. With this metric it is isometric to M_0.

Notice that if $\mathscr{P}_0$ and $\mathscr{P}_1$ are boundary patterns for (M, P) identical except for the fact that $\partial_0 \mathscr{P}_0$ is smaller than $\partial_0 \mathscr{P}_1$, then any geometric realization of $(M, \mathscr{P}_0)$ is automatically one for $(M, \mathscr{P}_1)$. Let $\rho' : \Gamma(M, \mathscr{P}_0) \to \text{PGL}_2(\mathbb{C})$ be the representation associated to the realization of $(M, \mathscr{P}_0)$. The representation $\rho : \Gamma(M, \mathscr{P}_1) \to \text{PGL}_2(\mathbb{C})$ coming from the induced realization of $(M, \mathscr{P}_1)$ is conjugate to the restriction of ρ' to $\Gamma(M, \mathscr{P}_1) \subset \Gamma(M, \mathscr{P}_0)$. The processes of enlarging $\partial_0 \mathscr{P}_0$ in effect "frees" more of the boundary. It is allowed to be locally convex instead of being piecewise geodesic.

We need a version of Selberg's theorem that gives sufficient conditions for $\Gamma(M, \mathscr{P})$ to have a torsion-free subgroup of finite index in the case in which we have no geometric realization in hyperbolic space. The reason is that we shall need such a subgroup in order to construct the geometric realization. The following lemma will suffice for our purposes.

V. Uniformization Theorem for Three-Dimensional Manifolds

LEMMA 14.1. *Let M be a 3-manifold and T a proper subsurface of ∂M. Let $\mathscr{P}$ be a boundary pattern for (M, T). Suppose that the only elements of finite order in $\Gamma(M, \mathscr{P})$ are (a) the reflections $r_{\tilde{f}}$ in the faces of $\tilde{\mathscr{P}} - \partial_0 \tilde{\mathscr{P}}$ and (b) powers of products of reflections $(r_{\tilde{f}} r_{\tilde{f}'})^h$ in adjacent faces sharing an edge of nonzero angle. Then $\Gamma(M, \mathscr{P})$ has a torsion-free subgroup of finite index.*

Proof. We have the surjection $\pi: \Gamma(M, \mathscr{P}) \to R_{\mathscr{P}}$. We shall construct a representation of $R_{\mathscr{P}}$ in a linear group and apply Selberg's theorem to the image. Let V be the (finite dimensional) vector space with basis the faces f of $\mathscr{P} - \partial_0 \mathscr{P}$. Define an inner product by

$$f \cdot f' = \begin{cases} 1 & \text{if } f = f', \\ 0 & \text{if } f \text{ and } f' \text{ are not adjacent,} \\ \cos(\theta_{\mathscr{P}}(e)) & \text{if } f \text{ and } f' \text{ share the edge } e. \end{cases}$$

(Notice that it is at this point that it becomes important that f and f' share at most one edge.) We define a representation $\psi: R_{\mathscr{P}} \to \text{aut}(V, \cdot)$. The reflection r_f is sent to reflection of V in the hyperplane perpendicular to f. One sees easily that this is a well-defined homomorphism. By the assumption on $\Gamma(M, \mathscr{P})$, any element of finite order in $\Gamma(M, \mathscr{P})$ maps via $\psi \circ \pi$ to an element of the same order in $\text{aut}(V, \cdot)$. By Selberg's theorem $\text{Im}(\psi)$ has a torsion-free subgroup of finite index. The preimage of this subgroup in $\Gamma(M, \mathscr{P})$ is also torsion-free and of finite index. ∎

Let (M, P) be a pared manifold with boundary pattern $\mathscr{P}$. The cell structure of $\mathscr{P}$, when restricted to $\partial_0 M - \partial_0 \mathscr{P}$ induces a cell structure on $\partial(\partial_0 \mathscr{P})$, which we now define (see Fig. 14.5). The vertices of this cell structure are the vertices of the original cell structure lying on $\partial(\partial_0 \mathscr{P})$ that are vertices of two faces in the complement of $\partial_0 \mathscr{P}$. Associated to each of these vertices is the angle $\theta_{\mathscr{P}}(e)$, where e is the edge incident to the vertex that separates the two faces outside $\partial_0 \mathscr{P}$. We denote the induced cell structure and angles by $\partial \mathscr{P}$.

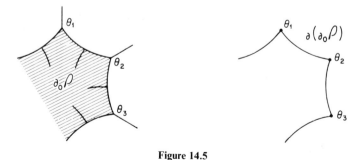

Figure 14.5

This pair is a surface with boundary pattern on its boundary. Hence, there is an associated Coxeter group associated to each component. Let Z be a component of $\partial_0 \mathscr{P}$. Take the universal cover $\tilde{Z}$ of Z. Let $\partial \tilde{\mathscr{P}} | \tilde{Z}$ be the induced cell structure and angles. Let $R_{\tilde{Z}}$ be the group generated by $\{r_{\tilde{d}}\}$ where $\tilde{d}$ ranges over the one cells of the cell structure on $\partial \tilde{Z}$. These generators are subject to the relations

(a) $(r_{\tilde{d}})^2 = 1$ and
(b) $(r_{\tilde{d}} r_{\tilde{d}'})^l = 1$, if $\tilde{d}$ and $\tilde{d}'$ share a vertex v with angle π/l associated to it.

The group $\Gamma(Z, \partial \mathscr{P} | Z)$ is $R_{\tilde{Z}} \rtimes \pi_1(Z)$.

A subgroup of $\Gamma(Z, \partial \mathscr{P} | Z)$ is *peripheral in Z* if it is conjugate in $\Gamma(Z, \partial \mathscr{P} | Z)$ to a subgroup of $\Gamma(N_\infty \cap Z, \partial \mathscr{P} | N_\infty \cap Z)$, where N is a regular neighborhood of $P \cup \theta_{\mathscr{P}}^{-1}(0)$ in M.

The inclusion of Z into M induces a natural homomorphism $\Gamma(Z, \partial \mathscr{P} | Z) \to \Gamma(M, \mathscr{P})$, which sends peripheral subgroups in $\Gamma(Z, \partial \mathscr{P} | Z)$ to peripheral subgroups in $\Gamma(M, \mathscr{P})$. We say that $(Z, \partial \mathscr{P} | Z)$ has *no accidental parabolics* if any subgroup whose image in $\Gamma(M, \mathscr{P})$ is peripheral is already peripheral in $\Gamma(Z, \partial \mathscr{P} | Z)$.

An isomorphism between surfaces with boundary patterns is a homeomorphism that preserves the tesselation of the boundaries and the "angles" between the edges of the tesselation. If $\varphi: (Z_0, \mathscr{P}_0) \to (Z_1, \mathscr{P}_1)$ is such an isomorphism, then it induces $\varphi_\#: \Gamma(Z_0, \mathscr{P}_0) \xrightarrow{\cong} \Gamma(Z_1, \mathscr{P}_1)$.

Let $\mathscr{P}$ be a boundary pattern for (M, T). Suppose that $\partial_0 \mathscr{P} \cong Z_0 \amalg Z_1$. Suppose that an isomorphism $\varphi: (Z_0, \partial \mathscr{P} | Z_0) \to (Z_1, \partial \mathscr{P} | Z_1)$ is given. We can form the topological space $M' = M/\varphi$, where we glue Z_0 to Z_1. The boundary of this manifold M' is $\overline{\partial M - (Z_0 \amalg Z_1)}/(\varphi | \partial Z_0)$. Let T' be the image of T in $\partial M'$. The boundary pattern $\mathscr{P}$ in $\partial M - T$, when restricted to $\partial M - (T \cup \partial_0 \mathscr{P})$, glues up under φ to form a boundary pattern for $\partial M' - T'$. The vertices of this decomposition are the images of the vertices of D_τ, which do not lie in $\partial_0 \mathscr{P}$ (nor on $\partial(\partial_0 \mathscr{P})$). The edges and faces of the decomposition come from D_τ. We take all edges of D_τ except those that are labeled 0 and that have both end points in $\partial_0 \mathscr{P}$. We glue two of these edges of D_τ together to make an edge (or part of an edge) of D'_τ whenever φ glues an end point of one to an end point of the other. Likewise, we glue two faces together whenever φ identifies an edge of one with an edge of the other. The function $\theta_\mathscr{P}$ induces a function $\theta_{\mathscr{P}'}: $ (edges $D'_\tau) \to \{\pi/2, \pi/3, \ldots, 0\}$. In general, there is no reason to expect this decomposition of $\partial M' - T'$ into faces to be dual to a triangulation. For example, it could happen that an edge of D'_τ is a simply closed curve without vertices. However, let us suppose that D'_τ is dual to a triangulation. Then we say that $(M, T, \mathscr{P}, \varphi)$ is *gluing data* and that (M', T') with boundary pattern $\mathscr{P}'$ ($\partial_0 \mathscr{P}' = \varnothing$) is the *result of gluing*. If M has two components (M_0, T_0) and (M_1, T_1), with $Z_i \subset \partial M_i$, then

$$\Gamma(M', \mathscr{P}') \cong \Gamma(M_0, \mathscr{P} | M_0) *_{\varphi_*} \Gamma(M_1, \mathscr{P} | M_1).$$

V. Uniformization Theorem for Three-Dimensional Manifolds

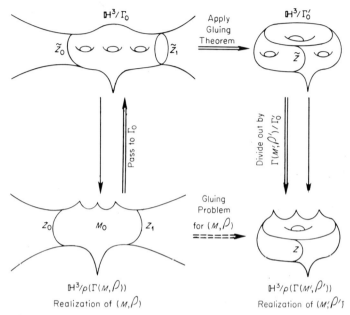

Figure 14.6 Diagram for the Proof of Theorem 14.2

If M is connected, then $\Gamma(M', \mathscr{P}')$ is the HNN construction

$$\Gamma(M, \mathscr{P}) *_{\varphi_*}.$$

The main result concerning gluing geometric realizations is the following theorem which generalizes the Gluing Theorem (see Fig. 14.6).

THEOREM 14.2. *Let (M, P) be a pared manifold with boundary pattern $\mathscr{P}$ that has a geometric realization. Let $\partial_0 \mathscr{P} = Z_0 \amalg Z_1$. Suppose that $(M, P, \mathscr{P}, \varphi)$ is gluing data and that the result of gluing is (M', P') with boundary pattern $\mathscr{P}'$. The latter has a geometric realization under the following hypotheses:*

(1) *The groups $\Gamma(Z_i, \partial \mathscr{P} | Z_i)$ map injectively into $\Gamma(M, \mathscr{P})$ and have no accidental parabolics, for $i = 1$ and 2;*

(2) *Every abelian, noncyclic subgroup of $\Gamma(M', \mathscr{P}')$ is peripheral.*

(3) *If p_0 is a nontrivial peripheral subgroup and α is an element of $\Gamma(M', \mathscr{P}')$ such that $\alpha p_0 \alpha^{-1}$ is also a peripheral subgroup, then $\langle \alpha, p_0 \rangle$ in $\Gamma(M', \mathscr{P}')$ is peripheral.*

(N.B.: Hypotheses (2) and (3) in the statement of Theorem 14.2 are the group theoretic analogues of Conditions (i) and (ii) in the definition of a pared manifold (Definition 4.8).)

Proof. Since (M, P) with boundary pattern $\mathscr{P}$ has a geometric realization, the only elements in $\Gamma(M, \mathscr{P})$ of finite order are reflections and power of products of reflections in adjacent edges. By the foregoing description of $\Gamma(M', \mathscr{P}')$ in terms of $\Gamma(M, \mathscr{P})$ and the Kurosh subgroup theorem, the same is true for $\Gamma(M', \mathscr{P}')$ provided that $\Gamma(Z_i, \partial \mathscr{P}|Z_i) \to \Gamma(M, \mathscr{P})$ is an injection for $i = 1$ and $i = 2$. This condition, of course, is required by hypothesis. According to Lemma 14.1, then $\Gamma(M', \mathscr{P}')$ has a torsion-free subgroup of finite index Γ'_0. This we can take to be normal. Let $\tilde{M}'$ be the manifold corresponding to Γ'_0. There is an action of $\Gamma(M', \mathscr{P}')/\Gamma'_0$ on $\tilde{M}'$ with quotient M'. There is a corresponding covering $\tilde{M}$ of M. The manifold $\tilde{M}'$ is obtained by gluing together $\tilde{M}$ along the surfaces $\tilde{Z}_0$ and $\tilde{Z}_1$ (where $\tilde{Z}_i$ is the preimage in $\tilde{M}$ of Z_i in M.) The manifold $\tilde{M}$ has a geometrically finite hyperbolic structure induced from the geometric realization. Furthermore, $\partial \tilde{M} = \tilde{Z}_0 \amalg \tilde{Z}_1$.

Suppose for the moment that we can find a hyperbolic structure on $\tilde{M}$ that glues together in the sense of Section 8. This would produce a complete hyperbolic structure of finite volume on $\tilde{M}'$. The group $\Gamma(M', \mathscr{P}')/\Gamma'_0$ of homeomorphisms of $\tilde{M}'$ is then represented as a group of isometries of $\tilde{M}'$. It is easy to see that the quotient of $\tilde{M}'$ by this action is a hyperbolic structure on $\tilde{M}'$ that realizes the boundary pattern $\mathscr{P}'$. Clearly, it has finite volume. Thus (M', P') with boundary pattern $\mathscr{P}'$ has a geometric realization.

It remains to study the gluing problem for $\tilde{M}$ and the homeomorphism $\tilde{\varphi}: \tilde{Z}_0 \to \tilde{Z}_1$. The whole point of introducing the Coxeter groups is that we have arrived at the type of gluing problem studied in Sections 7–12; that is, $\partial \tilde{M} - \tilde{P} = \tilde{Z}_0 \amalg \tilde{Z}_1$. According to the gluing theorem of Section 7, we can solve the gluing problem in order to find a complete hyperbolic structure of finite volume on $\tilde{M}/\varphi$ provided that

(a) $\tilde{M}$ has a geometrically finite hyperbolic structure,
(b) the glued-up pair $(\tilde{M}', \tilde{P}')$ is a pared manifold, and
(c) the image $\tilde{Z}$ of $\tilde{Z}_0$ in $\tilde{M}'$ is a superincompressible surface.

Condition (a) is immediate. To ensure condition (b) we must establish that $\tilde{M}'$ is irreducible and that conditions (i) and (ii) of Definition 4.8 hold for $(\tilde{M}', \tilde{P}')$. Hypotheses (2) and (3) in the statement of this theorem automatically hold for $\pi_1(\tilde{M}')$ (with peripheral being with respect to $\tilde{P}'$). The reason is that $\pi_1(\tilde{M}')$ is a subgroup of finite index in $\Gamma(M', \mathscr{P}')$. From this one deduces easily that conditions (i) and (ii) of Definition 4.8 hold for $(\tilde{M}', \tilde{P}')$. Each component of $\tilde{M}$ is a convex hyperbolic manifold and hence is irreducible. It follows immediately that $\tilde{M}'$, which is obtained from $\tilde{M}$ by gluing $\tilde{Z}_0$ to $\tilde{Z}_1$, is also irreducible.

Finally, we must show that $\tilde{Z} \hookrightarrow \tilde{M}'$ is a superincompressible surface; since $\Gamma(Z_i, \partial \mathscr{P}|Z_i) \to \Gamma(M, \mathscr{P})$ is injective, $\pi_1(\tilde{Z}_i) \to \pi_1(\tilde{M})$ is also injective, for $i = 1$ and 2. Thus $\tilde{Z} \subset \tilde{M}'$ is an incompressible surface. If it has a compressing annulus into $\partial \tilde{M}' = \tilde{P}'$, then one of the $\tilde{Z}_i$ would have a compressing annulus into $\tilde{P}$. This contradicts hypothesis (1).

This completes the proof of Theorem 14.2. ∎

Now we take up the question of when hypothesis (1) of Theorem 14.2 is satisfied. We find a simple, geometric condition that is sufficient.

PROPOSITION 14.3. *Let (M, P) be a pared manifold with boundary pattern $\mathscr{P}$. Let Z be a component of $\partial_0 \mathscr{P}$. The natural map $\Gamma(Z, \partial \mathscr{P}|Z) \to \Gamma(M, \mathscr{P})$ is an injection without accidental parabolics if*

(a) *$\pi_1(Z) \to \pi_1(M)$ is an injection without accidental parabolics;*
(b) *if $\tilde{Z}$ is the preimage of Z in the universal cover $\tilde{M}$ of M, then any face of $\tilde{\mathscr{P}}$ outside $\tilde{Z}$ that meets a component of $\tilde{Z}$ meets it in a connected set;*
(c) *if $\tilde{f}_1$ and $\tilde{f}_2$ are adjacent faces of $\tilde{\mathscr{P}}$ both lying outside $\tilde{Z}$ and both meeting a given component of $\tilde{Z}$, then $\tilde{f}_1$ and $\tilde{f}_2$ share a vertex on that component (Fig. 14.7).*

Proof. We shall use the equivariant Dehn's lemma and annulus theorem of Meeks and Yau [13]. However we can avoid the use of these results and give more directly naïve geometric arguments. Let us suppose for simplicity that (M, P) with boundary pattern $\mathscr{P}$ has a geometric realization. This is the only case that is required for the rest of the proof. By assuming this we avoid having to define the universal covering of a manifold with boundary pattern.

Let N be a convex hyperbolic manifold of finite volume realizing $(M, \mathscr{P})$ and let $\rho: \Gamma(M, \mathscr{P}) \to \mathrm{PGL}_2(\mathbf{C})$ be the associated representation. Let $\Gamma_0 \subset \Gamma(M, \mathscr{P})$ be a torsion-free, normal subgroup of finite index. Let $\tilde{N}$ be $\mathbb{H}^3/\Gamma_0$. Then $\Gamma(M, \mathscr{P})/\Gamma_0$ acts on $\tilde{N}$ with quotient N. Clearly, to prove Proposition 14.3 it suffices to show that for any component $\tilde{Z}$ of the induced covering of Z, we have $\tilde{Z} \to \tilde{N}$, an incompressible surface without accidental parabolics. Suppose that $\pi_1(\tilde{Z}) \to \pi_1(\tilde{N})$ has a nontrivial kernel. Then there is a disk $(D, \partial D) \subset (\tilde{N}, \tilde{Z})$ so that for each $\gamma \in \Gamma(M, \mathscr{P})/\Gamma_0$, $\gamma D \cap D = \varnothing$ or $\gamma D = D$. The stabilizer of D in $\Gamma(M, \mathscr{P})/\Gamma_0$ must be trivial, $\mathbf{Z}/2\mathbf{Z}$, or dihedral.

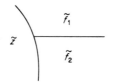

Figure 14.7

Figure 14.8

Accordingly, the image of D in N is (a) an embedded disk whose boundary is contained in Z; (b) an embedded disk meeting Z is an arc and meeting $\partial N - Z$ in an arc that crosses no edge of $\partial \mathcal{P}$ outside of Z (Fig. 14.8); or (c) an embedded disk meeting Z in an arc and meeting $\partial N - Z$ in an arc that crosses exactly one edge of ∂ outside of Z (Fig. 14.9).

Figure 14.9

The hypotheses imply that all such disks can occur in only the most trivial of ways. One sees easily from this that any such disk in $\tilde{N}$ must be parallel into $\tilde{Z}$ in $\tilde{N}$.

If $\tilde{Z} \subset \tilde{N}$ has accidental parabolics, then there is an essential annulus embedded in $\tilde{N}$ that connects a loop in $\tilde{Z}$ to a loop in $\tilde{P}$. Again we can choose this annulus A so that for all $\gamma \in \Gamma(M, \mathcal{P})/\Gamma_0$, $\gamma A \cap A = \emptyset$, or $\gamma A = A$. The image of A in N is either (a) an essential annulus for Z in N, or (b) a disk

Figure 14.10

as in Fig. 14.10. Again, the hypotheses tell us that there are no essential annuli for Z in N and that any disk as in (b) must actually be as shown in Fig. 14.11. It is easy to see that these lead to inessential annuli in $\tilde{N}$. ∎

Figure 14.11

V. Uniformization Theorem for Three-Dimensional Manifolds

15. Patterns of Circles

In this section we construct families of kleinian groups with torsion. If τ is a triangulation, then we let $E(\tau)$ denote the set of edges (one-simplexes) of τ.

THEOREM 15.1. (Andre'ev's Theorem [3]). *Let τ be a triangulation of S^2 not simplicially equivalent to $\partial \Delta^3$, and let $\theta: E(\tau) \to [0, \pi/2]$ be given. Suppose that*

(a) *if e_1, e_2, and $e_3 \in E(\tau)$ form a closed loop on S^2 and if $\sum_{i=1}^{3} \theta(e_i) \geq \pi$, then (e_1, e_2, e_3) are the three faces of a two-simplex of τ,*

(b) *if $e_1, e_2, e_3, e_4 \in E(\tau)$ form a closed loop on S^2 and if $\theta(e_i) = \pi/2$ for all $i = 1, \ldots, 4$, then e_1, e_2, e_3, e_4 form the boundary of the union of two adjacent two-simplexes of τ (see Fig. 15.1).*

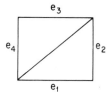

Figure 15.1

Then there is a realization of τ as a geodesic triangulation and a circle centered at each vertex of this triangulation so that the two circles associated to the vertices of an edge $e \in E(\tau)$ meet at an angle of $\theta(e)$. This triangulation and family of circles is unique up to conformal transformation of S^2.

COROLLARY 15.2. *Let (τ, θ) be as above and let D_τ be the dual cell decomposition to τ in S^2. The pair (D_τ, θ) determines a boundary pattern $\mathscr{P}$ on S^2, and there is a convex hyperbolic structure on the 3-ball with geodesic faces which realizes $(B^3, \mathscr{P})$, provided that any time (e_1, e_2, e_3) are edges forming the boundary of a two-simplex in τ either*

(a) $\sum_{i=1}^{3} \theta(e_i) > \pi$ *or*
(b) $\{\theta(e_1), \theta(e_2), \theta(e_3)\} = \{0, \pi/2, \pi/2\}$ *up to cyclic permutation.*

This structure is unique up to isometry.

The idea for deducing Corollary 15.2 from Theorem 15.1 is to realize the pattern of circles entered at the vertices of τ that meet at the given angles on the 2-sphere at infinity for the Poincaré model of $\mathbb{H}^3$. We let X be the region of $\mathbb{H}^3$ exterior to the open half-spaces that meet infinity in the interiors of these circles. The boundary of X is made up of geodesic faces, which we claim realizes $\mathscr{P}$. Conditions (a) and (b) insure that (D_τ, θ) is a boundary pattern.

Let us turn now to an analogue of Andre'ev's theorem of surfaces of negative Euler characteristic.

THEOREM 15.3. *Let S be the complement of a finite set of points in a connected closed surface $\hat{S}$. Let τ be a triangulation of $\hat{S}$ so that every point of $\hat{S} - S$ is a vertex.*
Let θ: edges(τ) $\to$ [0, $\pi/2$]. Suppose that
(1) $\chi(S) < 0$;
(2) *if e_1, e_2, e_3 are edges of τ forming a null homotopic loop in $\hat{S}$ and if $\sum_{i=1}^{3}(e_i) \geq \pi$, then e_1, e_2, and e_3 are the edges of a 2-simplex in τ;*
(3) *if e_1, e_2, e_3, e_4 are edges of τ forming a null-homotopic loop in $\hat{S}$ and if $\theta(e_i) = \pi/2$ for $i = 1, \ldots, 4$, then e_1, e_2, e_3 and e_4 are the four edges forming the boundary of two adjacent 2-simplices of τ.*

Then there is a complete hyperbolic structure on S and a geodesic triangulation τ' of S (with some vertices at infinity) so that

(a) *there is a homeomorphism $\varphi: S \to S$ properly homotopic to the identity so that $\varphi(\tau) = \tau'$;*
(b) *on the universal covering $\tilde{S}$ of S, the triangulation τ' lifts to $\tilde{\tau}'$; there is an equivariant family of hyperbolic circles on S centered at the vertices of τ'.[7] Two circles associated with the vertices of an edge $\tilde{e}$ meet at an angle $\theta(\tilde{e})$.*

The solution is unique up to isometry homotopic to the identity.

COROLLARY 15.4. *Let S, τ, and θ be as in Theorem 15.3. Suppose that for all triples of edges (e_1, e_2, e_3) that form the boundary of a 2-simplex of τ either $\sum_{i=1}^{3} \theta(e_i) > \pi$ or*

$$\{\theta(e_1), \theta(e_2), \theta(e_3)\} = \{0, \pi/2, \pi/2\}$$

up to a cyclic permutation. Then there is a convex hyperbolic structure of finite volume on $S \times I$ that realizes the dual boundary pattern on $S \times \{1\}$ and that is totally geodesic on $S \times \{0\}$. This structure is unique up to isometry.

Idea of the Proof of Corollary 15.4 from Theorem 15.3. Given (S, τ, θ), we use Corollary 15.4 to find a fuchsian group $\rho: \pi_1(S) \subset \text{PSL}_2(\mathbf{R})$ that realizes the hyperbolic structure on S in which the pattern of circles exists. We then realize the pattern of circles equivariantly on the upper-hemisphere of infinity in the Poincaré model. We cut out a region X bounded below by the equatorial plane and above by the equivariant family of planes bounded by the family of circles.[8]

---

[7] A hyperbolic circle centered at infinity is a horocycle.
[8] The circles at infinity are realized by circles tangent to the equatorial plane.

V. Uniformization Theorem for Three-Dimensional Manifolds

The subspace X is invariant under $\pi_1(S)$, and $Y = X/(\pi_1(S))$ is a convex hyperbolic 3-manifold of finite volume realizing the boundary pattern on $S \times \{1\}$ and totally geodesic on $S \times \{0\}$.

Proofs of Theorems 15.2 and 15.3 are presented in [19].

16. The Inductive Step in the Proofs of Theorems A' and B'

We are now, at last, ready to consider the inductive step. We have a pared manifold (M', P'). We have a superincompressible surface $(Z, \partial Z) \subset (M', \partial M')$ that satisfies Condition 4.9. We cut M' open along Z to obtain (M, P). Inside $\partial_0 M$ are the two copies of Z; Z_0, and Z_1. Both are superincompressible, and $\varphi: Z_0 \to Z_1$ is the gluing homeomorphism. By induction on the length of a hierarchy, we can assume that (M, P) has a convex hyperbolic structure of finite volume. We wish to find such a structure for (M', P').

The first step is to build a boundary pattern for the pared manifold (M', P'). Let $\hat{S}$ be $\overline{\partial_0 M'}$, with each boundary circle collapsed to a point. Then $(\partial_0 M')$ is embedded in $\hat{S}$ as the complement of a finite set of points. Choose a triangulation τ_0 of $\hat{S}$ so that each point outside $\partial_0 M'$ is a vertex. Let D_0 be the decomposition of $\partial_0 M'$, with faces, which is dual to τ_0. Replace each vertex of D_0 by a small triangle, as in Fig. 16.1. Label the original edges of D_0 with angles $\pi/7$ and the edges of the new triangles by $\pi/2$. This produces a decomposition and the numbers necessary to form a boundary pattern $\mathscr{P}'$ for (M', P') with $\partial_0 \mathscr{P}' = \emptyset$.

LEMMA 16.1. (M', P') *is a pared manifold with boundary pattern* $\mathscr{P}'$ *that satisfies the following*:

(a) *If l is a loop in $\partial M' - P'$ that crosses exactly three edges e_1, e_2, and e_3 of the decomposition, then either $\sum_{i=1}^{3} \theta(e_i) < \pi$ or l bounds a disk on $\partial M' - P'$ containing only one vertex of the decomposition.*

(b) *If e_1, e_2, and e_3 are three edges of the decomposition meeting at a vertex, then either $\sum_{i=1}^{3} \theta(e_i) > \pi$ or $\{\theta(e_1), \theta(e_2), \theta(e_3)\} = \{0, \pi/2, \pi/2\}$ up to cyclic permutation.*

Figure 16.1

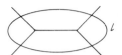

Figure 16.2

(c) *If l is a loop in $\partial M' - P'$ crossing exactly four edges, e_1, e_2, e_3, e_4, of the decomposition, and if $\theta(e_i) = \pi/2$ for $i = 1,\ldots, 4$, then l bounds a disk in $\partial M' - P'$ as in Fig. 16.2.*

One checks all these conditions in a straightforward fashion.

Now we wish to enhance the boundary pattern $\mathscr{P}'$ for (M', P') so that a boundary pattern $\mathscr{P}$ for (M, P) is produced. The first step is to deform $Z \subset M'$ by an isotopy until it crosses the minimal number of edges of $\mathscr{P}'$. Then cut $(M', \mathscr{P}')$ open along Z. Next we must extend the boundary pattern $\mathscr{P}'$ to the result of the cutting (M, P). There are two types of regions for which we must extend the boundary pattern: over the two copies Z_0 and Z_1 of Z in $\partial_0 M$ and over the annuli of P' that are cut open into a collection of disks in $\partial_0 M$. For the regions of the second type we extend as indicated in Fig. 16.3. On the two copies of Z, Z_0 and Z_1 in $\partial_0 M$, we can construct an equivariant cell structure that is dual to a triangulation and meets the boundary as indicated in Fig. 16.4. Then replace each interior vertex by a small triangle. Label the edges on ∂Z_i by $\pi/2$, label the edges of the original structure on int Z_i by $\pi/7$, and those of the small triangles by $\pi/2$.

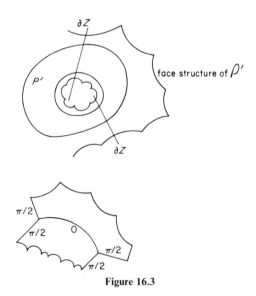

Figure 16.3

V. Uniformization Theorem for Three-Dimensional Manifolds

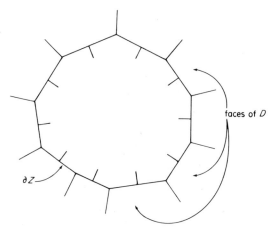

Figure 16.4

We call any such pattern $\mathscr{P}$ an *extension* of $\mathscr{P}'$ to (M, P). (We always let $\partial_0 \mathscr{P} = \varnothing$.) Let $\mathscr{P}_0$ be the same pattern but with $\partial_0 \mathscr{P}_0 = Z_0 \amalg Z_1$.

In order to glue up a geometric realization for $(M, \mathscr{P}_0)$ to form one for $(M', \mathscr{P}')$, we need to verify the three hypotheses of Theorem 14.2. The following proposition deals with hypotheses (2) and (3) of Theorem 14.2.

PROPOSITION 16.2. *Let (M', P') be a pared manifold with boundary pattern $\mathscr{P}'$ that satisfies Lemma 16.1(a), (b), and (c). Then*

(i) *any abelian, noncyclic subgroup of $\Gamma(M', \mathscr{P}')$ is peripheral;*
(ii) *if $p_0 \subset \Gamma(M', \mathscr{P}')$ is a nontrivial peripheral subgroup and if $\alpha p_0 \alpha^{-1}$ is also peripheral for some $\alpha \in \Gamma(M', \mathscr{P}')$, then $\langle \alpha, p_0 \rangle$ is peripheral.*

Idea of the Proof. The proof follows along the lines of the proof of Proposition 14.3. One passes to a torsion-free normal subgroup of finite index (using Lemma 14.1). There is an associated manifold pair $(\tilde{M}', \tilde{P}')$ on which the quotient group Q acts. The quotient space is M'. Conditions (i) and (ii) above are equivalent to the fact that $(\tilde{M}', \tilde{P}')$ is a pared manifold. It is easy to see that any essential torus in $\tilde{M}'$ is parallel into $\tilde{P}'$. Thus to complete the proof that $(\tilde{M}', \tilde{P}')$ is pared we need only show that there are no essential annuli in $(\tilde{M}', \tilde{P}')$. Again invoking [13], if there is an essential annulus, then there is a family of disjoint essential annuli invariant under Q. One rules out all possibilities for the quotient of such a family from conditions (a)–(c) of Lemma 16.1.

This shows that hypotheses (2) and (3) of Theorem 14.2 hold for $(M', \mathscr{P}')$. The following lemma concerns hypotheses (1) of Theorem 14.2.

LEMMA 16.3. *If (M', P') is a pared manifold with boundary pattern $\mathscr{P}'$ satisfying Lemma 16.1(a)–(c) and if Z is a superincompressible surface satisfying Condition 4.9, with Z cutting the minimal number of edges of $\mathscr{P}'$ among surfaces in its isotopy class, then*

(a) *(M, P) is a pared manifold with boundary pattern satisfying Lemma 16.1(a)–(c) and*
(b) *both Z_0 and Z_1 in $\partial_0 \mathscr{P}_0$ satisfy Proposition 14.3(a)–(c).*

One checks all these conditions in a straightforward manner. The fact that ∂Z meets a minimum number of edges of $\mathscr{P}'$, up to isotopy, implies conditions (b) and (c) of Proposition 14.3.

This completes the proof of Proposition 16.2. ∎

COROLLARY 16.4. *With notation and assumptions as in Lemma 16.3, if $(M, \mathscr{P})$ has a geometric realization, then so does $(M', \mathscr{P}')$.*

Proof. If $(M, \mathscr{P})$ has a geometric realization, then so does $(M, \mathscr{P}_0)$. According to Theorem 14.2, Propositions 14.3 and 16.2, and Lemma 16.3, if $(M, \mathscr{P}_0)$ has a realization, then so does $(M', \mathscr{P}')$. ∎

Thus our task is to find a geometric realization of $(M, \mathscr{P})$. We know from the inductive hypothesis that the underlying pared manifold (M, P) of $(M, \mathscr{P})$ has a convex hyperbolic structure of finite volume. The problem is to arrange that its boundary be piecewise geodesic and realize the boundary pattern $\mathscr{P}$.

We now find a geometric realization for $(M, \mathscr{P})$, one component at a time. We choose a component of $(M, \mathscr{P})$ which, for simplicity of notation, we continue to call $(M, \mathscr{P})$.

If M is a 3-ball, Lemma 16.1 and Corollary 15.2 tell us that there is a geometric realization of $(M, \mathscr{P})$.

If M is a solid torus, we cut it open along an embedded disk whose boundary is nontrivial. We perform an isotopy of this disk until it satisfies Condition 4.9 and until it crosses a minimal number of edges of $\mathscr{P}$. The resulting manifold is the 3-ball. We extend $\mathscr{P}$ over all S^2 as described above. By Lemmas 16.1 and 16.3 and Corollary 15.2, this pattern is realizable. By Corollary 16.4, so is $(M, \mathscr{P})$.

If M is homeomorphic to $T^2 \times I$, we cut it open along an essential annulus connecting the two boundary components. We arrange that the annulus meet a minimal number of edges of $\mathscr{P}$, and extend the boundary pattern as before. The cut open pared manifold with boundary pattern is of the type just considered. Thus, by the argument just given, it has a geometric realization. By Corollary 16.4, so does $(M, \mathscr{P})$.

According to Lemma 4.10, there is only one other possibility for $(M, \mathscr{P})$: each component of $\partial_0 M$ has negative Euler characteristic. If there are

V. Uniformization Theorem for Three-Dimensional Manifolds

compressing disks for $\partial_0 M$ in M, or essential annuli from $\partial_0 M$ into P, then we cut M open along a maximal family of such disjoint surfaces (each satisfying Condition 4.9). As above, we can extend the boundary patterns over the resulting pieces. If all of these have hyperbolic structures, then by Corollary 16.4 so will $(M, \mathscr{P})$. The result of cutting M open will be a collection of pared manifolds of the types considered above together with those that satisfy the following properties:

16.5

(a) Each component of $\partial_0 M$ has negative Euler characteristic.
(b) Each component of $\partial_0 M$ is incompressible in M.
(c) Each component of $\partial_0 M$ contains no accidental parabolics.

This shows that to complete the proof that $(M, \mathscr{P})$ has a geometric realization we can assume that (M, P) is pared and that (M, P) satisfies 16.5(a)–(c).

We have one additional piece of information: the underlying pared manifold (M, P) of $(M, \mathscr{P})$ is a component obtained by cutting open a pared manifold (L, R) that has a convex hyperbolic structure of finite volume. This implies that (M, P) has a convex hyperbolic structure of finite volume. The reason is that $\pi_1(M)$ is a finitely generated subgroup of $\pi_1(L)$. Thus the geometrically finite representation $\rho: \pi_1(L) \to \mathrm{PGL}_2(\mathbf{C})$ remains geometrically finite when restricted to $\pi_1(M)$ by Proposition 7.1. Thus M is realized as a convex hyperbolic manifold of finite volume. The parabolic locus for this structure is easily seen to be $P \subset M$. (Actually, one does not need to invoke Proposition 7.1; there is a direct construction of a convex structure on M from the one on L.)

The construction of the geometric realization for $(M, \mathscr{P})$ is completed by the following theorem.

THEOREM 16.6. *Let (M, P) be a pared manifold as in 16.5 with boundary pattern $\mathscr{P}$ satisfying 16.1. Suppose that (M, P) has a convex hyperbolic structure of finite volume. Then $(M, \mathscr{P})$ has a geometric realization as a convex hyperbolic manifold.*

Proof. Since every component of $\partial_0 M$ has negative Euler characteristic, and since $\mathscr{P}$ satisfies Lemma 16.1(a)–(c), Theorem 15.4 implies that there is a convex hyperbolic structure of finite volume on $(\partial_0 M) \times I$ that realizes the boundary pattern $\mathscr{P}$ on $(\partial_0 M) \times \{1\}$. The end $(\partial_0 M) \times \{0\}$ is totally geodesic, and hence each component of $(\partial_0 M) \times \{0\}$ corresponds to a fuchsian group.

By Theorem 7.2, the components of $\partial_0 M$ correspond to quasi-fuchsian groups in the convex hyperbolic structure on (M, P). We wish to glue the

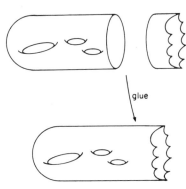

Figure 16.5

two structures together along surfaces which are quasi-fuchsian (see Fig. 16.5).

Clearly, this gluing problem satisfies the hypothesis of Theorem 14.2. The geometric realization of the result of gluing is exactly a convex hyperbolic structure of finite volume for $M - (P \cup \theta_{\mathscr{P}}^{-1}(0))$, which realizes the boundary pattern $\mathscr{P}$. ∎

This completes the study of all possibilities for $(M, \mathscr{P})$. We have shown that in all cases $(M, \mathscr{P})$ has a geometric realization. According to Corollary 16.4 this means that $(M', \mathscr{P}')$ also has a geometric realization. A fortiori the pared manifold (M', P') has a convex hyperbolic structure of finite volume. This completes the inductive step in the proofs of Theorems A′ and B′. The first step in the induction requires only that we find a convex hyperbolic structure of finite volume on a disjoint union of balls. Since this is easy to do, the proofs of Theorems A′ and B′ are complete.

References

[1] Ahlfors, L., Finitely generated kleinian groups, *Amer. J. Math.* **86** (1964), 413–429.
[2] Ahlfors, L., Fundamental polyhedrons and limit point sets of kleinian groups, *Proc. Nat. Acad. Sci. U.S.A.* **55** (1966), 251–254.
[3] Andre'ev, E. M., On convex polyhedra in Lobacevskii spaces, *Math. USSR-Sb.* **10** (1970), 413–440 (English translation).
[4] Bers, L., Quasiconformal mappings and Teichmüller's theorem, in "Analytic Functions" (R. Nevanlinna *et al.*, eds.), pp. 89–119. Princeton Univ. Press, Princeton, New Jersey, 1960.
[5] Borel, A., "Introduction aux groupes arithmétiques." Hermann, Paris, 1969.
[6] Hempel, J., "3-manifolds," *Ann. of Math. Studies* Vol. 86. Princeton Univ. Press, Princeton, New Jersey, 1976.

V. Uniformization Theorem for Three-Dimensional Manifolds

[7] Jaco, W., Finitely presented subgroups of 3-manifold groups, *Inv. Math.* **13** (1971), 335–346.

[8] Jaco, W., and Shalen, P., Seifert fibered spaces in 3-manifolds, *Mem. Amer. Math. Soc.* **220** (1979).

[9] Johannson, K., "Homotopy Equivalences of 3-manifolds with Boundary," Lecture Notes in Mathematics, Vol. 761. Springer–Verlag, Berlin and New York, 1979.

[10] Kurosh, A. G., "Theory of Groups," Chelsea, Bronx, New York, 1960.

[11] Margulis, G. A., Arithmeticity of nonuniform lattices, *Funkcional. Anal. Priložen.* **7** (1973), 88–89.

[12] Maskit, B., On Klein's combination theorem III, *in* "Advances in the theory of Riemann Surfaces," (L. Ahlfors *et al.*, eds.) *Ann. of Math Studies* Vol. 66, pp. 297–316. Princeton Univ. Press, Princeton, New Jersey, 1971.

[13] Meeks, W. and Yau, S.-T., The classical Plateau problem and the topology of three-dimensional manifolds, *Topology* **21** (1982), 409–440.

[14] Mostow, G., Quasi-conformal mappings in n-space and the rigidity of hyperbolic space forms, *Publ. IHES* **34** (1968), 53–104.

[15] Spivak, M., "Differential Geometry," Vol. 2. Publish or Perish, Boston, Massachusetts 1970.

[16] Sullivan, D., "Travaux de Thurston sur les groupes quasi-fuchsiens et les variétés hyperboliques de dimension 3 fibrées sur S^1," Lecture Notes in Mathematics, Vol. 842. Springer-Verlag, Berlin and New York, 1980.

[17] Sullivan, D., On the ergodic theorem at infinity of an arbitrary discrete group of hyperbolic motions, *Proc. Stony Brook Conf. Kleinian Groups Riemannian Surfaces, June 1978*.

[18] Thurston, W., Hyperbolic structures on acylindrical 3-manifolds, *Ann. of Math.*, to appear.

[19] Thurston, W., The geometry and topology of 3-manifolds. Lecture notes, Dept. of Math., Princeton Univ., Princeton, New Jersey, 1977.

[20] Thurston, W., Notes on geometric topology. Lecture notes. Dept. of Math., Princeton Univ., Princeton, New Jersey, 1978.

[21] Waldhausen, F., On irreducible 3-manifolds which are sufficiently large, *Ann. of Math.* **87** (1968), 56–88.

CHAPTER VI

Finitely Generated Subgroups of GL_2

Hyman Bass

Department of Mathematics
Columbia University
New York, New York

1. The GL_2-Subgroup Theorem

The GL_2-subgroup theorem is the following theorem, whose proof we shall outline below.

THEOREM. *Let Γ be a finitely generated subgroup of $GL_2(\mathbf{C})$. Then one of the following cases occurs:*

(a) *There is an epimorphism $f: \Gamma \to \mathbf{Z}$ such that $f(u) = 0$ for all unipotent elements $u \in \Gamma$.*

(b) *Γ is an amalgamated free product $\Gamma_0 *_\Lambda \Gamma_1$ with $\Gamma_0 \neq \Lambda \neq \Gamma_1$ and such that every finitely generated unipotent subgroup of Γ is contained in a conjugate of Γ_0 or of Γ_1.*

(c) *Γ is conjugate to a group of triangular matrices $\begin{pmatrix} a & b \\ 0 & d \end{pmatrix}$ with a and d roots of unity.*

(d) *Γ is conjugate to a subgroup of $GL_2(A)$, where A is a ring of algebraic integers.*

Remarks. (1) This theorem is proved in Bass [B, Theorem 6.5], (where one allows an arbitrary field in place of $\mathbf{C}$). The condition on unipotents in case (b) was suggested by Shalen. It is not contained in the statement of Bass [B, Theorem 6.5], but it is easily deduced from the proof there.

(2) The essential elements of the proof are taken from Serre's course [S]. They are

(I) a structure theorem for groups acting on trees, the arboreal presentation theorem, and

(II) exhibition of a natural tree on which GL_2 acts over a local field.

If we exclude cases (a) and (b) of the GL_2-subgroup theorem, then by applying (I) and (II) to Γ, one deduces that the traces of the elements of Γ are all algebraic integers. This argument is essentially contained in Serre [S, p. 83, Proposition 22]; we reproduce it here. The proof that integrality of traces implies case (c) or (d) is contained in Bass [B]; this elementary argument is also reproduced here. The arboreal presentation theorem is summarized in Section 2. The tree in (II) is described in Section 3. In Section 4 we assemble these ingredients and show how to deduce the GL_2-subgroup theorem.

(3) The GL_2-subgroup theorem was proved in order to establish the following conjecture of Shalen. Shalen persistently solicited this proof and correctly suggested that arboreal methods should yield it. We therefore owe him the main impulse behind this work, for which we express our gratitude.

COROLLARY 1 (Shalen's Conjecture). *Let Γ be a finitely generated subgroup of $GL_2(\mathbf{C})$. Assume that*

(1) *there is a subset X of $\mathbf{C}$ of transcendence degree ≥ 2 over $\mathbf{Q}$ such that $e(x) = \begin{pmatrix} 1 & x \\ 0 & 1 \end{pmatrix} \in \Gamma$ for all $x \in X$, and such that $e(X)$ and the commutator subgroup $[\Gamma, \Gamma]$ generate a subgroup of finite index in Γ;*

(2) *there is a matrix $s = \begin{pmatrix} a & b \\ c & d \end{pmatrix}$ in Γ with $c \neq 0$.*

Then case (b) of the GL_2-subgroup theorem holds.

Case (a) is clearly eliminated by (1). If $x \in X$, $x \neq 0$, then $e(x)$ leaves invariant a unique line L in $\mathbf{C}^2$, and L is not invariant under s in (2). This eliminates case (c). To eliminate case (d) and so prove the corollary, consider the standard basis $e_{11}, e_{12}, e_{21}, e_{22}$ of $M_2(\mathbf{C})$. Suppose that A is a ring of algebraic integers and that $t\Gamma t^{-1} \subset GL_2(A)$ for some $t \in GL_2(\mathbf{C})$. Let $f_{ij} = t^{-1} e_{ij} t$. Then the A-algebra $A\Gamma$ of A-linear combinations of elements of Γ is contained in $\bigoplus_{1 \leq i,j \leq 2} A f_{ij}$. For $x \in X$, the element $xe_{12} = e(x) - I$ belongs to $A\Gamma$. Choose $x \neq 0$ in X and set $y = xe_{12} \in A\Gamma$. Then $Xx^{-1}y = Xe_{12} \subset A\Gamma$. Write $y = \sum_{i,j} a_{ij} f_{ij}$ with $a_{ij} \in A$. For $x' \in X$ we have $x'x^{-1}y = \sum_{i,j} x'x^{-1} a_{ij} f_{ij} \in A\Gamma$, so $x'x^{-1} a_{ij} \in A$ ($1 \leq i, j \leq 2$). Choose an (i,j) for which

VI. Finitely Generated Subgroups of GL_2

$a_{ij} \neq 0$. Then the element $z = x^{-1}a_{ij}$ is nonzero, and $Xz \subset A$. It follows that $X \subset \mathbf{Q}(A, z)$, and so X has transcendence degree ≤ 1 over $\mathbf{Q}$, contradicting (1). This proves the corollary.

COROLLARY 2 (Bass [B, Corollary 6.7]). *Let Γ be a finitely generated group. Assume that*

(1) *The commutator quotient of Γ is finite;*
(2) *Γ is not a nontrivial amalgamated free product.*

Let $\rho: \Gamma \to GL_2(\mathbf{C})$ be a homomorphism. Then $\rho(\Gamma)$ satisfies the conditions of case (c) or case (d) of the GL_2-subgroup theorem.

It is easily seen that $\rho(\Gamma)$ also satisfies conditions (1) and (2), and these conditions clearly eliminate cases (a) and (b), respectively, of the GL_2-subgroup theorem. Whence the corollary.

COROLLARY 3 (cf. Bass [B, Proposition 5.3]). *Let Γ be as in Corollary 2. Then Γ has only finitely many isomorphism classes of irreducible representations of dimension ≤ 2 over $\mathbf{C}$.*

Condition (1) is equivalent to the finiteness of the set $\text{Hom}(\Gamma, \mathbf{C}^\times)$ of one-dimensional representations. The classes of irreducible two-dimensional representations $\rho: \Gamma \to GL_2(\mathbf{C})$ can be parametrized by an algebraic variety V defined over $\mathbf{Q}$. Such ρ's, by Corollary 2, must occur only in case (d) of the GL_2-subgroup theorem. It follows that *every* point of V is algebraic over $\mathbf{Q}$. Applying this to generic points of the components of V, we see that $\dim V = 0$, i.e., V is finite.

Remarks. (1) Corollary 3 expresses a strong rigidity property of two-dimensional representations of Γ. I am grateful to William Thurston for suggesting that such a result should be deducible from the GL_2-subgroup theorem.

(2) Corollaries 2 and 3 are valid with any algebraically closed field F in place of $\mathbf{C}$. In case $\text{char}(F) = p > 0$, then in case (d) of the GL_2-subgroup theorem the ring A should be taken to be a finite field.

(3) In fact, Corollary 3 has a version over $\mathbf{Z}$, which yields a uniform bound on the number of irreducible two-dimensional representations of Γ in characteristic p for every p. This version fully uses the integrality of A in case (d), whereas the proof of Corollary 3 over $\mathbf{C}$ requires only that A be algebraic over $\mathbf{Q}$.

(4) One can generalize the above corollaries as follows: Let Γ be a finitely generated group and U a subset of Γ such that

(i) the subgroup generated by U and all commutators has finite index in Γ;

' (ii) Γ is not an amalgamated free product $\Gamma_0 *_\Lambda \Gamma_1$ with $\Gamma_0 \neq \Lambda \neq \Gamma_1$ such that each $u \in U$ lies in a conjugate of Γ_0 or of Γ_1.

Consider irreducible representations $\rho: \Gamma \to GL_2(\mathbf{C})$ such that $\rho(u)$ is unipotent for all $u \in U$. Then case (d) of the GL_2-subgroup theorem applies to $\rho(\Gamma)$. Further, there are only finitely many conjugacy classes of such ρ's.

The proofs are as above. One need only note that the conditions $\rho(u)$ is unipotent for $u \in U$ define a closed subvariety of the variety V in the proof of Corollary 3.

(5) Let $PSL_2(\mathbf{C}) = GL_2(\mathbf{C})/\mathbf{C}^\times \cdot I$. Then the GL_2-subgroup theorem applies, as stated, to a finitely generated subgroup Γ of $PSL_2(\mathbf{C})$. In fact, let $\tilde{\Gamma}$ denote the inverse image of Γ in $SL_2(\mathbf{C})$, so that $\Gamma = \tilde{\Gamma}/\{\pm I\}$. Each condition of the GL_2-subgroup theorem applies to Γ if and only if it applies to $\tilde{\Gamma}$. This is obvious for conditions (a), (c), and (d). As for (b), a decomposition $\Gamma = \Gamma_0 *_\Lambda \Gamma_1$ yields $\tilde{\Gamma} = \tilde{\Gamma}_0 *_{\tilde{\Lambda}} \tilde{\Gamma}_1$, where $\tilde{X}$ denotes the inverse image of X in $\tilde{\Gamma}$. Conversely, given a decomposition $\tilde{\Gamma} = \tilde{\Gamma}_0 *_{\tilde{\Lambda}} \tilde{\Gamma}_1$, the central element $-I$ must belong to $\tilde{\Lambda}$, so we obtain a decomposition of Γ by passage to quotients.

2. Arboreal Group Theory[1]

Let X be a graph that consists of vertices P and edges e. Each edge e is considered to be oriented, with an initial vertex P_e and a terminal vertex Q_e. The oppositely oriented edge is denoted $\bar{e}$; the unordered pair $\{e, \bar{e}\}$ represents a *geometric edge* of X.

Let Γ be a group acting on X (as graph automorphisms). We always assume that Γ acts without inversion (of edges), i.e., that $se \neq \bar{e}$ for all $s \in \Gamma$ and all edges e. Equivalently, the stabilizer of a geometric edge $\{e, \bar{e}\}$ coincides with the stabilizer Γ_e of e, and so is contained in the vertex stabilizers Γ_{P_e} and Γ_{Q_e}.

Suppose henceforth that X is a tree, i.e., that X is nonempty, connected, and simply connected. We propose to present Γ in terms of the stabilizers Γ_P, Γ_e, and the fundamental group of the *quotient graph* $X' = X/\Gamma$.

Choose a *maximal subtree* T' of X' containing all vertices of X'. Let B' be a set of edges e' representing the geometric edges of $X' - T'$. Contracting

---

[1] See Serre [S, Chapter I, Section 5].

VI. Finitely Generated Subgroups of GL_2

T' to a point transforms X' to a bouquet of circles, one for each $e' \in B'$. These circles define free generators of $\pi_1(X')$ $(=\pi_1(X', T'))$.

Since T' is simply connected, it can be lifted to a subtree T of X, which is a "tree of representatives of the vertices of X mod Γ." Lift B' to a set B of edges e of X having initial vertices P_e in T. For each $e \in B$, choose an element $t_e \in \Gamma$ that transforms Q_e into T.

With these notations we can now state the

ARBOREAL PRESENTATION THEOREM (cf. Serre [S, p. 76, Theorem 13]). *The group Γ admits the following presentation*:

Generators

G_v *the groups* Γ_P (P *a vertex of* T),
G_e *the elements* t_e ($e \in B$).

Relations

R_v *If e is an edge of T and $x \in \Gamma_e$, identify the occurrence of x in Γ_{P_e} with its occurrence in Γ_{Q_e}.*

R_e *If $e \in B$ and if $x \in \Gamma_e$ ($\subset \Gamma_{P_e}$), identify $t_e x t_e^{-1}$ with its occurrence in $t_e \Gamma_{Q_e} t_e^{-1} = \Gamma_{t_e Q_e}$. (Recall that P_e and $t_e Q_e$ are vertices of T.)*

Remarks. (1) We have tacitly included the given group structure in each Γ_P in this presentation. The generators G_v and relations R_v present the free product of the vertex groups Γ_P ($P \in T$), with the intervening edge groups Γ_e ($e \in T$) amalgamated, via the inclusions $\Gamma_{P_e} \hookleftarrow \Gamma_e \hookrightarrow \Gamma_{Q_e}$. Denote this amalgamated free product by $\pi(\Gamma, T)$. The obvious homomorphism $\pi(\Gamma, T) \to \Gamma$ is injective; we shall view it as an inclusion.

(2) Since $\Gamma \cdot T$ contains every vertex of X, it follows that the subgroup N of Γ generated by all vertex stabilizers is just the least normal subgroup of Γ containing $\pi(\Gamma, T)$. It is clear then from the arboreal presentation of Γ that we have an exact sequence

$$1 \to N \to \Gamma \to \pi_1(X', T') \to 1,$$

where $\pi_1(X', T')$ can be identified with the free group based on the elements t_e ($e \in B$).

(3) Suppose that Γ acts freely on the edges, so that $\Gamma_e = \{1\}$ for all e. Then all of the arboreal relations above become trivial, and we conclude that

$$\Gamma = \left(\underset{P \in T}{*} \Gamma_P \right) * \pi_1(X', T')$$

(4) In particular, if Γ acts freely on X ($\Gamma_P = \{1\}$ for all P), then we recover the well-known fact that Γ is the free group $\pi_1(X', T')$.

(5) On the other hand, suppose that $\pi_1(X', T') = \{1\}$, i.e., that X' is a tree, and so $T' = X'$. Then the tree $T \subset X$ is a fundamental domain for $X \mod \Gamma$, and we conclude that Γ is the amalgamated free product $\pi(\Gamma, T)$.

(6) The simplest case that mixes the preceding phenomena is that in which X' is a loop:

$$P' \bigcirc e'.$$

Then B consists of an edge

$$\underset{P}{\circ} \xrightarrow{e} \underset{Q}{\circ}.$$

whose initial vertex P constitutes all of T. Choose $t\,(=t_e)$ in Γ, so that $tQ = P$. Then $\Gamma_e \subset \Gamma_P$, and we have the monomorphism $\beta \colon \Gamma_e \to \Gamma_P$, $\beta(x) = txt^{-1}$. (Note that $t\Gamma_e t^{-1} \subset t\Gamma_Q t^{-1} = \Gamma_P$.) The arboreal presentation of Γ is just the HNN-*group* associated to $\Gamma_e \subset \Gamma_P$ and the monomorphism β.

(7) The tree T and edges B in X furnish liftings of each vertex and edge of X'. If x' is a vertex on edge of X' and x is its lifting to X, we shall write $G_{x'}$ for the stabilizer Γ_x. If e' is an edge of X', then $G_{e'}$ comes equipped with monomorphisms

$$G_{P_{e'}} \xleftarrow{\alpha_{e'}} G_{e'} \xrightarrow{\beta_{e'}} G_{Q_{e'}}.$$

When $e' \in T'$, these are inclusions. When $e' \in B'$, then $\alpha_{e'}$ is an inclusion and $\beta_{e'}(x) = t_e x t_e^{-1}$. The data just constructed constitute what is called a *graph of groups*; we denote it by (G, X'). Clearly, the arboreal presentation of Γ depends only on (G, X') and the maximal tree T' in X'. Thus we may denote the group so presented by $\pi_1(G, X', T')$.

There is the following *converse* to the arboreal presentation theorem: Let (G, X') be any graph of groups with X' connected and nonempty. Let T' be a maximal subtree of X'. Then there is a tree $X = X(G, X', T')$ on which $\Gamma = \pi_1(G, X', T')$ operates with quotient $X/\Gamma = X'$, and so that (G, X') arises by the construction above relative to suitable liftings T and B. See Serre [S, Chapter I, Section 5] for details.

3. The Tree of SL_2 Over a Local Field[2]

Let K be a field with a *discrete valuation* v; explicitly, v is a surjective homomorphism $K^\times \to \mathbf{Z}$ such that $v(x+y) \geq \mathrm{Inf}(v(x), v(y))$ for all $x, y \in K$, with the convention $v(0) = \infty$. It follows that the elements $x \in K$, for which $v(x) \geq 0$, form a ring A, the *valuation ring* of v. Choosing $p \in K$ with $v(p) = 1$,

---

[2] See Serre [S, Chapter II, Section 1].

VI. Finitely Generated Subgroups of GL_2

the nonzero ideals of A are of the form Ap^n ($n \geq 0$); $k = A/Ap$ is called the *residue class field*.

Let V be a two-dimensional vector space over K. A *lattice* in V is a finitely generated sub-A-module L of V that spans V over K. Then there is a K-basis e_1, e_2 of V such that $L = Ae_1 \oplus Ae_2$. Call lattices L and L' in V equivalent if $L' = xL$ for some $x \in K^\times$, and write (L) for the equivalence class of L. These classes (L) form the vertices of a combinatorial graph X, where we connect two vertices (L) and (L') by an edge if there is an $x \in K^\times$ such that $xL' \subset L$ and $L/xL' \cong A/pA$.

THEOREM (Serre [S, p. 98]). *X is a tree.*

The group $GL(V) \cong GL_2(K)$ operates on X: If $s \in GL(V)$ and L is a lattice in V, then so also is sL; moreover, s preserves equivalence and pairs of connected vertices. However, s might well invert some edges. To avoid this we pass to the group

$$GL(V)^0 = \ker(GL(V) \xrightarrow{\det} K^\times \xrightarrow{r} \mathbf{Z}).$$

Thus $s \in GL(V) \Leftrightarrow \det(s) \in A^\times$; in particular $SL(V) \subset GL(V)^0$. The group $GL(V)^0$ acts without inversion on X ([S, p. 105]). To apply arboreal group theory we must know something about the stabilizers for this action. Note that the scalars act trivially so that the action factors through $PGL(V)$.

LEMMA (Serre [S, p. 105]). *Let Γ be a subgroup of $GL(V)^0$, and let L be a lattice in V. The stabilizer $\Gamma_{(L)}$ of (L) coincides with the stabilizer $\Gamma_L = \Gamma \cap GL(L)$ of L.*

COROLLARY *Choose a basis of V to identify $GL(V)$ with $GL_2(K)$. A subgroup Γ of $GL_2(K)^0$ has a fixed point in X if and only if Γ is conjugate to a subgroup of $GL_2(A)$. Every finitely generated unipotent subgroup of $GL_2(K)$ has a fixed point in X.*

The first assertion is immediate from the lemma. The second assertion follows from the first since a unipotent subgroup is conjugate to a subgroup of $\begin{pmatrix} 1 & K \\ 0 & 1 \end{pmatrix}$, and if $S \subset K$ is finite, then

$$\begin{pmatrix} p^n & 0 \\ 0 & p^{-n} \end{pmatrix} \begin{pmatrix} 1 & S \\ 0 & 1 \end{pmatrix} \begin{pmatrix} p^{-n} & 0 \\ 0 & p^n \end{pmatrix} = \begin{pmatrix} 1 & p^{2n}S \\ 0 & 1 \end{pmatrix}$$

lies in $GL_2(A)$ for suitable n.

4. Proof of the GL_2-Subgroup Theorem

Let Γ be a finitely generated subgroup of $GL_2(\mathbf{C})$. We assume that case (a) of the theorem does not occur; in other words,

(1) the subgroup of Γ generated by commutators and unipotents has finite index in Γ.

It follows that

(2) $\det(\Gamma)$ is a finite subgroup of $\mathbf{C}^x$.

Let F denote the field generated by the matrix entries of elements of Γ. Since Γ is finitely generated, it follows that

(3) F is a finitely generated extension of $\mathbf{Q}$ and $\Gamma \subset GL_2(F)$.

Let V denote the set of discrete valuations of F. If $v \in V$, let A_v denote its valuation ring. From (3) and [B, Lemma 6.8] it follows that

(4) $A = \bigcap_{v \in V} A_v$ is the ring of algebraic integers in a finite extension of $\mathbf{Q}$.

Let $v \in V$. Let X_v be the tree of SL_2 over F relative to the valuation v as in Section 3. Condition (2) implies that $\det(\Gamma) \subset A_v^x$, i.e., $\Gamma \subset GL_2(F)^0$. From Section 3, lemma and corollary, we conclude that

(5) Γ acts without inversion on X_v. The vertex stabilizers are intersections of Γ with conjugates of $GL_2(A_v)$. Each finitely generated unipotent subgroup fixes some vertex.

Let N denote the subgroup of Γ generated by all vertex stabilizers. Then (Section 2, Remark (2)) Γ/N is isomorphic to the free group $\pi_1(X_v/\Gamma)$. Condition (1) implies that this free group must be trivial, i.e., that X_v/Γ is a tree. Thus there is a fundamental domain T for X mod Γ that is a tree, and Γ is the free product $\pi(\Gamma, T)$ of the vertex stabilizers along T, with the intervening edge stabilizers amalgamated (Section 2, Remark (5)). The group $\pi(\Gamma, T)$ is the filtered union of groups $\pi(\Gamma, S)$, where S varies over finite subtrees of T. Since Γ is finitely generated, $\Gamma = \pi(\Gamma, S)$ for some such S. Choose S minimal with this property. Every vertex stabilizer in Γ is conjugate to a subgroup of Γ_P for some $P \in S$. Let P be a terminal vertex of S. If S has other vertices, then P is the initial vertex of an edge e joining P to a tree S' containing all other vertices of S. We then have

$$\Gamma = \pi(\Gamma, S) = \Gamma_P *_{\Gamma_e} \pi(\Gamma, S'),$$

VI. Finitely Generated Subgroups of GL_2

and the minimality of S implies that this amalgamation is nontrivial. Further, every finitely generated unipotent subgroup of Γ is (in view of (5) and the remarks above) contained in a conjugate of Γ_P or of $\pi(\Gamma, S')$. Thus we have case (b) of the GL_2-subgroup theorem.

Now suppose that case (b) does not occur. Then the tree S above is a single vertex P, which is a fixed point of Γ, and so (see (5)) Γ is conjugate to a subgroup of $GL_2(A_v)$. Hence the set $\text{tr}(\Gamma)$ of traces of elements of Γ is contained in A_v. Moreover, this must happen for all $v \in V$, whence

$$(6) \qquad \text{tr}(\Gamma) \subset A = \bigcap_{v \in V} A_v.$$

We now conclude the proof of the GL_2-subgroup theorem by showing that conditions (1), (4), and (6) imply that case (c) or (d) of the GL_2-subgroup theorem occurs.

Case (c) occurs when Γ acts reducibly on $\mathbf{C}^2$, so Γ is triangularizable, and condition (1) then implies that the diagonal entries of elements of Γ are roots of unity.

It remains to establish case (d) when Γ acts irreducibly on $\mathbf{C}^2$. Thus the proof is completed by the next proposition.

PROPOSITION (cf. Bass [B, Corollary 2.5]). *Let Γ be a subgroup of $GL_n(\mathbf{C})$. Let A be the ring of integers in a finite extension E of $\mathbf{Q}$. Assume that*

(i) $\text{tr}(\Gamma) \subset A$ *and*
(ii) Γ *acts irreducibly on $\mathbf{C}^n$.*

Then there is a finite extension F of E, with algebraic integers B, such that Γ is conjugate to a subgroup of $GL_n(B)$.

Condition (ii) implies that Γ contains a basis $s_1, \ldots, s_{n^2}$ of $M_n(\mathbf{C})$ (cf. [B, Lemma 1.1]). Let $t_1, \ldots, t_{n^2}$ be the dual basis relative to the trace form $\text{tr}(t_i s_j) = \delta_{ij}$, $1 \le i, j \le n^2$. If $s \in M_n(\mathbf{C})$, then $s = \sum_i \text{tr}(ss_i) t_i$. If $s \in \Gamma$, then $ss_i \in \Gamma$ for all i, so $\text{tr}(ss_i) \in A$ by (i). Let $A\Gamma$ be the A-algebra in $M_n(\mathbf{C})$ of A-linear combinations of elements of Γ. We see from above that

$$\bigoplus_{1 \le i \le n^2} As_i \subset A\Gamma \subset \bigoplus_{1 \le i \le n^2} At_i.$$

It follows that $A\Gamma$ is a finitely generated A-module and an A-*form* of $M_n(\mathbf{C})$, i.e., $\mathbf{C} \otimes_A A\Gamma \to M_n(\mathbf{C})$ is an isomorphism. The same observations with E (the field of fractions of A) in place of A shows that $E\Gamma$ is an E-form of $M_n(\mathbf{C})$ and hence a central simple E-algebra. Replacing E by a finite extension and A by the corresponding ring of integers, we may assume that $E\Gamma \cong M_n(E)$. It follows easily that $E\Gamma$ is conjugate in $M_n(\mathbf{C})$ to $M_n(E)$. Therefore, we may further assume that $E\Gamma = M_n(E)$. Then $A\Gamma$ is an A-order in $M_n(E)$, and so it leaves invariant an A-*lattice* $P \subset E^n$ (i.e., P is a finitely generated A-module containing an E-basis of E^n). If $P \cong A^n$, then an A-basis for P furnishes a coordinate system with $\Gamma \subset \text{GL}_n(A)$. In general, there is a finite extension F of E, with integers B such that the B-lattice $B \otimes_A P$ in F^n is a free B-module. Then, as above, we can conjugate Γ into $\text{GL}_n(B)$.

References

[B] Bass, H., Groups of integral representation type, *Pacific J. Math.* **86** (1980), 15–51.
[S] Serre, J.-P., Arbres, amalgames, SL_2, *Astérisque* **42** (1977).

PART **C**

THE CASE OF AN INCOMPRESSIBLE SURFACE

CHAPTER VII

Incompressible Surfaces in Branched Coverings

C. McA. Gordon[†]   and   R. A. Litherland[‡]

Department of Mathematics
University of Texas at Austin
Austin, Texas

Department of Pure Mathematics and
Mathematical Statistics
University of Cambridge
Cambridge, England

1. Introduction

Our primary aim is to show that if $\tilde{\Sigma}$ is the n-fold cyclic branched covering, $n > 1$, of a prime knot K in a homotopy 3-sphere Σ, such that $\Sigma - K$ is irreducible and contains a nonperipheral incompressible surface, then $\tilde{\Sigma}$ is not a homotopy sphere. This will be achieved by using the equivariant loop theorem (see Chapter VIII by Yau and Meeks and Chapter IX by Meeks and Yau) to show that $\tilde{\Sigma}$ contains an incompressible surface of positive genus. Actually, we shall work in the more general context of a regular branched covering of a link in an arbitrary closed, orientable 3-manifold,

---

[†] This author's work was supported partially by National Science Foundation grant MCS 78-02995.

[‡] Present address: Department of Mathematics, Louisiana State University, Baton Rouge, Louisiana.

as this is needed for the study of noncyclic finite group actions on homotopy 3-spheres (see Chapter X by Davis and Morgan).

Our main result is Theorem 1, which asserts that, under certain circumstances, an incompressible surface in the complement of a link gives rise to one in any regular branched covering of the link. The special case of incompressible tori is considered in Theorem 2; here the proof actually *uses* the Smith conjecture. Interpreting Theorem 2 in terms of hyperbolic structures, using Thurston's uniformization theorem for Haken manifolds (see Thurston [Th] and Chapter V, this volume by Morgan) we obtain Corollary 2.1, which states that if a regular branched covering of a link is hyperbolic, then so is the complement of the link.

These results are stated in Section 2, which also contains the necessary definitions and terminology. The proofs of Theorems 1 and 2 and Corollary 2.1 are given in Section 3. Finally, in Section 4, we give an elementary proof of the equivariant loop theorem for involutions (Theorem 3). A similar proof has been given by Kim and Tollefson [K–To].

2. Terminology and Statement of Results

We shall work in the smooth category, and all our manifolds, including surfaces, will be assumed to be orientable. 3-manifolds will generally be assumed to be connected.

A *link* in a closed 3-manifold is a closed one-dimensional submanifold.

Let $\tilde{M}$ be a closed, oriented 3-manifold, and G a finite nontrivial group acting effectively on $\tilde{M}$ as a group of orientation-preserving diffeomorphisms. Suppose that the union of the exceptional orbits is a link $\tilde{L}$. (Equivalently, all the isotropy groups G_x, $x \in \tilde{M}$, are cyclic.) Let M be the quotient manifold $\tilde{M}/G$, $p: \tilde{M} \to M$ the quotient map, and $L = p(\tilde{L})$. Then L is a link in M, and we say that $\tilde{M}$ is a *regular branched covering of* (M, L), *with group* G.

If L_i is a link in M_i, $i = 1, 2$, then a *connected sum* $(M_1, L_1) \# (M_2, L_2)$ of (M_1, L_1) and (M_2, L_2) is formed by choosing $B_i \subset M_i$ such that $(B_i, B_i \cap L_i)$ is diffeomorphic to the standard pair (B^3, B^1) and identifying the boundaries of the deleted pairs $(M_i^*, L_i^*) = (M_i - \text{int } B_i, L_i - \text{int } B_i \cap L_i)$, $i = 1, 2$, by some diffeomorphism.

A link is *trivial* if it consists of one component and bounds a disk. A link is *prime* if it is not a connected sum of two nontrivial links. An extension of Haken's finiteness theorem ([Hak], [Ja, Theorem III.20]) applied to annuli implies that every link is a finite connected sum of prime links.

Let X be a 3-manifold and $F \subset X$ be a closed surface such that either $F \subset \text{int } X$ or $F \subset \partial X$. A *compressing disk* for F is a 2-disk $D \subset X$ such that

VII. Incompressible Surfaces in Branched Coverings

$D \cap (\partial X \cup F) = \partial D \subset F$, and ∂D is essential on F. If F has a compressing disk, it is *compressible*; otherwise, it is *incompressible*.

We shall say that a closed 3-manifold is *sufficiently large* if it contains a closed incompressible surface of genus > 0.

Let L be a link in a closed 3-manifold M. A closed surface $F \subset M - L$ is *peripheral* if each component of F is parallel in $M - L$ to the boundary of a tubular neighborhood (in M) of some component of L. Recall that a 3-manifold is *irreducible* if every 2-sphere in it bounds a 3-ball. We shall say that L is *sufficiently large* if $M - L$ is irreducible and contains a closed nonperipheral incompressible surface.

Our main result follows.

THEOREM 1. *Let L be a prime, sufficiently large link in a closed 3-manifold M, and let $\tilde{M}$ be a regular branched covering of (M, L). Then either $\tilde{M}$ is sufficiently large or M and $\tilde{M}$ both contain a nonseparating 2-sphere.*

The special case of this that is needed in the proof of the Smith conjecture is

COROLLARY 1.1. *Let $\tilde{\Sigma}$ be the n-fold cyclic branched covering, $n > 1$, of a prime, sufficiently large knot in a homotopy 3-sphere Σ. Then $\tilde{\Sigma}$ is not simply connected.*

Examples of sufficiently large knots in S^3 are those with nontrivial companions, the incompressible surface in question being a torus. Here, the arguments of [G, Section 4] are enough to show that $\tilde{\Sigma}$ is sufficiently large in many specific cases, for instance that of doubled knots.

Since a 2-bridge link is prime [Sc] and has a lens space as its 2-fold branched covering, Theorem 1 implies

COROLLARY 1.2. *The complement of a 2-bridge link contains no closed, nonperipheral incompressible surface.*

This is also proved in Hatcher and Thurston [Hat–Th], which, in addition, determines all incompressible surfaces with boundary in a 2-bridge knot exterior.

Theorem 1 may be combined with the Smith conjecture to give the following theorem. Recall that a 3-manifold is *atoroidal* if it contains no nonperipheral incompressible torus.

THEOREM 2. *Let L be a link in a closed 3-manifold M, and let $\tilde{M}$ be a regular branched covering of (M, L). If $\tilde{M}$ is irreducible and atoroidal, then $M - L$ is irreducible and atoroidal.*

Let us say briefly that a 3-manifold is *hyperbolic* if it has a complete hyperbolic structure with finite volume. Then Thurston's main theorem on the existence of hyperbolic structures on Haken manifolds ([Th], Chapter V), together with Theorem 2, yields

COROLLARY 2.1. *Let L be a link in a closed 3-manifold M, and let $\tilde{M}$ be a regular branched covering of (M, L). If $\tilde{M}$ is hyperbolic, then $M - L$ is hyperbolic.*

We remark that the assumption in Theorem 1 that $M - L$ is irreducible is only for convenience; it is not hard to take account of any 2-spheres in $M - L$ that do not bound 3-balls. Similarly, if L is not prime, so that $(M, L) \cong (M_1, L_1) \# (M_2, L_2)$, then $\tilde{M}$ can be analyzed in terms of $\tilde{M}_1$ and $\tilde{M}_2$. (See the proof of Theorem 2 in Section 3.)

Again, in the setting of Theorem 1, note that if M contains a nonseparating 2-sphere, then $\tilde{M}$ is not necessarily sufficiently large as we have defined it. For example, let $M = S^1 \times S^2$ and K be the Whitehead double of the core $S^1 \times *$ of $S^1 \times S^2$. It turns out that the 2-fold branched covering of $(S^1 \times S^2, K)$ is $S^1 \times S^2 \# RP^3$. It can also be shown that K is prime, and that $S^1 \times S^2 - K$ contains a nonperipheral incompressible torus. On the other hand, it is easy to see that $S^1 \times S^2 \# RP^3$ is not sufficiently large.

Finally, Culler and Shalen [C–Sh] have pointed out that Theorem 1 continues to apply, with the same proof, if the notion of a sufficiently large link is extended to allow an incompressible surface with boundary in the link exterior which is not boundary-parallel and whose boundary components are meridians of the link.

3. Proofs of Theorems 1 and 2

Recall from Section 2 the definition of a compressing disk. The proof of Theorem 1 is an application of the following equivariant version of the loop theorem—Dehn's lemma, due to Meeks and Yau (Chapters VIII and IX) (for the case $G = \mathbf{Z}/2$, see [K–To] or Section 4).

THEOREM (Meeks and Yau). *Let X be a 3-manifold, and G a finite group acting on X. Suppose that F is a compressible component of ∂X. Then there exists a compressing disk D for F such that for all $g \in G$, either $g(D) = D$ or $g(D) \cap D = \emptyset$.*

We remark that although we are dealing only with orientable manifolds, the theorem is true whether or not X is orientable.

VII. Incompressible Surfaces in Branched Coverings

Now suppose that the action is effective and orientation-preserving and that the union E of the exceptional orbits is a proper one-dimensional submanifold of X. Let D be a compressing disk as above, and suppose that $H = \{g \in G : g(D) = D\} \neq \{1\}$. If H preserves the orientation of D, then, since it acts freely on ∂D, it is cyclic and (by two-dimensional Smith theory, or an elementary Euler characteristic argument) $D \cap E$ consists of a single point. An examination of the action in a neighborhood of this point shows that it is a point of transverse intersection. If H contains an element that reverses the orientation of D, then by our assumption on E we must have $H \cong \mathbf{Z}/2$ and $D \cap E$ a proper arc in D. A small normal translate D' of D then satisfies $g(D') \cap D' = \varnothing$ for all $g \in G - \{1\}$.

In general, any disk D such that, for all $g \in G$, either $g(D) = D$ or $g(D) \cap D = \varnothing$, and such that D meets E transversely in a single point, will be called a *nice equivariant disk*.

If F is a closed surface in int X and D a disk in int X such that $D \cap F = \partial D$, the inclusion map $D = D \times \{0\} \to \text{int } X$ extends to an embedding $f : D \times [-1, 1] \to \text{int } X$ with image $U(D)$, say, such that $U(D) \cap F = f(\partial D \times [-1, 1])$. We then say that the surface $F' = (F - f(\partial D \times (-1, 1))) \cup f(D \times \{-1, 1\})$, with corners rounded, is obtained from F by *surgery along* D.

If, in addition, F is G-invariant and D is a nice equivariant disk, we may choose $U(D)$ so that $g(U(D)) = U(D)$ (resp. $g(U(D)) \cap U(D) = \varnothing$) whenever $g(D) = D$ (resp. $g(D) \cap D = \varnothing$), and such that $(U(D), U(D) \cap E) \cong (D \times [-1, 1], 0 \times [-1, 1])$, where 0 is the center of D. We may then do surgery along each disk in $G(D) = \cup\{g(D) : g \in G\}$, producing a G-invariant surface F' and two points of transverse intersection of F' with E for each component of $G(D)$.

Proof of Theorem 1. Let F be a closed, connected, nonperipheral, incompressible surface in $M - L$. Suppose that $\tilde{M}$ is not sufficiently large, so that any closed surface of genus > 0 in $\tilde{M}$ is compressible. Let $\tilde{F} = p^{-1}(F)$, a (possibly disconnected) G-invariant surface in $\tilde{M}$, where G is the group of the branched covering.

CLAIM A. *There is a sequence $\tilde{F} = \tilde{F}_0, \tilde{F}_1, \ldots, \tilde{F}_m$ of G-invariant surfaces in $\tilde{M}$ such that*

(a) $\tilde{F}_{i+1}$ *is obtained from $\tilde{F}_i$ by surgery along (D_i) for some nice equivariant compressing disk D_i for $\tilde{F}_i$, $0 \leq i \leq m - 1$;*
(b) *each component of $\tilde{F}_m$ is a 2-sphere.*

Proof of Claim A. Suppose that we have constructed $\tilde{F}_0, \tilde{F}_1 \ldots, \tilde{F}_k$, for some $k \geq 0$ such that (a) holds for $0 \leq i \leq k - 1$, but some component of $\tilde{F}_k$ is not a 2-sphere. Since $\tilde{M}$ is not sufficiently large, $\tilde{F}_k$ has compressing disks. We begin with the following observation.

CLAIM B. *Any compressing disk D for $\tilde{F}_k$ must meet $\tilde{L}$.*

Proof of Claim B. If $D \cap \tilde{L} = \varnothing$, we could move D off $G(U(D_{k-1}))$ by an isotopy keeping $D \cap \tilde{F}_k = \partial D$ and preserving the property that $D \cap \tilde{L} = \varnothing$. We assert that D is now a compressing disk for $\tilde{F}_{k-1}$. For certainly $D \cap \tilde{F}_{k-1} = \partial D$, and if ∂D were to bound a disk $\Delta \subset \tilde{F}_{k-1}$, then either $G(\partial D_{k-1}) \cap \Delta = \varnothing$, or $g(\partial D_{k-1}) \subset \Delta$ for some $g \in G$. But the former implies that $\Delta \subset \tilde{F}_k$, contradicting the essentiality of ∂D on $\tilde{F}_k$, whereas the latter contradicts the essentiality of ∂D_{k-1} on $\tilde{F}_{k-1}$.

Repeating this procedure, we would eventually obtain a compressing disk for $\tilde{F}_0 = \tilde{F}$ in $\tilde{M} - \tilde{L}$. But since $p|\tilde{F}': \tilde{F}' \to F$ induces an injection of fundamental groups for each component $\tilde{F}'$ of $\tilde{F}$, this contradicts the incompressibility of F in $M - L$.

This completes the proof of Claim B.

Now cut $(\tilde{M}, \tilde{L})$ open along $\tilde{F}_k$ to obtain $(\tilde{M}', \tilde{L}')$, say. The action of G on $(\tilde{M}, \tilde{F}_k)$ induces an action of G on $\tilde{M}'$. Since $\tilde{F}_k$ is compressible in $\tilde{M}$, $\partial \tilde{M}'$ is compressible in $\tilde{M}'$. Also, it follows from Claim B that any compressing disk for $\partial \tilde{M}'$ in $\tilde{M}'$ must meet L'. The equivalent loop theorem stated at the beginning of this section, together with the subsequent discussion, therefore guarantees the existence of a nice equivariant compressing disk D'_k for $\partial \tilde{M}'$. Let D_k be the image of D'_k under the regluing map $\tilde{M}' \to \tilde{M}$. We claim that D_k is a nice equivariant compressing disk for $\tilde{F}_k$. This could only fail if, for some $g \in G$ such that $g(D_k) \neq D_k$, we had $g(\partial D_k) \cap \partial D_k \neq \varnothing$, so that D_k and $g(D_k)$ would have to be locally on opposite sides of some component of $\tilde{F}_k$. But this is precluded by the fact that G preserves an orientation of the normal bundle of $\tilde{F}_k$, since an orientation of $F_k = p(\tilde{F}_k)$ lifts to a G-invariant orientation of $\tilde{F}_k$. We may therefore construct $\tilde{F}_{k+1}$.

It is straightforward to see that this process must terminate. For example, one may define the complexity $c(F)$ of a surface F by $c(F) = \Sigma(1 - \chi(F'))$ summed over all components F' of F which are not 2-spheres. This clearly has the following properties. Firstly, $c(F) \geq 0$, with equality if and only if F is a union of 2-spheres, and secondly, if F_2 is obtained from F_1 by (abstract) surgery along a simple closed curve C, then $c(F_2) \leq c(F_1)$, with equality if and only if C is inessential on F_1.

Now regard the surgery on $\tilde{F}_k$ along $G(D_k)$ as a sequence of surgeries along its constituent disks. Since the first of these surgeries is necessarily along an essential simple closed curve, we have $c(\tilde{F}_{k+1}) < c(\tilde{F}_k)$. It follows that the process just described must indeed eventually terminate, giving $\tilde{F}_m$, say, a union of 2-spheres as desired.

VII. Incompressible Surfaces in Branched Coverings

This completes the proof of Claim A.

Since no component of $\tilde{F}$ is a 2-sphere, we see that $m \geq 1$ and that every component of $\tilde{F}_m$ meets $\tilde{L}$. Let $\tilde{S}$ be a component of $\tilde{F}_m$, and write $S = p(\tilde{S})$. Then $\tilde{S}$ is a regular branched covering of $(S, S \cap L)$ with group $H = \{g \in G : g(\tilde{S}) = \tilde{S}\} \neq \{1\}$. Let n be the order of H, and b the cardinality of $S \cap L$ (the number of branch points on S), or equivalently, the number of exceptional orbits on $\tilde{S}$. Let k_i be the number of points in the ith exceptional orbit, $i = 1, \ldots, b$. Counting cells shows that the Euler characteristics of $\tilde{S}$ and S are related by the formula

$$\chi(\tilde{S}) = n\chi(S) - \sum_{i=1}^{b} (n - k_i).$$

Therefore, S is a 2-sphere and

$$2(n - 1) = \sum_{i=1}^{b} (n - k_i).$$

But $k_i | n$ and $k_i < n$, hence $1 \leq k_i \leq n/2$, $i = 1, \ldots, b$, and we obtain the inequalities

$$b(n/2) \leq 2(n - 1) \leq b(n - 1),$$

which show that $b = 2$ or 3. (In fact, any finite group action on S^2 is equivalent to an orthogonal action, so that if $b = 2$, then H is cyclic, and if $b = 3$, then H is dihedral, tetrahedral, octahedral, or icosahedral; see, for example, Wolf [W, p. 86].)

Let $F_m = p(\tilde{F}_m)$, $F_{m-1} = p(\tilde{F}_{m-1})$, and $D = p(D_{m-1})$. Then F_m is obtained from F_{m-1} by surgery along D. Let F'_{m-1} be the component of F_{m-1} containing ∂D, and let S be a component of F_m with $S \cap F'_{m-1} \neq \emptyset$. There are two cases to consider: either S separates M or it does not. We consider them independently.

Case 1. S separates M. Then $S \cap L$ must consist of two points. If ∂D were separating on F'_{m-1}, then it would bound a disk $\Delta \subset F'_{m-1}$, namely, $S \cap F'_{m-1}$ plus a boundary collar, with $\Delta \cap L =$ one point. We would then have $\partial D_{m-1} = \partial \tilde{\Delta}$, where $\tilde{\Delta} \subset \tilde{F}_{m-1}$ is a (cyclic) branched covering of $(\Delta, \Delta \cap L)$, contradicting the essentiality of ∂D_{m-1} on $\tilde{F}_{m-1}$. Hence ∂D is nonseparating on F'_{m-1}, showing that F'_{m-1} is a torus. Note also that $F'_{m-1} \cap L = \emptyset$. Therefore $F'_{m-1} = F_{m-1} = F$, and $S = F_m = F_1$ is obtained from F by surgery along D.

By hypothesis, S separates (M, L), into (M_1^*, L_1^*) and (M_2^*, L_2^*), say, corresponding to a connected sum decomposition $(M, L) \cong (M_1, L_1) \# (M_2, L_2)$. We may suppose that the numbering is such that, if A denotes a tubular neighborhood of the arc component of L_2^*, then F is obtained from S by the 0-surgery corresponding to the attachment of the 1-handle A to M_1^*.

If L_1 were a trivial link in M_1, then L_1^* would lie in a collar of S in M_1^*, and hence, since $M - L$ is irreducible by assumption, M_1^* would be a 3-ball. Then $M_1^* \cup A$ would be a solid torus, with core the component of L containing L_1^*, contradicting the fact that F is nonperipheral. On the other hand, if L_2 were trivial, then F would be compressible in $M - L$. Since neither L_1 nor L_2 is trivial, L is not prime, contrary to the hypothesis of the theorem. Thus the assumption that $\tilde{M}$ is not sufficiently large has in this case led to a contradiction.

Case 2. Some component S of F_m fails to separate M. Let C be an oriented, simple, closed curve in $M - L$ that intersects S transversely in a single point. Then some power of C lifts to an oriented simple closed curve $\tilde{C}$ in $\tilde{M} - \tilde{L}$ which meets an arbitrary component $\tilde{S}$ of $p^{-1}(S)$ transversely in a nonzero number of points. Since the action of G preserves orientations of $\tilde{M}$ and $\tilde{F}_m$, all the intersections of $\tilde{C}$ with $\tilde{S}$ have the same sign. Hence $\tilde{S}$ is also nonseparating.

This completes the proof of Theorem 1. ∎

An examination of the proof of Theorem 1 shows that we have actually proved that if $\tilde{M}$ is a regular branched covering of a sufficiently large prime link $L \subset M$, such that there is no nonseparating 2-sphere $S \subset M$ with $S \cap L = 2$ points and no 2-sphere $S \subset M$ with $S \cap L = 3$ points, then $\tilde{M}$ is sufficiently large. Note also that if G is cyclic, the last condition can be omitted.

Proof of Theorem 2. There are three steps.

(1) *$\tilde{M}$ irreducible implies $M - L$ irreducible.* Suppose that $M - L$ contains a 2-sphere that does not bound a 3-ball in $M - L$. Then it contains one S, say, that separates M, into M_1^* and M_2^*, say, corresponding to a connected sum decomposition $M \cong M_1 \# M_2$. Let L_i be the (possibly empty) sublink $L \cap M_i^*$ of L, $i = 1, 2$. Then $\tilde{M} \cong \tilde{M}_1^* \cup \tilde{M}_2^*$, the union being along the (2-sphere) boundary components of $\tilde{M}_i^*$, where $\tilde{M}_i^*$ is a regular branched (possibly disconnected) covering of (M_i^*, L_i), $i = 1, 2$.

If, for $i = 1$ and 2, each component of $\tilde{M}_i^*$ has more than one boundary component (this is necessarily the case if L_i is nonempty), then $\tilde{M}$ contains a nonseparating 2-sphere, contrary to hypothesis.

If L_1, say, is empty and some component of $\tilde{M}_1^*$ has connected boundary, then that component must be homeomorphic to M_1^*. Since $\tilde{M}$ is irreducible, M_1^* is a 3-ball, contrary to our hypothesis on S.

The next step uses the Smith conjecture.

(2) *$\tilde{M}$ irreducible implies L prime.* On the contrary, suppose that there is a 2-sphere S in M, such that $S \cap L$ consists of 2 points, separating (M, L) into (M_1^*, L_1^*), (M_2^*, L_2^*), corresponding to a nontrivial connected sum

decomposition $(M, L) \cong (M_1, L_1) \# (M_2, L_2)$. Then $\tilde{M} \cong \tilde{M}_1^* \cup \tilde{M}_2^*$, the union being along the (2-sphere) boundary components of $\tilde{M}_i^*$, $i = 1, 2$.

If, for $i = 1$ and 2, each component of $\tilde{M}_i^*$ has disconnected boundary, then $\tilde{M}$ contains a nonseparating 2-sphere.

If some component N of $\tilde{M}_i^*$, say, has a single boundary component, then N must be an n-fold cyclic branched covering of (M_1^*, L_1^*), for some $n > 1$. Since N is a 3-ball, by the irreducibility of $\tilde{M}$, L_1 is connected (for example, by Smith theory), and the Smith conjecture implies that L_1 is trivial (and M_1 is S^3), contrary to hypothesis.

(3) $\tilde{M}$ *irreducible and atoroidal implies $M - L$ atoroidal.* Suppose that $T \subset M - L$ is a nonperipheral incompressible torus. Then $\tilde{T} = p^{-1}(T)$ is a collection of tori. Since $\tilde{M}$ is atoroidal, $\tilde{T}$ is compressible. Let $\tilde{D}$ be a nice equivariant compressing disk for $\tilde{T}$, as in the proof of Theorem 1, and let $D = p(\tilde{D})$. Then D is a compressing disk for T. (For if ∂D bounds a disk Δ on T, then Δ lifts to a disk $\tilde{\Delta} \subset \tilde{T}$ bounded by $\partial \tilde{D}$.) Surgery on T along D yields a 2-sphere S meeting L in 2 points. If S were nonseparating, then, as in the proof of Theorem 1, $\tilde{M}$ would also contain a nonseparating 2-sphere. Hence, again as in the proof of Theorem 1, S decomposes L as a nontrivial connected sum of links, contrary to (2) above. This contradicts the existence of T, showing that $M - L$ is atoroidal. ■

Proof of Corollary 2.1. Suppose that $\tilde{M}$ is hyperbolic. Then it is irreducible, since its universal covering is $\mathbb{H}^3$, which is homeomorphic to int B^3, and atoroidal, since $\pi_1(\tilde{M})$ contains no subgroup isomorphic to $\mathbb{Z} \times \mathbb{Z}$. It follows from Theorem 2 that $M - L$ is irreducible and atoroidal. Therefore, by the torus theorem [F1, F2, Ja-Sh, Jo], either $\pi_1(M - L)$ contains no nonperipheral $\mathbb{Z} \times \mathbb{Z}$ subgroup, or $M - L$ is a (very special) Seifert manifold. In the first case, $M - L$ is hyperbolic by Thurston [Th] (see Morgan (Chapter V)). In the second case, $\tilde{M} - \tilde{L}$ is a Seifert manifold, and hence by Heil [He] $\tilde{M}$ is either a Seifert manifold or a connected sum of lens spaces (including $S^1 \times S^2$), neither of which can be hyperbolic. ■

4. The Equivariant Loop Theorem for Involutions

Throughout this section we leave corner smoothing to the reader.

By an elementary cut and paste argument we shall prove the following theorem (see also [K-To]). We do not assume that X is orientable.

THEOREM 3. *Let X be a 3-manifold with an involution g, and F a compressible component of ∂X. Then there exists a compressing disk D for F such that either $g(D) \cap D = \varnothing$, or $g(D) = D$ and D meets the fixed point set of g transversely.*

Let $\mathscr{F}$ be the fixed point set of g. By the loop theorem, there is a compressing disk D for F.

LEMMA 1. *We can take D to be transverse to both $g(D)$ and $\mathscr{F}$.*

Proof. Suppose first that $\mathscr{F}$ is empty, and consider any surface G properly embedded in X. Let $X' = X/\langle g \rangle$, and let $f : G \to X'$ be the immersion obtained by restricting the projection $X \to X'$. Approximate f, in the C^1-topology, by a general position immersion $f_1 : G \to X'$. (We are grateful to A. J. Casson for suggesting the use of the C^1-topology.) Lift f_1 to a map $i_1 : G \to X$ approximating the inclusion $i : G \to X$. Since the embeddings $G \to X$ form an open set in the C^1-topology, we can ensure that i_1 is an embedding, and then $i_1(G)$ meets $gi_1(G)$ transversely. Note that if, for some closed subset A of ∂G, A already meets $g(A)$ transversely in ∂X, we may assume that $i_1 | A = i | A$.

Next, suppose that $\mathscr{F}$ is one-dimensional. Let T be an invariant tubular neighborhood of $\mathscr{F}$. Put D transverse to $\mathscr{F}$, and consider a single point x of $D \cap \mathscr{F}$. Next x, we may assume that D meets T in a fiber disk, which is invariant under g. By perturbing this disk slightly we can make $D \cap g(D)$ transverse near x (see Fig. 1). Now let $X_1 = X - \text{int } T$, and $G = D \cap X_1$.

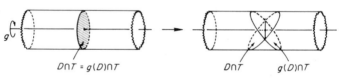

Figure 1

G is a disk with holes properly embedded in X_1, and g acts on X_1 without fixed points. By the above, we may assume G meets $g(G)$ transversely, without moving $G \cap \partial T$. This establishes the lemma when $\mathscr{F}$ is one-dimensional. The other cases (which are not needed for our present purposes) are similar, and we omit the details. (Note: one needs to observe that any closed curve of intersection of D with a two-dimensional component of $\mathscr{F}$ preserves orientation in X.) ∎

Proof of Theorem 3. By Lemma 1, $D \cap g(D)$ consists of simple closed curves and arcs with endpoints on ∂D. We now attempt to reduce the number of components of $D \cap g(D)$. First suppose that there exists a simple closed curve of intersection. Let C be one such curve which is innermost on $g(D)$; i.e., if E is the disk bounded by C on $g(D)$, then $E \cap D = C$. Let C bound the disk G on D. (It may be that int $G \cap g(D) \ne \varnothing$.) Let E' be a nearby parallel

VII. Incompressible Surfaces in Branched Coverings

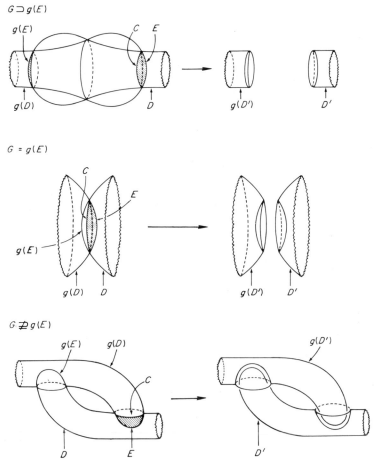

Figure 2

copy of E, with $E' \cap D = \partial E' \subset D - G$. Let $\partial E'$ bound G' on D, and set $D' = (D - G') \cup E'$ (see Fig. 2). Then $\partial D' = \partial D$, $D' \cap g(D') \subseteq D \cap g(D)$, and $C \not\subseteq D' \cap g(D')$. In this way we can eliminate all closed curves from $D \cap g(D)$.

Now suppose that $D \cap g(D)$ consists only of arcs. Let A be one such arc which is extremal on $g(D)$; i.e., for one of the disks into which A divides $g(D)$, say E, we have $E \cap D = A$. Let B be the arc in $\partial g(D)$ such that $\partial E = A \cup B$. Let the two disks into which A divides D be G_1 and G_2, with $G_2 \supseteq g(E)$, and let the corresponding arcs of ∂D be C_1 and C_2. At least one of the curves $B \cup C_1$, $B \cup C_2$ must be essential on F (since $C_1 \cup C_2 = \partial D$ is); choose $j = 1$ or 2 so that $B \cup C_j$ is. There are two cases.

Case 1. $A \neq g(A)$ *or* $j = 1$. Let E' be a nearby parallel copy of E, whose boundary is the union of parallel copies A', B' of A, B in D, F, respectively, and such that $E' \cap D = A' \subset G_j$. A' divides D into two disks; let G'_j be the disk with $G'_j \subset G_j$. Set $D' = G'_j \cup E'$ (see Fig. 3). Then $\partial D'$ is essential on F, $D' \cap g(D') \subseteq D \cap g(D)$ and $A \not\subseteq D' \cap g(D')$.

Case 2. $A = g(A)$ *and* $j = 2$. In this case $D' = E \cup g(E) = E \cup G_2$ is a compressing disk for F with $g(D') = D'$ (see Fig. 3).

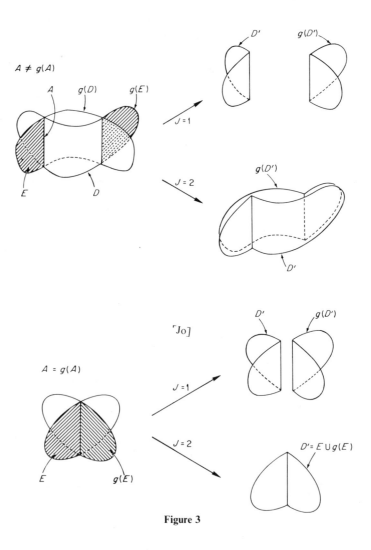

Figure 3

VII. Incompressible Surfaces in Branched Coverings

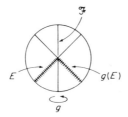

Figure 4

In Case 1 we repeat the procedure with another arc; the process terminates either once we have eliminated all arcs of $D \cap g(D)$ by Case 1 or when Case 2 first occurs. It remains only to check that our final disk is transverse to $\mathscr{F}$. Clearly, the transversality of D and $\mathscr{F}$ is preserved each time we remove a component of $D \cap g(D)$. Suppose that we are in Case 2. Let x be any point of $\mathscr{F} \cap A$. We can find a transverse disk Δ to A at x such that $\mathscr{F}$ meets Δ in an arc and g acts on Δ by reflection in this arc. The intersections of D, $g(D)$, and $\mathscr{F}$ with Δ must then be as in Fig. 4, showing that $\mathscr{F}$ is transverse to $E \cup g(E)$. ∎

References

[C–Sh]   Culler, M., and Shalen, P., Varieties of group representations and splittings of 3-manifolds, *Ann. of Math.* **117** (1983), 109–146.

[F1]   Feustel, C. D., On the torus theorem and its applications, *Trans. Amer. Math. Soc.* **217** (1976), 1–43.

[F2]   Feustel, C. D., On the torus theorem for closed 3-manifolds, *Trans. Amer. Math. Soc.* **217** (1976), 45–57.

[G]   Gordon, C. McA., Uncountably many stably trivial strings in codimension two, *Quart. J. Math. Oxford Ser.* (2) **28** (1977), 369–379.

[Hak]   Haken, W., Theorie der Normalflachen, *Acta Math.* **105** (1961), 245–375.

[Hat–Th]   Hatcher, A., and Thurston, W., Incompressible surfaces in two-bridge knot complements, to appear.

[He]   Heil, W., Elementary surgery on Seifert fiber spaces, *Yokohama Math. J.* **22** (1974), 135–139.

[Ja]   Jaco, W., Lectures on three-manifold topology, *Proc. C.B.M.S. Regional Conf. Three-Manifold Topology, Blacksburg, Virginia,* 1977.

[Ja–Sh]   Jaco, W., and Shalen, P., Seifert-fibered spaces in 3-manifolds, *Mem. Amer. Math. Soc.* **21**, No. 220, (1979).

[Jo]   Johannson, K., "Homotopy Equivalences of 3-manifolds with Boundary." Lecture Notes in Mathematics, Vol. 761. Springer-Verlag, Berlin and New York, 1979.

[K–To] Kim, P. K., and Tollefson, J. L., Splitting the PL involutions of nonprime 3-manifolds, *Mich. Math. J.* **27** (1980), 259–274.
[Sc] Schubert, H., Über eine numerische Knoteninvariante, *Math. Z.* **61** (1954), 245–288.
[Th] Thurston, W., The geometry and topology of 3-manifolds. Lecture notes, Dept. of Math., Princeton Univ., Princeton, New Jersey, 1979.
[W] Wolf, J. A., "Spaces of Constant Curvature." McGraw-Hill, New York, 1967.

CHAPTER VIII

The Equivariant Loop Theorem for Three-Dimensional Manifolds and a Review of the Existence Theorems for Minimal Surfaces

Shing-Tung Yau[*]   and   William H. Meeks, III

Department of Mathematics
Stanford University
Stanford, California

Instituto Mathemática Pura e Aplicada
Rio de Janeiro, Brazil

The details of this chapter appeared in Meeks and Yau [4, 5]. The equivariant loop theorem that is needed in settling the Smith conjecture can be described as follows.

Let G be finite group acting smoothly on a compact three-dimensional manifold M. Let $S = \amalg_j S_j$ be a union of components of ∂M such that $g(S) = S$ for all $g \in G$. For all j, we have the inclusion $i_j : S_j \to M$. Let $K_j \subset \pi_1(S_j)$ be the kernel of $(i_j)_* : \pi_1(S_j) \to \pi_1(M)$. The equivariant version of the loop theorem says that there are a finite number of properly embedded disks $D_1, \ldots, D_k$ in M which satisfy the following properties:

(1) K_j is the normal subgroup of $\pi_1(S_j)$ generated by the boundary circles ∂D_i, which lie in S_j.

---

[*] This author's work was supported by National Science Foundation grant MCS 79-12938. Present address: Institute for Advanced Study, School of Mathematics, Princeton University, Princeton, New Jersey.

(2) For any $g \in G$ and $1 \leq i, i' \leq k$, either $D_i \cap g(D_{i'}) = \emptyset$ or $D_i = g(D_{i'})$.

The major point here is that our solution to the loop theorem respects the action of the group G in a suitable manner. The classical proof of Papakyriakopoulos, Whitehead and Shapiro, and Stallings does not seem to be readily generalizable to cover this case.

Our proof can be sketched in the following manner. We put a metric on M so that the group G acts isometrically and so that ∂M is convex with respect to the outward normal. Then with respect to this metric, we demonstrate the existence of an immersed disk D_1 in M whose boundary ∂D_1 represents a nontrivial element in $\pi_1(S)$ and whose area is minimal among all such disks. Partially using the classical topological methods developed in Papakyriakopoulos [8], Shapiro and Whitehead [11], and Stallings, [12] and partially using the properties of minimal surfaces, we show that D_1 is embedded.

If the smallest normal subgroup of $\pi_1(S)$ containing $[\partial D_1]$ is not equal to $\pi(S)$, then we minimize the area of all disks whose boundary curve is an element in $\pi_1(S)$, which does not belong to this group. In this way, we construct another embedded minimal disk D_2. Continuing this process, we obtain embedded disks $D_1, D_2, \ldots$. This process has to stop because there is only a finite number of pairwise disjoint Jordan curves that are not isotopic to each other.

Having constructed $D_1, \ldots, D_k$, we can prove (2) by using the minimality of the disks. The point is that when two minimal disks intersect nontrivially along a Jordan arc or a closed Jordan curve they must intersect transversally except at finite number of points. One can then prove that this is in contradiction to the minimality of the area by cutting the disks and deforming along their intersection curve. In this way we prove that two minimal disks with the properties described above are either equal or disjoint. Since g is an isometry, it is clear that if D is a minimal disk, then $g(D)$ has similar properties. Property (2) follows easily from this remark.

Up to now, the argument sketched above was described in detail in Meeks and Yau [4]. For the rest of this chapter, we review some of the existence theorems for minimal surfaces that may be useful to the study of the topology of three-dimensional manifolds. For that reason, we generalize some of these classical theorems to a somewhat more general category.

1. Morrey's Solution for the Plateau Problem in a General Riemannian Manifold

Let M be a complete m-dimensional manifold that is homogeneously regular in the following sense of Morrey [7]: There exists a constant $C > 0$ such that for each point $x \in M$, there exists a bi-Lipschitz homeomorphism

VIII. Equivariant Loop Theorem and Existence Theorems

of a neighborhood of x onto the unit ball in $\mathbf{R}^m$ with Lipschitz constants less than C. By Nash's isometric embedding theorem, we can assume that M is a properly embedded submanifold of a higher-dimensional euclidean space $\mathbf{R}^n$. Let Σ be a compact Riemann surface with boundary $\partial\Sigma$. Let $f: \Sigma \to M$ be a smooth map and $\mathscr{F}$ be the family of maps $g: \Sigma \to M$ so that the energy of g is

$$E(g) = \frac{1}{2} \int_\Sigma |\nabla g|^2 < \infty \tag{1.1}$$

and $g|\partial\Sigma = f|\partial\Sigma$.

THEOREM 1 (Morrey). *There exists a map $f_0 \in \mathscr{F}$ such that $E(f_0) = \inf\{E(g) | g \in \mathscr{F}\}$, and any such f_0 is smooth.*

Proof. Let g_i be a sequence in $\mathscr{F}$ so that $\lim_{i \to \infty} E(g_i) = \inf\{E(g)|g \in \mathscr{F}\}$. Then, since $g_i|\partial\Sigma = f$ for each i and $E(g_i)$ has an upper bound independent of i, a subsequence of g_i converges weakly in the Hilbert space of vector valued mappings g from Σ into R^n with $g|\partial\Sigma = f$ and $\|g\|^2 = \int_\Sigma |g|^2 + \int_\Sigma |\nabla g|^2 < \infty$. We may assume the subsequence is $\{g_i\}$ itself and the weak limit is g_0. It is easy to check that $g_0 \in \mathscr{F}$. As $E(g_0) \leq \lim_{i \to \infty} E(g_i)$, $E(g_0) = \inf\{E(g)|g \in \mathscr{F}\}$. It remains to check that g_0 is smooth.

Let x be a point in the interior of Σ. Let $B_x(r)$ be disks of radius r around x. Then we assert that for some constant $a > 0$,

$$\int_{B_x(r)} |\nabla g_0|^2 \leq ar \int_{\partial B_x(r)} |\nabla g_0|^2 \tag{1.2}$$

for r smaller than some positive constant independent of a.

If the length of $g_0(\partial B_x(r))$ is greater than $1/c$, where c is the Lipschitz constant that appears in the definition of homogeneous regularity of M, then

$$2\pi r \int_{\partial B_x(r)} |\nabla g_0|^2 \geq \left(\int_{\partial B_x(r)} |\nabla g_0|\right)^2 \geq \left(\frac{1}{c}\right)^2 \tag{1.3}$$

and (1.2) follows by choosing $a = 2\pi c^2 \int_\Sigma |\nabla g_0|^2$.

If the length of $g_0(\partial B_x(r))$ is smaller than $1/c$, we can assume that the image of $\partial B_x(r)$ under g_0 lies in the unit ball in the coordinate system that appeared in the definition of homogeneous regularity. Then we define a map from $B_x(r)$ into this unit ball by requiring each component of the map to be harmonic (with respect to the coordinate system) and its restriction on $\partial B_x(r)$ to be given by $g_0|\partial B_x(r)$. Call this map h. We can define a new map $\tilde{g}_0 \in \mathscr{F}$ by requiring that $\tilde{g}_0 = g_0$ on $\Sigma \setminus B_x(r)$ and $\tilde{g}_0 = h$ on $B_x(r)$.

Since g_0 minimizes the energy in $\mathscr{F}$, it is clear that

$$\int_{B_x(r)} |\nabla g_0|^2 \leq \int_{B_x(r)} |\nabla h|^2. \tag{1.4}$$

Because $g_0|\partial B_x(r) = h|\partial B_x(r)$, it suffices to prove (1.2) by demonstrating that

$$\int_{B(x,r)} |\nabla h|^2 \leq ar \int_{\partial B(x,r)} |\nabla_{(1/r)(\partial/\partial\theta)} h|^2. \tag{1.5}$$

However, this last inequality follows easily by expanding each coordinate of h in Fourier series.

Therefore, inequality (1.2) is proved and we can rewrite it as

$$\frac{d}{dr} \log\left(\int_{B_x(r)} |\nabla g_0|^2\right) \geq \frac{1}{a} \frac{d}{dr} \log r. \tag{1.6}$$

By integrating, we find that

$$\int_{B_x(r)} |\nabla g_0|^2 \leq r^{1/a}\left(R^{-1/a} \int_{B_x(R)} |\nabla g_0|^2\right), \tag{1.7}$$

where $0 \leq r < R$ and $B_x(R)$ is a fixed ball around x.

As $\int_{B_x(R)} |\nabla g_0|^2$ is bounded, (1.7) measures how $\int_{B_x(r)} |\nabla g_0|^2$ decays when $r \to 0$. A calculus lemma of Morrey [7] then shows that g_0 is Hölder continuous at x with constants that depend only on a and $R^{-1/a} \int_{B_x(R)} |\nabla g_0|^2$.

Therefore g_0 is continuous in the interior of Σ. By using the fact that $g_0|\partial\Sigma$ is Lipschitz, one can use an argument similar to that given before to prove that g_0 is Hölder continuous near $\partial\Sigma$.

The Hölder continuity of g_0 guarantees that the image of a suitable disk $B_x(r)$ of any point $x \in \Sigma$ lies in a coordinate neighborhood of M. Using the coordinate system and the fact that g_0 minimizes energy in $\mathscr{F}$, g_0 must satisfy the variational equation

$$\frac{\partial}{\partial x}\left(\sum_i h_{ij}(g_0) \frac{\partial g_0^i}{\partial x}\right) + \frac{\partial}{\partial y}\left(\sum_i h_{ij}(g_0) \frac{\partial g_0^i}{\partial y}\right) = 0, \tag{1.8}$$

where $\sum_{i,j} h_{ij} dx^i dx^j$ is the metric tensor of M.

Since we are dealing with vector-valued functions, a difference-quotient argument (see Morrey [7]) can be used to prove the higher differentiability of g_0. This completes the proof of Theorem 1. ∎

In order to state the second theorem, one needs to introduce some terminology. Let Σ and Σ' be two not necessarily connected surfaces. Let $f: \partial\Sigma \to M$ and $f': \partial\Sigma' \to M$ be smooth diffeomorphisms, each mapping onto the same disjoint union of Jordan curves in M. Then we say that $(\Sigma, f) > (\Sigma', f')$ if Σ' can be obtained from Σ by surgery on a collection of disjoint, simple, closed curves in the interior of Σ (i.e., cut along the curve and sew

VIII. Equivariant Loop Theorem and Existence Theorems

back disks in neighborhoods defined as "homogeneously regular manifolds.") and f' is the same as f up to reparametrization of $\partial\Sigma$. (If Σ is oriented, we can require this reparametrization to preserve the orientation.)

For each pair (Σ, f), let $\mathscr{F}(\Sigma, f)$ be the family of all maps $g: \Sigma \to M$ so that for some conformal structure on Σ, $\int_\Sigma |\nabla g|^2 < \infty$ and $g|\partial\Sigma = f$ up to reparametrization. Let $A(\Sigma, f)$ be the infimum of all $\frac{1}{2}\int_\Sigma |\nabla g|^2$, where ∇ is taken with respect to some conformal structure over Σ and $g|\partial\Sigma$ is equal to f up to reparametrization. Using the existence of isothermal coordinates on a surface, one verifies that $A(\Sigma, f)$ is simply the infimum of the area of all possible maps $g: \Sigma \to M$ so that $g|\partial\Sigma$ is equal to f up to reparametrization.

THEOREM 2. *Suppose for each $(\Sigma', f') < (\Sigma, f)$, that $A(\Sigma', f') > A(\Sigma, f)$. Then there exists a conformal structure over Σ and a smooth conformal map $g_0: \Sigma \to M$ whose area is equal to $A(\Sigma, f)$.*

Proof. First, we fix a conformal structure on Σ and we minimize the energy $E(g)$ over all maps $g: \Sigma \to M$, so that $g|\partial\Sigma = f$ up to a reparametrization. Let g_i be a minimizing sequence. By Theorem 1, we can assume that each g_i is harmonic, and for the latter purpose we may assume that the choice of the conformal structure gives rise to $\lim_{i \to \infty} E(g_i) < A(\Sigma', f')$ whenever $(\Sigma', f') < (\Sigma, f)$.

Each g_i satisfies the equation for harmonicity. Hence the (standard) arguments in Theorem 1 of Meeks and Yau [5] show that we may assume g_i converges smoothly on compact subsets of the interior of Σ_0 to a smooth harmonic map g'_0.

We are now going to prove that g'_0 is continuous in a neighborhood of $\partial\Sigma$ and is equal to $f|\partial\Sigma$ up to reparametrization. There are two cases to be discussed. If Σ is not the disk, we proceed as follows.

Let x be an arbitrary point on a component σ of $\partial\Sigma$. Then, by using the argument of Lebesgue (see the proof of Theorem 1 in [5]), we can find a number $0 < r^2 < r_i < r$ so that the length of $g_i(\partial B_x(r) \cap \Sigma)$ is not greater than $\sqrt{2\pi E(g_i)} (\log r)^{-1/2}$. Since $E(g_i)$ is uniformly bounded, the last number is arbitrarily small when r is small enough. When r is small, the arc $B_x(r) \cap \sigma$ is either mapped to an arc of $f(\sigma)$ with length small, compared with r, or mapped to the complement of such an arc on $f(\sigma)$. If the former case occurs for all $x \in \sigma$, then g_i is equicontinuous on σ and the replacement arguments in Theorem 1 show that g'_0 is Hölder continuous near $\partial\Sigma$. If the latter case occurs, both $g_j(\partial B_x(r) \cap \Sigma)$ and $g_j(\sigma)\backslash g_j(B_x(r) \cap \sigma)$ have small length and hence bound a disk with small area. We can form a new surface Σ' by putting two new disks with $g_j(B_x(r) \cap \Sigma)$ and $g_j(\Sigma)\backslash g_j(B_x(r) \cap \Sigma)$ together, respectively. In this way, we form a new pair (Σ', f') with an area close to the area of g_j. This gives a contradiction to $A(\Sigma, f) < A(\Sigma'; f')$.

In case Σ is a disk, one has to fix the points 1, -1, and $\sqrt{-1}$ on the unit circle and then fix the points $g_j(1)$, $g_j(-1)$, and $g_j(\sqrt{-1})$ in $f(\sigma)$. One can make this assumption because any three distinct points on the unit circle can be mapped to the other three by a conformal automorphism of Σ and the total energy is invariant under conformal parametrization. With this assumption, the maps g_j always map small arcs to small arcs and hence are equicontinuous. Therefore, g'_0 is Hölder continuous in a neighborhood of $\partial\Sigma$.

Now we change the conformal structures over Σ. The treatment here is the same as the one in Schoen and Yau [10]. We only treat the case when Σ is not the disk or the annulus. By doubling Σ and putting the Poincaré metric on the doubled Riemann surface, we can assume that Σ admits a Poincaré metric whose boundary consists of geodesics. The space of conformal structures over Σ can be identified with the space of these metrics. For each fixed conformal structure ω, which satisfies the assumption mentioned in the beginning of the proof, we can choose a map g_ω by the procedure mentioned above and the energy of this map defines a lower semicontinuous function over the space of conformal structures. However, the last space is noncompact. Hence, in order to prove that the lower semicontinuous function has a minimum, we demonstrate that it is proper. (Strictly speaking, we have to study on the Teichmüller space instead of the moduli space. The method in [10] can be used to overcome this problem.) This follows because a sequence of conformal structure tends to infinity iff for each of these conformal structures there exists an embedded closed geodesic or a geodesic arc joining the boundaries whose length with respect to the Poincaré metric tends to zero. If the length of the image of these curves under g_ω is bounded away from zero, then the arguments in [10] demonstrate that the energy of g_ω tends to infinity. Otherwise, the length of the image of the geodesics tends to zero (these geodesics must be closed geodesics because the curves in $f(\partial\Sigma)$ are fixed in M). Hence, eventually the image of the geodesics bound a disk with small area and the arguments used above show that we will violate the condition $A(\Sigma, f) < A(\Sigma', f')$.

In conclusion, we have found a conformal structure on Σ and a map $g_0: \Sigma \to M$ such that $E(g_0) = A(\Sigma, f)$ and $g_0|\partial\Sigma = f$ up to reparametrization. Furthermore, the arguments also showed that g_0 is smooth in the interior of Σ and Hölder continuous in a neighborhood of $\partial\Sigma$. The arguments of Hildebrandt [2] then show that g_0 is in fact smooth in a neighborhood of $\partial\Sigma$. This finishes the proof of Theorem 2. ∎

Remark. In Hildebrandt [2] a proof of the theorem of Lewy and Morrey was also given. The proof states that if M is real analytic and if the image curves $f(\partial\Sigma)$ are real analytic, then the minimal surface constructed in Theorem 2 must be real analytic. The proof consists of estimating the derivatives of g_0 carefully and proving the convergence of the Taylor series of g_0.

VIII. Equivariant Loop Theorem and Existence Theorems

2. The Existence Theorem for Manifolds with Boundary

In this section we extend Theorem 2 to the case in which M is allowed to have boundary ∂M. We say M is homogeneously regular if M is a subdomain of another homogeneously regular manifold N that has no boundary.

THEOREM 3. *Let M be a three-dimensional, homogeneously regular manifold whose boundary ∂M has nonnegative mean curvature with respect to the outward normal. Let Σ be a compact surface with boundary and $f : \Sigma \to M$ be a smooth map so that $f : \partial \Sigma \to M$ is an embedding. Suppose that $A(\Sigma, f) < A(\Sigma', f')$ for all $(\Sigma', f') < (\Sigma, f)$. (See the definitions in Section 1.) Then there exists a conformal structure over Σ and a conformal map $g : \Sigma \to M$ so that $g|\partial\Sigma$ is equal to f up to reparametrization of $\partial\Sigma$ and the area of g is not greater than the area of any map with the same property.*

Proof. First we notice that in case M is compact and ∂M has nonnegative mean curvature with respect to the outward normal, then Theorem 3 is valid. Furthermore, the same theorem remains valid if M is the intersection of a finite number of compact domains of the above form. This was carried out in Meeks and Yau [6].

In general, let Ω_i be an increasing sequence of compact, smooth domains in M such that $M = \bigcup_{i=1}^\infty \Omega_i$, $\partial M = \bigcup_{i=1}^\infty (\partial M \cap \Omega_i)$, and $\bigcup_{x \in \Omega_i} B_x(1) \subset \Omega_{i+1}$ for all i. Then for each i, we can change the metric in $\Omega_i \setminus \Omega_{i-1}$ so that $\partial \Omega_i$ has nonnegative mean curvature with respect to the outward normal. By the remark in the last paragraph, we can then find a conformal structure ω_i over Σ and a conformal map $g_i : \Sigma_i \to \Omega_i$ that minimizes area with respect to the changed metric on Ω_i.

We claim that for i large enough, we may assume that $\omega_i = \omega_{i+1} = \cdots$ and $g_i = g_{i+1} = \cdots$. This claim clearly implies the theorem.

For each $j \leq i - 2$ and $x \in g_i(\Sigma) \cap \Omega_j$, let $B_x(1)$ be the ball with center x and radius 1 in N. Then, as $B_x(1) \subset \Omega_{i-1}$ and the metric on Ω_{i-1} is unchanged, there is a bi-Lipschitz diffeomorphism φ of $B_x(1)$ to R^n whose Lipschitz constant is not greater than a constant C. (This comes from the definition of homogeneous regularity of N.)

We are going to bound the area of $g_i(\Sigma) \cap B_x(1)$ from below by a positive constant that is independent of x and i. Hence, we may assume that $g_i(\Sigma) \cap B_x(1)$ minimizes area in $B_x(1)$ instead of minimizing area in $B_x(1) \cap \Omega_i$. For almost every $0 < r \leq 1$, we may assume that $B_x(r) \cap g_i(\Sigma)$ is a disjoint union of Jorden curves $\sigma_1(r), \ldots, \sigma_k(r)$. Let $D_i(r)$ be the minimal disk in $\mathbf{R}^n$ whose boundary is given by $\varphi(\sigma_i(r))$. Then the area of $\varphi^{-1}(D_i(r))$ is not greater than $C^2 A_e(D_i(r))$, where $A_e(D_i(r))$ is the euclidean area of $D_i(r)$. Since $g_i(\Sigma) \cap B_x(1)$ minimizes area in $B_x(1)$, the area of $g_0(\Sigma) \cap B_x(r)$ is not greater than the sum of the areas of $\varphi^{-1}(D_i(r))$, and hence not greater than

$$C^2 \sum_{i=1}^{k} A_e(D_i(r)).$$

By the isoperimetric inequality for minimal surfaces in $\mathbf{R}^n$, $(1/4\pi)A_e(D_i(r))$ is not greater then the square of the euclidean length of $\varphi(\sigma_i(r))$. Hence the area $A(g_i(\Sigma) \cap B_x(r))$ is not greater than $4\pi C^4 l^2(g_i(\Sigma) \cap \partial B_x(r))$, where $l(g_i(\Sigma) \cap \partial B_x(r))$ is the length of $g_i(\Sigma) \cap \partial B_x(r)$.

Since $|\nabla r| \leq 1$ on $g_i(\Sigma)$ when r is restricted to $g_i(\Sigma)$, the coarea formula [1] shows that

$$4\pi C^4 \left(\frac{d}{dr}(A(g_i(\Sigma) \cap B_x(r)))\right)^2 \geq A(g_i(\Sigma) \cap B_x(r)). \tag{2.1}$$

By integrating this inequality, we obtain

$$A(g_i(\Sigma) \cap B_x(r)) \geq r^2/\sqrt{\pi}\, C^2 \tag{2.2}$$

for $0 < r \leq 1$.

In particular, $A(g_i(\Sigma) \cap B_x(1)) \geq (\sqrt{\pi}\, C^2)^{-1}$. Without loss of generality, we may assume $\partial(g_i(\Sigma)) \subset \Omega_1$. If $g_i(\Sigma) \cap (\Omega_j \backslash \Omega_{j-1}) \neq \varnothing$, then we can find points $x_1, \ldots, x_{[(j-1)/2]}$ with $x_k \in g_i(\Sigma) \cap (\Omega_{2k-1} \backslash \Omega_{2k-2})$ (Here $[(j-1)/2]$ is the largest integer less than $(j-1)/2$). The balls $B_{x_1}(1), B_{x_2}(1), \ldots$ are disjoint and hence the area of $g_i(\Sigma)$ is not less than $\Sigma_k A(g_i(\Sigma) \cap B_{x_k}(1)) \geq ((j-1)/2)$ $(\sqrt{\pi}\, C^2)^{-1}$. This implies $j \leq 2 + 2\sqrt{\pi}\, C^2 A(\Sigma, f)$ and that $g_j(\Sigma)$ lies in a fixed compact set of M. This finishes the proof of Theorem 3. ∎

3. Existence of Closed Minimal Surfaces

In this section we shall record the existence of incompressible, closed, minimal surfaces in a three-dimensional homogeneous regular manifold M (possibly with boundary). If M is compact, without boundary, this was proved independently by Sacks and Uhlenbeck [9] and Schoen and Yau [10].

THEOREM 4. *Let Σ be a compact surface without boundary. Let $f: \Sigma \to M$ be a smooth map, so that $f_*: \pi_1(\Sigma) \to \pi_1(M)$ is injective and f is not homotopic to the sum of two spheres and a map which can be deformed off of every compact subset of M. If ∂M has nonnegative mean curvature with respect to the outword normal, then there exists a conformal structure on Σ and a conformal map $g: \Sigma \to M$, so that $g_*|\pi_1(\Sigma) = f_*|\pi_1(\Sigma)$ and the area of g is not greater then the area of any map from Σ to M that is homotopic to g.*

VIII. Equivariant Loop Theorem and Existence Theorems

Proof. First, we treat the following special case. Assume that M has no boundary and that for any point $x \in M$ there is a contractible neighborhood Ω_x of x with $d(x, \partial\Omega_x) \to \infty$ when x tends to infinity. (In this case, we do not need the assumption that f cannot be homotopic to infinity.)

We minimize, for each conformal structure over Σ, the energies of maps whose action on $\pi_1(\Sigma)$ is the same as f. Thus let ω be any conformal structure over Σ. Then we can find a sequence of smooth maps $g_1, g_2, \ldots$ whose action on $\pi_1(\Sigma)$ is the same as f and $\lim_{i \to \infty} E(g_i)$ is the infinmum of the energies of such maps. By following the same procedure as in [10], we can produce a harmonic map g_ω with minimal energy in our class if we can prove that for some $y \in M$, there exists $\varepsilon > 0$ and $L > 0$ such that the set $\{x \in \Sigma \mid d(g_i(x), y) \leq L\}$ has measure greater than ε. (This is true because after isometric embedding of M into R^n and applying the Poincaré inequality for vector-valued functions over Σ, we can estimate the L^2-norm of g_i in terms of ε, L, and the energy of g_i. Then we can take a weak convergent subsequence of g_i and proceed as in [10].)

If the last statement were wrong, then by passing to a sequence of g_i, we may assume that for some $\varepsilon_i \to 0$ and $L_i \to \infty$ the measure of

$$\{x \in \Sigma \mid d(g_i(x), y) \leq L_i\}$$

is less than ε_i. Assume $\Sigma \neq \mathbf{R}P^2$ and fix an annulus region $[0, a] \times S^1$ in Σ so that $f(\{0\} \times S^1)$ is homotopically nontrivial in M. By Fubini's theorem, we can find $[0, a] \times \{\tau\}$ in $[0, a] \times S^1$ so that, except for a set of measure δ_i in $[0, a] \times \{\tau\}$ with $\delta_i \to 0$, $d(g_i(t, \tau), y) > L_i$. By assumption, there exists contractible domain $\Omega_{(t, \tau)}$ containing $g_i(t, \tau)$ so that $d(g_i(t, \tau), \partial\Omega_{(t, \tau)})$ tends to infinity uniformly as $d(g_i(t, \tau), y) > L_i$ and $i \to \infty$. Since $g_i(\{t\} \times S^1)$ is homotopically nontrivial, its length $L(g_i(\{t\} \times S^1))$ tends to infinity uniformly also. Therefore $\int_0^a L(g_i(\{t\} \times S^1)) \, dt \to \infty$. Because the energy of g_i over $[0, a] \times S^1$ is dominated from below by $\{\int_0^a L(g_i(\{t\} \times S^1)) \, dt\}^2$ up to a constant independent of i, $E(g_i) \to \infty$. This is a contradiction.

If Σ is $\mathbf{R}P^2$, one can proceed as follows. Let $U(\Sigma)$ be the unit tangent bundle of Σ and let $\mathscr{C}$ be the two-dimensional surface that parametrizes the set of all closed geodesics in Σ. Since the measure of $\{x \in \Sigma \mid d(g_i(x), y) \leq L_i\}$ is less than ε_i, the measure of the closed geodesics that pass through

$$\{x \in \Sigma \mid d(g_i(x), y) \leq L_i\}$$

is small compared with ε_i. We can consider $|\nabla g_i|^2$ as a function over $U(\Sigma)$, which fibers over $\mathscr{C}$. Hence, $E(g_i)$ can be obtained by integrating $|\nabla g_i|^2$ over the closed geodesics first and then over $\mathscr{C}$. As above, if there is a point x on the closed geodesic so that $d(g_i(x), y) > L_i$, then the integral is dominated from below by L_i^2. This also gives a contradiction and we have proved the existence of g_ω.

As in [10], we have to change the conformal structures ω and minimize $E(g_\omega)$ to achieve a map with minimal area. This can be done in exactly the same way as in [10]. This proves the theorem in the special case described above.

By using the method of [4] and [6], we see that Theorem 4 also holds if M is compact with nonnegative mean curvature with respect to the outward normal.

For the general case, we proceed as in Theorem 3. We construct an increasing sequence of compact domains Ω_i in M so that $M = \bigcup_{i=1}^{\infty} \Omega_i$ and $\bigcup_{x \in \Omega_i} B_x(1) \subset \Omega_{i+1}$. For each Ω_i, we construct a metric so that $\partial \Omega_i$ has nonnegative mean curvature with respect to the outward normal and the metric coincide with the original one on Ω_{i-1}. We may also assume that $f(\Sigma) \subset \Omega_1$.

We can then minimize area in Ω_i and obtain $g_i: \Sigma \to \Omega_i$, which is homotopic to f up to the connected sum of two spheres. By the topological assumption on f, we may find a point $x_i \in \Sigma$ so that $\lim_{i \to \infty} g_i(x_i)$ exists. This fact and the arguments provided in the proof of Theorem 3 then imply that $g_i(\Sigma)$ stays in a fixed compact set of M. This finishes the proof of Theorem 4. ∎

COROLLARY. *Let Ω be a domain in $\mathbf{R}^3$ that has nonnegative mean curvature with respect to the outward normal. Then there exists no compact, incompressible surface with nontrivial fundamental group in Ω.*

Proof. Let Σ be the compact, incompressible surface with nontrivial fundamental group in Ω. Then we can choose a two-dimensional sphere S (with positive mean curvature) in $\mathbf{R}^3$ that encloses Σ. The domain $\Omega \cap S$ becomes a compact manifold with nonnegative mean curvature with respect to the outward normal. Hence we can minimize the area of Σ within $\Omega \cap S$ and obtain a compact minimal surface in $\mathbf{R}^3$. (The previous theorem applies, by smoothing, even if $\Omega \cap S$ has corners [6].) Since $\mathbf{R}^3$ has no compact minimal surface, this is a contradiction. ∎

By using some topological arguments, one can then derive the following.

COROLLARY. *Let Σ be a properly embedded minimal cylinder in $\mathbf{R}^3$. Then Σ is isotopic to the catenoid.*

4. Existence of the Free Boundary Value Problem for Minimal Surfaces

By using the arguments of the above sections and [5], we can generalize Theorem 1 of [5] in the following way.

Let M be a three-dimensional, homogeneously regular manifold whose boundary ∂M has nonnegative mean curvature with respect to the outward normal. Let Σ be a compact surface with boundary and let $\mathscr{F}_\Sigma$ be the family of smooth maps $f: \Sigma \to M$ so that $f(\partial\Sigma) \subset \partial M$, so that f is not homotopic rel $\partial\Sigma$ to a map whose image is in ∂M and so that there is a fixed compact set $K \subset M$ that meets the image of every map $g: \Sigma \to M$ homotopic, as a map of pairs, to f. Let A be the infimum of the area of the maps in $\mathscr{F}_\Sigma$. We say that $\Sigma' < \Sigma$ if Σ' can be obtained by surgery along a simple closed curve or a Jordan arc of Σ which disconnects Σ.

THEOREM 5. *Let M be a three-dimensional, homogeneously regular manifold whose boundary has nonnegative mean curvature with respect to the outward normal. Let Σ be a compact surface with $A_\Sigma < A_{\Sigma'}$ for any $\Sigma' < \Sigma$. Then we can find a conformal structure over Σ and a smooth conformal map $f \in \mathscr{F}_\Sigma$ so that the area of f is equal to A_Σ.*

References

[1] Federer, H. "Geometric Measure Theory." Springer-Verlag, Berlin and New York, 1969.
[2] Hildebrandt, S., Boundary behavior of minimal surfaces, *Arch. Rational Mech. Anal.* **35** (1969), 47–82.
[3] Lemaire, L., Applications harmoniques de surfaces riemanniennes, *J. Differential Geom.* **131** (1978), 51–87.
[4] Meeks, W. H., III, and Yau, S. T., The classical Plateau problem and the topology of three-dimensional manifolds, *Topology* **21** (1982), 409–440.
[5] Meeks, W. H., III, and Yau, S. T., Topology of three-dimensional manifolds and the embedding problems in minimal surface theory, *Ann. of Math.* **112** (1980), 441–484.
[6] Meeks, W. H., III, and Yau, S. T., The existence of embedded minimal surfaces and the problem of uniqueness, *Math. Z.* **179** (1982), 151–168.
[7] Morrey, C. B., "Multiple Integrals in the Calculus of Variations." Springer-Verlag, Berlin and New York, 1966.
[8] Papakyriakopoulos, C. D., On Dehn's lemma and the asphericity of knots, *Ann. of Math.* **66** (1957), 1–26.
[9] Sacks, J., and Uhlenbeck, K., The existence of minimal immersions of the 2-sphere, *Ann. of Math.* **113** (1981), 1–24.
[10] Schoen, R., and Yau, S. T., Existence of incompressible minimal surfaces and the topology of three-dimensional manifolds with non-negative scalar curvature, *Ann. of Math.* **110** (1979), 127–142.
[11] Shapiro, A., and Whitehead, J. H. C., A proof and extension of Dehn's lemma, *Bull. Amer. Math. Soc.* **64** (1958), 174–178.
[12] Stallings, J. R., "Group Theory and Three Dimensional Manifolds." Yale Univ. Press, New Haven, Connecticut, 1971.

PART **D**

GENERALIZATIONS

CHAPTER IX

Group Actions on R^3

William H. Meeks, III
Instituto de Matemática Pura e Aplicada
Rio de Janeiro, Brazil

and   Shing-Tung Yau [†]
Department of Mathematics
Stanford University
Stanford, California

The Smith conjecture has many equivalent forms and each of these forms has various consequences and generalizations. Our point of view is that the Smith conjecture is a structure theorem about symmetries of the product of a compact surface with an interval. Here the interval may be closed or open. For example, the usual Smith conjecture is equivalent to proving the smooth Z_n actions on $S^2 \times [0, 1]$ are conjugate to actions that preserve the product structure. Thus this generalized Smith conjecture represents the belief that all the symmetries of the product of a compact surface with an interval actually arise from the symmetries of the surface extended trivially to the product structure.

In case the compact surface Ω has boundary, we make the additional assumption that symmetries of the three-dimensional manifold $M = \Omega \times [0, 1]$ preserve the ends $\Omega \times \{0, 1\}$. In this form the usual Smith conjecture can be restated as follows. Let $M = D \times [0, 1]$, where D is the unit disk.

[†] Present address: Institute for Advanced Study, School of Mathematics, Princeton University, Princeton, New Jersey.

Then if $f: M \to M$, where $f^n = \mathrm{id}$, $f(D \times \{0\}) = D \times \{0\}$ and $f(D \times \{1\}) = D \times \{1\}$, then f is conjugate to a diffeomorphism $g: M \to M$, where $g(D \times \{t\}) = D \times \{t\}$ and the conjugating diffeomorphism $L: M \to M$ preserves the ends of M. More generally, let $M = \Omega \times [0, 1]$, where Ω is a compact surface (possibly with boundary). Suppose that $f: M \to M$ is a diffeomorphism that preserves the ends of M. We shall say that f preserves a product structure on M if there is a diffeomorphism $h: M \to M$ so that h preserves ends and hfh^{-1} preserves the original product structure.

In this chapter we shall show how to apply minimal surfaces to study the generalized Smith conjecture on $S^2 \times I$, where I is an open or closed interval. This analysis leads to a classification of compact groups that can act smoothly on $\mathbf{R}^3$. Furthermore, we shall prove that if the compact group is not algebraically isomorphic to the icosahedral group, then the action of the compact group on $\mathbf{R}^3$ is linear. John Morgan informed us that he and Michael Davis have similar results in the case S^3.

In [M-Y 3] the authors proved a geodesic version of the loop theorem and a new equivariant version of the loop theorem. The new equivariant loop theorem is easier to apply in the study of group actions, and it will be used to prove the theorems in this section. We now state these theorems.

THEOREM 1 (The Geodesic Loop Theorem). *Suppose that M is a compact three-dimensional manifold with boundary and let Z denote the collection of all loops on ∂M that are homotopically nontrivial in ∂M but homotopically trivial in M. Then with respect to any riemannian metric on ∂M*

(1) *There exists a collection of geodesics $\Gamma = \{\gamma_1, \ldots, \gamma_n\}$ contained in Z so that Z is the smallest normal subgroup of $\pi_1(\partial M)$ generated by the conjugacy classes represented by $\gamma_1, \ldots, \gamma_n$. Furthermore, γ_1 is a geodesic of least length in Z and γ_i is a geodesic of least length in the complement of the normal subgroup generated by the conjugacy classes represented by $\gamma_1, \ldots, \gamma_{i-1}$.*

(2) *Any such collection Γ is a pairwise disjoint collection of embedded geodesics.*

(3) *If $\Gamma' = \{\alpha_1, \alpha_2, \ldots, \alpha_k\}$ is another such collection of geodesics and α_i intersects some γ_j, then $\alpha_i = \gamma_j$.*

We shall call a collection of geodesics Γ that arise in the above theorem a *short generating set* for Z. The disjointness property (3), coupled with the equivariant Dehn's lemma in [M-Y 3], proves the following equivariant version of the loop theorem.

THEOREM 2 (Equivariant Loop Theorem). *Let M and Z be as above and suppose that M is orientable. Suppose that G is a group of orientation-preserving isometries of a riemannian metric on M and let Γ be a short generating* set for Z in this metric. Then $G\Gamma = \{g(\gamma_i)|g \in G, \gamma \in \Gamma\}$ is a pairwise disjoint collection of geodesics that bound a pairwise disjoint collection C of disks. Furthermore, if G acts freely on the union of Γ, then we may assume that G preserves the union of C.

With the above theorems we are now in a position to examine the compact groups that act on $S^2 \times [0, 1]$. First, we state the following lemma of a relative version of the Smith conjecture. The proof, which uses the solution of the Smith conjecture, will be left to the reader. Throughout the remainder of this section all diffeomorphisms will be orientation-preserving unless otherwise stated.

LEMMA 1. *Let $M = D \times [0, 1]$ and F be a foliation of $\partial D \times [0, 1]$ by circles S_t^1. If $f : M \to M$ is a diffeomorphism of finite order so that $f(S_t^1) = S_t^1$ for all t, then there is a foliation of M by disks D_t for $t \in [0, 1]$ such that*

(1) $\partial D_t = S_t^1$ and
(2) $f(D_t) = D_t$.

DEFINITION. The term Ω_k denotes a compact planar domain with k boundary curves, $M_k = \Omega_k \times [0, 1]$.

LEMMA 2. *Suppose that $f : M_2 \to M_2$ is a diffeomorphism of order two which preserves ends and that f interchanges the two circles of $\partial \Omega_2 \times \{0\}$. Then f preserves a product structure on M_2.*

Proof. First, consider the usual linear action R of M_2 given in this lemma, which is induced by rotation around the Z-axis by $180°$ (see Fig. 1). The reader should note that in the product structure chosen for M_2, the ends $\Omega_2 \times \{0, 1\}$ consist of interior and exterior annuli of the solid cylinder. Note that there exists a disk D_1 such that D_1 and $D_2 = R(D_1)$ are disjoint and $D_1 \cup D_2$ disconnects the ends of M_2 into disks V_1, V_2 and W_1, W_2, respectively. We now prove the existence of similar disks for f rather than R.

It is not difficult to check that there is an f-invariant metric on ∂M_2, so that any curve γ of least length in the kernel K of $i_* : \pi_1(\partial M_2) \to \pi_1(M_2)$ is disjoint from the fixed point set of f and also that $\gamma \cap (\partial \Omega_2 \times [0, 1])$ are two embedded intervals. (In fact, there exists such an invariant metric that is a product metric on one component of $\partial \Omega_2 \times [0, 1]$ and is flat on the other boundary component. The metric on the second boundary component is induced by f from the first metric. We claim that if we choose the curves $\{x\} \times [0, 1]$ to be long enough in the product metric of the first component, the geodesic of least length in K can have only one interval component in

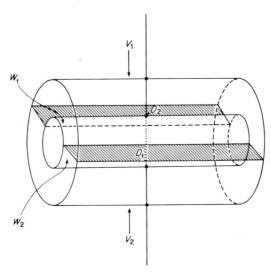

Figure 1

each component of $\partial\Omega_2 \times [0, 1]$. To see this, let σ be a closed curve in K whose intersection with the first component of $\partial\Omega_2 \times [0, 1]$ has the form $\{x\} \times [0, 1]$ and whose intersection with the second component of $\partial\Omega_2 \times [0, 1]$ is the arc induced from the first one by f. Let σ' be the part of σ that lies in the complement of $\partial\Omega_2 \times [0, 1]$ and let l be its length. If we pick the product metric on the first component so that $\{x\} \times [0, 1]$ has length greater than l, then any closed geodesic in K whose intersection with either the first component or the second component more than once has length not less than three times the length of $\{x\} \times [0, 1]$. By comparison with the length of σ, we obtain a contradiction and we have proved the claim. To guarantee that the geodesic avoids the fixed point set of f, we may just blow up the metric in a small neighborhood of each fixed point of f.)

By the geodesic loop theorem, γ and $f(\gamma)$ are equal or disjoint. By hypothesis, f preserves the ends of M_2 and so cannot act freely on γ. Because f has no fixed points on γ, f cannot leave γ invariant, which shows that γ and $f(\gamma)$ must be disjoint. Therefore, γ and $f(\gamma)$ disconnect $\Omega_2 \times \{0\}$ into two disks V_1 and V_2. The equivariant loop theorem stated in this section implies that there exists a disk D_1 such that $\partial D_1 = \gamma$ and $D_2 = f(D_1)$ is disjoint from D_1. The disks D_1 and D_2 disconnect M_2 into two balls B_1 and B_2, where $V_i \subset B_i$. By hypothesis, f interchanges the circles $\partial\Omega_2 \times \{0\}$. The Lefschetz fixed point theorem implies that $f|\Omega_3 \times \{0\}$ has a fixed point. Because the balls B_i have fixed points in their interiors, they are left invariant by f. Let W_i be the intersection of B_i with $\Omega \times \{1\}$ for $i = 1, 2$.

IX. Group Actions on R³

Now choose product foliations F_i of $\partial B_2 - (\mathring{V}_i \cup \mathring{W}_i)$ for $i = 1, 2$ that agree on the intersection $\partial B_1 \cap \partial B_2$ and that are preserved by f. Because B_1 and B_2 are invariant under the restriction of f, Lemma 1 implies that the foliations F_1 and F_2 are the boundary curves of a product foliation of M_2 preserved by f. Fitting together the associated invariant product structures along $\partial B_1 \cap \partial B_2$ gives rise to a product structure on M_2 that is preserved by f. This proves the lemma. ∎

LEMMA 3. *If $f: M_3 \to M_3$ is a diffeomorphism of order two or three that preserves the ends, then f preserves a product structure on M_3.*

Proof. First consider the case in which f has order two. The usual linear action R on M_3 of order two given in this lemma is induced by rotating M_3 around the Z-axis by 180° (see Fig. 2).

Note that there exists a disk D_1 such that D_1 and $D_2 = R(D_1)$ are disjoint and $D_1 \cup D_2$ disconnects the ends of M_3 into disks C, E and annuli A_1, A_2, B_1, B_2. Furthermore, D_1 and D_2 each intersect $\partial \Omega_3 \times [0, 1]$ only along the invariant component and in two intervals. We now prove the existence of similar disks for f rather than R.

First, note that since f has order two and acts on $\partial \Omega_3 \times \{0\}$, f leaves invariant one or three of these circles. However, if f leaves invariant all three circles, then $f | \Omega_3 \times \{0\}$ would induce a diffeomorphism of order two on the disk with two or more fixed points. As this is impossible by the Lefschetz fixed point theorem, f leaves invariant exactly one circle. As in Lemma 2, there is an f-invariant metric on ∂M_3, so that any curve γ of least length in

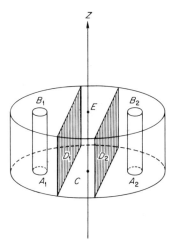

Figure 2

the kernel K of $i_*: \pi_1(\partial M_3) \to \pi_1(M_3)$ is disjoint from the fixed point set f and $\gamma \cap (\partial\Omega_3 \times [0, 1])$ is two intervals on the component of $\partial\Omega_3 \times [0, 1]$ invariant under f. (There is an invariant metric that is a product metric on two components X_1, X_2 of $\partial\Omega_3 \times [0, 1]$ and is flat on the other component $X_3 = f(X_2)$. In fact, choose a metric so that the length of $\{x\} \times [0, 1]$ on X_1 is sufficiently long and the length of $\{y\} \times [0, 1]$ on X_2 is much longer. The arguments in Lemma 2 now show that γ has the correct properties.)

By the geodesic loop theorem, γ and $f(\gamma)$ are equal or disjoint. The same argument as in the proof of Lemma 2 shows that γ and $f(\gamma)$ are disjoint. Therefore, γ and $f(\gamma)$ disconnect $\Omega_3 \times \{0\}$ into three parts A_1, A_2, and C, where $f(A_1) = A_2$ and C is an invariant disk. The equivariant loop theorem implies that there exists a disk D_1 such that $\partial D_1 = \gamma$ and $D_2 = f(D_1)$ is disjoint from D_1. The disks D_1 and D_2 disconnect M_3 into a ball B and two regions R_1, R_2 diffeomorphic to $A_1 \times [0, 1]$. Let $B_i = R_i \cap (\Omega_3 \times \{1\})$ and $E = B \cap (\Omega_3 \times \{1\})$.

Now choose a product foliation F_1 of R_1 with end leaves A_1, B_1 and such that $F_1 \cap D_1$ is a product foliation by intervals. Let F_2 be the foliation on R_2 induced from F_1 by f. The partial foliations on $\partial B \cap (D_1 \cup D_2)$ induced by $F_1 \cup F_2$ can be extended to a product foliation on $\partial B \cap (\partial\Omega_3 \times [0, 1])$ by circles S_t^1 with $f(S_t^1) = S_t^1$. Lemma 1 implies that there is a foliation F_3 of disks of B with $\partial D_t = S_t^1$ and $f(D_t) = D_t$. Fitting together the foliations F_1, F_2, and F_3 along ∂B gives rise to a product structure for M_3 that is preserved by f. This proves the first case in Lemma 3. The proof of the second case in Lemma 3 is similar and will be left to the reader. ∎

Remark. The technique used in the above lemmas generalizes to characterize finite group actions which preserve the ends of M_n as being conjugate to actions preserving the product structure, and the similar statement holds for an M that is a product of an interval with any compact surface with nonempty boundary.

THEOREM 3. *Suppose that G is a compact Lie group that acts smoothly and preserves orientation on the ball $B = \{(x, y, z) \in \mathbf{R}^3 | x^2 + y^2 + z^2 \leq 1\}$ and that G is not isomorphic to the alternating group A_5. Then G is conjugate to a linear action on B.*

PROOF. Since the group G acts on the boundary sphere of B and compact group actions on S^2 are standard, G is isomorphic to a compact subgroup of $SO(3)$. The nonfinite compact subgroups of $SO(3)$ are $SO(3)$, S^1, and a Z_2 extension of S^1. In the case of $SO(3)$, the orbit space is an interval and in the other two cases the orbit space is a disk. In the nonfinite cases it is straightforward to check that the action of G on B is conjugate to the linear action.

Because A_5 is the only nonsolvable finite subgroup of $SO(3)$, our hypotheses imply that the group G is solvable. Standard Smith theory implies

IX. Group Actions on R³

that G has a global fixed point and the fixed point set fix(g) of any diffeomorphism $g \in G$ is an interval with boundary in the boundary sphere of B. The solution of the Smith conjecture shows that the interval fix(g) is unknotted and that $g: B \to B$ is conjugate to a rotation. Hence, for G cyclic, the theorem is known. The proof of the theorem will depend on the analysis of the remaining possibilities. The other possible groups that act on S^2 are listed in [W, pp. 83–85] and are the dihedral groups D_n, the tetrahedral group T, and the octahedral group.

Since the above solvable groups G have a global fixed point p and the action of G is linear near p, the action will be conjugate to the linear action if the restricted action on the complement of a small invariant open ball centered at p is standard. In other words, if a G action of $M = S^2 \times [0, 1]$ is conjugate to an action that preserves the product structure, then G acts on the compact three-dimensional ball standardly. Because product structures on $S^2 \times [0, 1]$ differ by diffeomorphisms, we need only find some product structure that is preserved by G. To do this we shall use the solvability of G.

The finite groups G that we are considering have a normal series $H_1 = \{e\} \subset H_2 \subset \cdots \subset H_n = G$, where H_k/H_{k-1} is cyclic and this cyclic group is equal to Z_2 or Z_3 when $k = 2$. In fact, for D_{2n}, $k = 3$, for the tetrahedral group, $k = 4$, and for the octahedral group, $k = 5$. This normal series of G gives rise to a family of cyclic branched covering spaces

$$X_1 \underset{p_2}{\to} X_2 \underset{p_3}{\to} \cdots \to X_{n-1} \underset{p_n}{\to} X_n,$$

where $X_k = M/H_k$ and H_{k+1}/H_k are the covering transformations for $p_{k+1}: X_k \to X_{k+1}$.

After averaging a riemannian metric on X_1, G acts as a group of isometries. By the solution of the Smith conjecture, X_2 is diffeomorphic to $S^2 \times [0, 1]$. Furthermore, the branch locus for $p_2: X_1 \to X_2$ consists of two intervals I_1 and I_2 in X_2 such that the small open ε-neighborhood N of $I_1 \cup I_2$ is invariant under H_3. Also, $X_2 - N$ is naturally diffeomorphic to $M_2 = \Omega_2 \times [0, 1]$. By solvability of the group G, M_2 is invariant under the generator f of the group of the covering transformations H_3/H_2 and f preserves the ends of M_2.

By suitable choice of the normal series, we may assume that f has order two at this stage and f interchanges the circles $\partial \Omega_2 \times \{0, 1\}$. Thus by Lemma 2, f preserves a product structure on M_2. This implies that the complement C of the branch set of the induced action of H_3/H_2 on M_2 has an invariant product structure. The quotient of this manifold C is naturally diffeomorphic to $M_3 = \Omega_3 \times [0, 1]$. Furthermore, H_4/H_3 induces an action on M_3 that preserves ends. Lemma 3 implies that H_4/H_3 preserves a product structure on M_3. Therefore, the complement of the ε-neighborhood of the branch set of H_4/H_3 has an invariant product structure.

Continuing the above arguments for the other groups H_k/H_{k-1}, we can show that the complement of the branch locus of $p_n \circ p_{n-1} \circ \cdots \circ p_2: M_{k-1} \to M_k$ has a product structure. This implies that $G = H_n$ acting on M preserves a product structure. As was noted earlier, this proves Theorem 3. ∎

LEMMA 4. *If G is a compact subgroup of* $\text{diff}^+(\mathbf{R}^3)$, *then there exists a sequence of balls* $B_1 \subset B_2 \subset \cdots \subset B_n \subset \cdots \subset \mathbf{R}^3$ *satisfying*

(1) $B_i \subset \text{int}(B_{i+1})$,
(2) $\bigcup_{i=1}^{\infty} B_i = \mathbf{R}^3$, *and*
(3) $g(B_i) = B_i$ *for all $g \in G$*.

Proof. The proof of the lemma for nonfinite groups is rather straightforward and will be left to the reader. Now suppose that G is finite. Smith's theory [B] for the powers g^k of elements g of G with prime order easily implies that the fixed point set fix(g) of g is diffeomorphic to $\mathbf{R}$ properly embedded in $\mathbf{R}^3$ and that the cyclic subgroup generated by g acts freely outside fix(g). The isotropy subgroups $I_p \subset G$ of any point $p \in \mathbf{R}^3$ has a natural representation from their differentials on $T_p\mathbf{R}^3$ in the group SO(3). From this representation, it is straightforward to put a differential structure on the orbit space $M = \mathbf{R}^3/G$ so that the projection map $p: \mathbf{R}^3 \to M$ is a smooth, branched immersion. Furthermore, if X is a smooth, compact, embedded surface in M that is in general position with respect to the branch locus of p, then $p^{-1}(X)$ is a smooth, embedded surface in $\mathbf{R}^3$. ∎

ASSERTION 1. *There exists a proper Morse function $H: M \to \mathbf{R}$ and an increasing sequence of numbers $r_i \in \mathbf{R}$ that satisfy the following properties.*

(1) $N_i = (H \cdot p)^{-1}[r_i, r_{i+1}]$ *is a smooth, compact three-dimensional submanifold of $\mathbf{R}^3$ that is invariant under f.*
(2) *There exists an embedded 2-sphere in N_i that intersects $\alpha[0, \infty]$ and $\alpha[-\infty, 0]$ transversely in an odd number of points, where $\alpha: \mathbf{R} \to \text{fix}(f)$ is a fixed diffeomorphism for some fixed, nontrivial $f \in G$.*
(3) $H \circ p \circ \alpha(0) < r_1$ *and* $\lim_{i \to \infty} r_i = \infty$.

Proof. Let $H: M \to \mathbf{R}$ be a proper Morse function that is in general position with respect to $p \circ \alpha: \mathbf{R} \to M$. Then choose $r_1 \in \mathbf{R}$ large enough so that $\alpha(0) \in \text{fix}(f)$ is in the open submanifold $(H \circ p)^{-1}(-\infty, r_1)$. Pick D_1 to be a smooth ball in $\mathbf{R}^3$ that contains $(H \circ p)^{-1}((-\infty, r_1])$ and such that ∂D_1 is in general position with respect to the embedded curve $\alpha(\mathbf{R})$. Pick an $r_2 > \max(H \circ p(D_1)) + 1$. Now choose D_2 to be a smooth ball in $\mathbf{R}^3$ that contains $(H \circ p^{-1}((-\infty, r_2])$ and let $r_3 > \max(H \circ p(D_2)) + 1$. Continuing this process, one has an infinite set of r_i with $r_i < r_{i+1}$, $H \circ p \circ \alpha(0) < r_1$

IX. Group Actions on R³

and $\lim_{i \to \infty} r_i = \infty$. As H is a Morse function, we may choose r_i to be regular values of H and $(H \circ p \circ \alpha)$. By choice of r_i, $H^{-1}(r_i)$ is a compact surface, possibly disconnected, that is transversal to $(p \circ \alpha)(\mathbf{R})$ in a finite number of points. This implies that $N_i = (H \circ p)^{-1} \times ([r_i, r_i])$ is a smooth, compact submanifold of $\mathbf{R}^3$ that is invariant under f.

We now show that each sphere $S_i = \partial D_i$ intersects each of the intervals $\alpha([0, \infty))$ and $\alpha((-\infty])$ in an odd number of points. In fact, by the construction of D_i, $\alpha([0, \infty))$ intersects S_i transversally in a finite number of points. Since only a compact part of $\alpha[0, \infty)$ lies in D_i, there is an integer $m_i > 0$ so that $\alpha([m_i, \infty))$ is disjoint from D_i. As $\alpha(0)$ is contained in D_i and $\alpha(m_i)$ is contained in the complement of D_i, the curve $\alpha(0, m_i]$ must intersect S_i an odd number of times. Hence, $\alpha([0, \infty))$ must intersect S_i transversely an odd number of times. The same proof shows that $\alpha((-\infty, 0])$ intersects S_i in an odd number of points.

ASSERTION 2.   There exists a sequence of closed, smooth balls

$$B_1 \subset B_2 \subset \cdots \subset B_n \subset \cdots \subset \mathbf{R}^3$$

that satisfy

(1) $B_i \subset \text{int}(B_{i+1})$,
(2) $\bigcup_{i=1}^{\infty} B_i = \mathbf{R}^3$, and
(3) $f(B_i) = B_i$.

Proof. By Assertion 1 there exists an embedded sphere S_i in N_i such that $\alpha([0, \infty))$ intersects S_i in an odd number of points. By construction, $\alpha([0, \infty]) \cap N_i$ is a finite number of embedded intervals whose boundaries are contained in ∂N_i. We may consider $\alpha[0, \infty) \cap N_i$ as representing an element $\bar{\alpha}$ in $H_1(N_i, \partial N_i, Z_2)$. Considering the sphere S_i as representing an element $\bar{S}_i$ in $H_2(N_i, Z_2)$, the above geometric intersection shows that $\bar{S}_i \cap \bar{\alpha}_i$ is nonzero, where $\cap$ is the intersection pairing on homology. In particular, the class $\bar{S}_i$ is nonzero, which implies that S_i represents a nontrivial element in the second homotopy group $\pi_2(N_i)$ of N_i.

Since the group G_i of diffeomorphisms of N_i generated by the restrictions $g | N_i$ of elements of G to N_i is finite, we may assume G_i acts as a group of isometries with respect to a convex metric on N_i. Now apply the equivariant sphere theorem to get a finite generating set $g_1, g_2, \ldots, g_n: S^2 \to N_i$ of embedded spheres such that if $(f | N_i)(g_k(S^2))$ intersects $g_k(S^2)$, then f leaves this sphere invariant. Since $g_1, g_2, \ldots, g_n$ represent a $\pi_1(N_i)$-generating set for $\pi_2(N_i)$, the homology class $\bar{S}_i$ of the sphere S_i can be expressed as a linear combination of the associated homology classes $\bar{g}_1, \ldots, \bar{g}_n$ in $H_2(N_i, Z_2)$. As the intersection number $\bar{S}_i \cap \bar{\alpha}$ is nonzero, $\bar{\alpha} \cap \bar{g}_k$ is also nonzero for some k. Assume that $k = 1$.

Since $\bar{\alpha} \cap \bar{g}_1$ is nonzero, $\alpha \cap g_1(S^2)$ is nonempty. This implies that f has a fixed point on $g_1(S^2)$. Therefore, $f(g_1(S^2))$ intersects $g_1(S^2)$ and f must leave the sphere $g_1(S^2)$ invariant. By Alexander's theorem, $g_1(S^2)$ disconnects $\mathbf{R}^3$ into a ball B_i and a noncompact component. As f leaves invariant the boundary sphere of the closed ball B_i, f leaves B_i invariant. Since f is orientation-preserving, $f|B_i: B_i \to B_i$ is orientation-preserving. This shows $f|g_1(S^2)$ is orientation-preserving as a diffeomorphism of the sphere $g_1(S^2)$. Since finite groups act standardly on two spheres, f has two fixed points on $g_1(S^2)$. The local behavior of f near its fixed point set shows that $\text{fix}(f) = \alpha(\mathbf{R})$ is transversal to $g_1(S^2)$ at the two intersection points. As the intersection number of $\alpha([0, \infty))$ and $g_1(S^2)$ is odd, $\alpha([0, \infty))$ intersects $g_1(S^2)$ transversely in exactly one point. This shows that $\alpha(0)$ must be in the compact component of $\mathbf{R}^3$ bounded by the sphere S_i and hence $\alpha(0)$ is contained in the ball B_i. Since N_i and N_{i+2} are disjoint, the boundary sphere S_i of B_i is disjoint from the boundary sphere S_{i+2} of the ball B_{i+2}. This fact, together with the fact that $\alpha(0) \in B_i$ for all i, shows that $B_i \subset \text{int}(B_{i+2})$.

We now show that the union of $\bigcup_{i=1}^{\infty} B_{2i} = \mathbf{R}^3$. (Hence, after reindexing, $B_2, B_4, \ldots, B_{2n}, \ldots$ will provide the required balls for Assertion 2.) Consider the compact, possibly disconnected, manifolds $M_i = H^{-1}((-\infty, r_i])$ for each i. Let C_i be the component of M_i that contains the point $\alpha(0)$. If the union $\cup_i C_i$ is not all of $\mathbf{R}^3$, there is a boundary point x_0 of $\cup C_i$. As C_i is contained in the interior of C_{i+1}, there is a sequence $x_i \in \partial C_i$ that converges to x_0. However, $H(x_i) = r_i$ and hence $H(x) = \lim_{i \to \infty} H(x_i) = \lim_{i \to \infty} r_i = \infty$, which is impossible. This shows that $\cup_i C_i = \mathbf{R}^3$.

The sphere $\partial B_{2i} = \bar{S}_{2i} \subset N_{2i}$ is disjoint from the manifold C_{2i-2} and the ball B_{2i} intersects C_{2i-2} at the point $\alpha(0)$. Since C_{2i-2} is connected and disjoint from the boundary sphere of B_{2i}, C_{2i-2} is a subset of B_{2i}. Thus $\bigcup_i C_{2i} = \mathbf{R}^3 \subset \bigcup_i B_{2i} = \mathbf{R}^3$, and therefore $\mathbf{R}^3 = \bigcup_{i=1}^{\infty} B_{2i}$. This equation completes the proof of the assertion.

ASSERTION 3. *There is an N such that for $i > N$ the ball B_i in Assertion 2 is invariant under all elements of G.*

Proof. By construction of the ball B_i and the proof of the previous assertion, the ball B_i is invariant under a diffeomorphism $g \in G$ if $\text{fix}(g)$ intersects B_i nontrivially. However, as $\text{fix}(g)$ is nonempty and $\bigcup_{i=1}^{\infty} B_i = \mathbf{R}^3$, $\text{fix}(g)$ must intersect some ball B_i for i large. As there are only a finite number of elements in G, the assertion holds. The lemma follows immediately from Assertion 3. ∎

LEMMA 5. *Suppose that M is a compact surface with boundary and let $X = M \times [0, 1]$. Let $Z_1 = M \times \{0\}$ and $Z_2 = \partial M \times [0, 1]$. If $f: Z_1 \cup Z_2 \to$*

IX. Group Actions on $\mathbf{R}^3$

$Z_1 \cup Z_2$ is a homeomorphism of finite order with $f(Z_i) = Z_i$ and $f|Z_i$ smooth for $i = 1, 2$, then f extends to a diffeomorphism of $M \times [0, 1]$.

Proof. First, suppose that $f(p, t) = (f(p, 0), t)$. In this case, f can be extended to X by defining $f(p, t) = (f(p, 0), t)$. This case is easy because f preserves the fixed-product structure on Z_2.

There is an elementary classical result that states there exists an isotopy between two diffeomorphisms of an annulus that preserve boundary curves and whose restriction to one of the boundary circles is preassigned. This classical result is equivalent to proving that the lemma holds when every component of M is an annulus. We now apply the fact that the lemma holds in this generality.

Let N be an invariant neighborhood of ∂M in M that is a collection of annular components. Let $Z_3 = (\partial N \setminus \partial M) \times [0, 1]$ and define $g: Z_1 \cup Z_2 \cup Z_3 \to X$ by

$$g(p, t) = \begin{cases} f(p, t) & \text{if } (p, t) \in Z_1 \cup c_2, \\ (f(p, 0), t) & \text{if } (p, t) \in Z_3. \end{cases}$$

By the extension property for M a collection of annular surfaces, it follows that g can be extended to $N \times [0, 1]$. As observed in the first case in the proof, g has natural extension as a homeomorphism to the rest of X. By making an appropriate choice of the extension of g to $N \times [0, 1]$, the required extension will be a diffeomorphism. This proves the lemma. ∎

THEOREM 4. *If G is a compact subgroup of $\mathrm{diff}^+(\mathbf{R}^3)$, then G is isomorphic to a compact subgroup of $SO(3)$. If G is not isomorphic to alternating group A_5 (also called the icosahedral group), then G is conjugate in $\mathrm{diff}^+(\mathbf{R}^3)$ to a subgroup of $SO(3)$.*

Proof. That G is isomorphic to a compact subgroup of $SO(3)$ follows immediately from Lemma 4 because G acts as a compact group of orientation-preserving diffeomorphism on the boundary sphere ∂B_i and all such actions are conjugate to linear actions. Now assume that G is not isomorphic to A_5.

Let $B_1 \subset B_2 \subset \cdots \subset B_n \subset \cdots \mathbf{R}^3$ be the equivariant balls given in Lemma 4. By the Schoenflies theorem for annular domains in $\mathbf{R}^3$ we may assume, after conjugation by some diffeomorphism, that the balls B_i are balls of radius i centered at the origin in $\mathbf{R}^3$. Suppose for the moment that there exist diffeomorphisms $\tilde{h}_n: B_n \to B_n$ so that $\tilde{h}_n G \tilde{h}_n^{-1}$ is a group of linear isometries of B_n, and the $\tilde{h}_n(B_{n-1})$ and $\tilde{h}_n|B_{n-1} = \tilde{h}_{n-1}$. Define $\tilde{h}: \mathbf{R}^3 \to \mathbf{R}^3$ by $\tilde{h}(x) = \tilde{h}_n(x)$, where $|x| \leq n$. Clearly, hGh^{-1} is contained in the subgroup $SO(3)$ of $\mathrm{diff}^+(\mathbf{R}^3)$. Hence, the theorem will follow from the existence of the diffeomorphisms $\tilde{h}_n: B_n \to B_n$. We now give a proof by induction of the existence of $\tilde{h}_n$. By Theorem 3, any action of G on the closed three-dimensional ball of radius 1 is conjugate to the linear action. Translated into algebraic terms on the associated branched covering spaces, this conjugacy can be expressed by stating that the commutative diagram

$$\begin{array}{ccc} B_1 & \xrightarrow{\tilde{h}_1} & B \\ \rho_1 \downarrow & & \downarrow \rho'_1 \\ B & \xrightarrow{h_1} & B' \end{array}$$

exists.

Here p_1 is the natural branched covering space arising from the action of G on B, p'_1 is the natural branched covering space arising from a fixed linear action of G on the ball B_1; h_1 is a diffeomorphism that maps the unknotted branch set of p, which is diffeomorphic to an unknotted interval to the branch set of p' in the ball B'; $\tilde{h}_1$ is a lift of h_1 to the branched covering spaces. Actually, once we have constructed a diffeomorphism h_1, a lifting $\tilde{h}_1$ exists if h_1 preserves the branch locus and the local monodromy. This fact can be proved using the theory of normal covering spaces in this case. If $\tilde{h}_1$ is some lifting, then clearly $\tilde{h}_1 g \tilde{h}_1^{-1}$ is a covering transformation for the branched covering space $p'_1: B_1 \to B'$. As the covering transformations of this branched covering space is G, $\tilde{h}_1$ is a diffeomorphism that conjugates G to a linear action.

Note that in the above discussion the conjugating diffeomorphism $\tilde{h}_1$ depends on h_1 and the choice of the lifting of this h_1 to the total spaces of the branched covering spaces. Now suppose by induction that the commutative diagram

$$\begin{array}{ccc} B_{n-1} & \xrightarrow{\tilde{h}_{n-1}} & B_{n-1} \\ \rho_{n-1} \downarrow & & \downarrow \rho'_{n-1} \\ B & \xrightarrow{h_{n-1}} & B \end{array}$$

exists.

We now want to extend the diagram above to get the commutative diagram

$$\begin{array}{ccc} B_n & \xrightarrow{\tilde{h}_n} & B_n \\ \rho_n \downarrow & & \downarrow \rho'_n \\ B & \xrightarrow{h_n} & B' \end{array}$$

From the previous discussion this extension problem is equivalent to showing that we can extend the diffeomorphism h_{n-1} to a diffeomorphism $h_n : B \to B'$ so that $h_n|B_{n-1} = h_{n-1}$ and the image of the branch set of p is the branch set of p'.

Let Y_n denote the branch set of p_n and $\overline{Y}_n$ the branch set of p'_n. To construct h_n, we first extend h_{n-1} to a diffeomorphism h'_n of some small regular neighborhood N_n of Y_n to a small regular neighborhood $\overline{N}_n$ of $\overline{Y}_n$ in such a way that $h'_n(Y_n) = \overline{Y}_n$ and $h'_n(N_n \cup \partial B) = \overline{N}_n \cap \partial B'$. As Y_n and $\overline{Y}_n$ are unknotted in their respective balls, $B\backslash\text{int}(N_n \cup B)$ and $B'\backslash\text{int}(\overline{N}_n \cup B')$ are products of a planar domain with an interval. Lemma 5 implies h'_n extends to the required diffeomorphism.

It follows that the lift $\tilde{h}_n$ of h_n completes the proof of the existence of the above diagram. Thus, by induction, we have proved the existence of the required diffeomorphism $\tilde{h}_n$. As noted earlier, this completes the proof of the theorem. ∎

References

[B]   Bredon, G., "Introduction to Compact Transformation Groups." Academic Press, New York, 1972.

[M]   Meeks, W. H., III, Lectures on Plateau's problem. Instituto de Matemática Pura e Aplicada, Rio de Janeiro, Brazil, 1978.

[MY-1] Meeks, W. H., III, and Yau, S. T., The classical Plateau problem and the topology of three-dimensional manifolds, *Topology* **21** (1982), 409–440.

[M-Y 2] Meeks, W. H., III, and Yau, S. T., Topology of three-dimensional manifolds and the embedding problems in minimal surface theory, *Ann. of Math.* **112** (1980), 441–485.

[M-Y 3] Meeks, W. H., III, and Yau, S. T., The equivariant Dehn's lemma and loop theorem, *Comm. Math. Helv.* **56** (1981), 225–239.

[M-Y 4] Meeks, W. H., III, and Yau, S. T., Generalized boundary conditions and the existence and uniqueness of minimal surfaces, *Comm. Math. Helv.* **56** (1981), 225–239.

[W]   Wolf, J. A., "Spaces of Constant Curvature." McGraw-Hill, New York, 1967.

CHAPTER X

Finite Group Actions on Homotopy 3-Spheres

Michael W. Davis and *John W. Morgan*

Department of Mathematics
Ohio State University
Columbus, Ohio

Department of Mathematics
Columbia University
New York, New York

The Smith conjecture states that certain types of smooth actions of a cyclic group on a homotopy 3-sphere are essentially linear. Our working hypothesis is that every smooth action of a finite group on a homotopy 3-sphere is essentially linear. By making use of the methods presented earlier in this volume, we have established this hypothesis in a substantial number of cases. Our results are summarized by the following theorem.

THEOREM A. *Let $\tilde{\Sigma}$ be a homotopy 3-sphere and let $G \times \tilde{\Sigma} \to \tilde{\Sigma}$ be an action of a finite group of orientation-preserving diffeomorphisms. If every isotropy group is cyclic and if at least one isotropy group has order greater than 5, then the action is essentially linear.*

It is an immediate consequence of the Schönflies theorem and the definition of "essentially linear" that any essentially linear action on S^3 is equivariantly diffeomorphic to a linear action. (The definition of essentially linear is given in Chapter 1.) Thus we have the following corollary of the above theorem.

THEOREM B. *Let $G \times S^3 \to S^3$ be an action of a finite group of orientation-preserving diffeomorphisms. If every isotropy group is cyclic and if at least one isotropy group has order greater than 5, then the given action is linear.*[1]

This chapter is divided into eight sections. The first introduces the notion of an orbifold. This is a term coined by Thurston to describe spaces that are locally the quotients of finite group actions. In Sections 2 and 3 we discuss two- and three-dimensional examples of these spaces in some detail. We find the language of orbifolds a convenient framework for dealing with quotient spaces of properly discontinuous group actions. Using them, one can treat all properly discontinuous actions the way one normally deals with free, properly discontinuous actions—by working with the quotient space.

We establish the connection between finite group actions and linear actions through the intermediary of Seifert-fibered orbifolds. Essentially, a Seifert-fibered orbifold comes about when one has a group acting on a Seifert-fibered manifold preserving the Seifert-fibered structure. The basic properties of these orbifolds are developed in Section 4. In Section 5 we prove an analogue of a theorem of Seifert and Threlfall. We show that any three-dimensional, Seifert-fibered orbifold with finite fundamental group is diffeomorphic to S^3/G for some finite group G in $SO(4)$ that normalizes the standard S^1 in $SO(4)$. This reduces the problem of showing a finite group action is linear to the problem of showing that its quotient orbifold can be given the structure of a Seifert fibration.

We prove Theorem A by showing that if $G \times \tilde{\Sigma} \to \tilde{\Sigma}$ is a finite group action on a homotopy 3-sphere, then the quotient $\tilde{\Sigma}/G$ is isomorphic to $X_1 \# \Sigma$, where X_1 is Seifert-fibered and Σ is a homotopy 3-sphere. Since X_1 is equivalent to the quotient of a linear action, it follows easily that the action we began with is essentially linear.

In Section 6 we reformulate the result along these lines. In Section 7 we study an important special case. In the special case we study a knot K in a homotopy sphere Σ with the property that the cyclic n-sheeted branched cover of Σ, branched over K, has finite fundamental group for some $n > 5$. We show that this implies that K is unknotted in Σ. This is very similar to the situation of the Smith conjecture. The difference is that in the Smith conjecture the cyclic branched cover is assumed to be simply connected (also, one doesn't restrict to $n > 5$). The general plan of attack that resolves the Smith conjecture, bolstered by some old results of Dickson's concerning subgroups

---

[1] By linear we mean that the action is equivariantly diffeomorphic to an action arising from a representation $G \hookrightarrow SO(4)$.

X. Finite Group Actions on Homotopy 3-Spheres

of $PSL_2(F)$, for F a finite field, is used in Section 7 to deal with the special case.

In Section 8 we show how to reduce the general case to the special case. The basic idea is to start with an action and restrict it to a certain type of normal subgroup. We show that if the quotient by the subgroup is Seifert-fibered, then so is the quotient by the full group. The main result needed to establish this is that a prime three-dimensional orbifold with boundary is Seifert-fibered if and only if its fundamental group contains a normal, infinite, cyclic subgroup. Using this reduction result, one deduces Theorem A from the special case considered in Section 7.

Lastly, there is an appendix that gives Dickson's classification of subgroups of $PSL_2(F)$ that contain an element whose order equals the characteristic of F.

The results of the first three sections of this chapter are not original. They are properly called folklore. We have adopted Thurston's terminology and point of view on the subject. The heart of the chapter is Section 7, where we generalize the argument proving the Smith conjecture to deal with the special case considered there.

1. Orbifolds

Suppose that H is a discrete group acting smoothly, effectively, and properly discontinuously on a manifold M^n. We wish to analyze the local structure of the orbit space M^n/H. For each $x \in M$ the isotropy group H_x is finite. Furthermore, each $x \in M$ has an H_x-invariant open neighborhood U_x such that $h(U_x) \cap U_x = \emptyset$ for all $h \in H - H_x$. It follows that the image of HU_x in M^n/H is naturally isomorphic to U_x/H_x.

Since the action is smooth, its differential induces a linear action $H_x \times T_x M \to T_x M$. Choose an H_x-invariant metric on M and an isometry $T_x M \cong \mathbf{R}^n$. This yields a faithful representation $\mu: H_x \hookrightarrow O(n)$, well defined up to conjugation. Let $[x]$ denote the image of x in M^n/H. The conjugacy class of $\mu(H_x) \subset O(n)$ is called the *local group type at* $[x]$. It is independent of all choices. If $D^n \subset \mathbf{R}^n$ is a disk about the origin of sufficiently small radius, then the exponential map $\mathbf{R}^n \cong TM_x \to M$ can be used to define an H_x-invariant, smooth embedding $\varphi: D^n \hookrightarrow M^n$, taking 0 to x. Clearly, we can arrange that the image of φ is contained in U_x. Assuming this, it induces an embedding $\bar\varphi: D^n/\mu(H_x) \hookrightarrow M^n/H$ taking $[0]$ to $[x]$. The map $\bar\varphi$ is called a *smooth orbifold chart for* M^n/H *centered at* $[x]$.

Suppose that G_1 and G_2 are conjugate subgroups of $O(n)$ and that $b: D^n/G_1 \to D^n/G_2$ is a map taking $[0]$ to $[0]$. The map b is called a *smooth isomorphism* if it can be lifted to a diffeomorphism $\tilde{b}: D^n \to D^n$ so that the following diagram commutes:

$$\begin{array}{ccc} D^n & \xrightarrow{\tilde{b}} & D^n \\ \downarrow & & \downarrow \\ D^n/G_1 & \xrightarrow{b} & D^n/G_2 \end{array}$$

The map $\tilde{b}$ is called a *lifting* of b. If $\tilde{b}'$ is another lifting of b, then there is an element $g \in G_1$ so that $\tilde{b}' = \tilde{b} \cdot g$. A map of orbit spaces $f: M/H \to M'/H'$ is a *smooth isomorphism* (or a *diffeomorphism*) if it is a homeomorphism and if for each $[x] \in M/H$ there are smooth orbifold charts centered at $[x]$ and $f[x]$ so that in these coordinates the map is a smooth isomorphism. Notice that if $f: M/H \to M'/H'$ is a smooth isomorphism, then it preserves the local group types. (However, it is not necessary that $M \cong M'$ or that $H \cong H'$ for M/H to be diffeomorphic to M'/H'.)

Now we generalize these concepts from spaces that are globally quotients of properly discontinuous group actions to those that locally have such descriptions. Such a notion was first suggested by Satake [8] under the name of *V-manifolds*. Recently, Thurston has made use of these spaces and has introduced the word *orbifolds*. We follow Thurston's point of view and terminology.

By a (*smooth*) *n-dimensional orbifold* we shall mean a paracompact Hausdorff space X together with a maximal atlas of local charts of the form $D^n/G \hookrightarrow X$, where G ranges over the conjugacy classes of finite subgroups of $O(n)$ and where the overlap maps are diffeomorphisms in the sense defined above. The notions of the *local group at a point* and of a *diffeomorphism of orbifolds* then have unambiguous meanings. A point in X is called a *manifold point* if its local group is trivial; otherwise, it is *exceptional*. An orbifold is a *manifold* if it has empty exceptional set.

It is clear that if M^n is a manifold and $H \times M^n \to M^n$ is an effective, properly discontinuous, smooth action, then M^m/H receives naturally the structure of an orbifold.

An *orientation* for D^n/G is an orientation for D^n in which G acts as a group of orientation-preserving maps. An *orientation* for an orbifold is a compatible system of orientations for the charts in the atlas defining the orbifold structure.

There is a similar notion of *orbifold with boundary*. Here one uses, in addition, charts of the form $(D^n)_+/G$, where $(D^n)_+ = \{(x_1, \ldots, x_n) | \Sigma x_i^2 \leq r$ and $x_n \geq 0\}$ and $G \subset O(n)$ leaves the half-space $\{x_n \geq 0\}$ invariant. The boundary of an n-dimensional orbifold is an $(n-1)$-dimensional orbifold without boundary.

X. Finite Group Actions on Homotopy 3-Spheres

An orbifold is *compact* if its underlying topological space is compact. An orbifold is *closed* if it is compact and has empty boundary. (N.B.: The underlying topological manifold can have a boundary even if the orbifold is closed.)

If X is an orbifold whose underlying space is Q and if $H \times Q \to Q$ is a group action, then the action is said to be a (*smooth*), *properly discontinuous action on the orbifold* X if

(1) $H \times Q \to Q$ is properly discontinuous, and
(2) each $h \in H$ acts via a smooth isomorphism on X.

Suppose that $H \times X \to X$ is a properly discontinuous action on the orbifold X. We shall give the quotient space the structure of an orbifold. Let Q be the underlying space of X, and let x be a point in Q/H. Choose $\tilde{x} \in Q$ a point that projects to x. Choose an orbifold chart for X centered at $\tilde{x}$, $\varphi: D^n/G \hookrightarrow X$, so that $h(\varphi(D^n/G)) \cap \varphi(D^n/G) = \emptyset$ for all $h \in H - H_{\tilde{x}}$. Choose an open set $U \subset D^n/G$ with $[0] \in U$ and with $h(\varphi(U)) \subset \varphi(D^n/G)$ for all $h \in H_x$. Then $\varphi^{-1} \circ h \circ \varphi: U \hookrightarrow D^n/G$ is a diffeomorphism onto its image for all $h \in H_x$. Thus there is a lifting $(\varphi^{-1} \circ h \circ \varphi)\tilde{\,}: \tilde{U} \hookrightarrow D^n$, where $\tilde{U}$ is the preimage of U in D^n. This lifting can be varied by any element of G. Let $\tilde{H}_{\tilde{x}}$ be the group of germs at $0 \in D^n$ of all liftings of all $\varphi^{-1} \circ h \circ \varphi$ for $h \in H_{\tilde{x}}$. There is an exact sequence

$$1 \to G \to \tilde{H}_{\tilde{x}} \to H_{\tilde{x}} \to 1$$

In particular, $\tilde{H}_{\tilde{x}}$ is a finite group of germs of diffeomorphisms. Each element $\alpha \in \tilde{H}_{\tilde{x}}$ is represented by a smooth embedding $\varphi_\alpha: \tilde{U} \hookrightarrow D^n$. Pull back the standard metric on D^n under each of the φ_α. Average the resulting finite collection of metrics on $\tilde{U}$. This produces a riemannian metric on $\tilde{U}$, in which the elements of $\tilde{H}_{\tilde{x}}$ are germs of isometries. It follows that there is a disk D' of radius ε centered at $0 \in \tilde{U}$ so that $D'/\tilde{H}_{\tilde{x}}$ is embedded in $(D^n/G)/H_{\tilde{x}}$, which, in turn, is embedded by φ in Q/H. The composite is an orbifold chart for Q/H centered at x. The collection of all charts constructed in this manner defines the orbifold structure which we denote X/H. It is called the *quotient of the action of* H *on* X.

Local models for the map $X \to X/H$ are $\pi: D^n/G \to D^n/G'$, where $G \subset G' \subset O(n)$ and π is the natural projection.

We turn now to the question of covering spaces for orbifolds. Let $\{U_\alpha\}_{\alpha \in I}$ be a family of orbifolds. A mapping $p: \amalg_{\alpha \in I} U_\alpha \to D^n/G$ *evenly covers* D^n/G if for each $\alpha \in I$ there is an isomorphism $\varphi: D^n/G_\alpha \cong U_\alpha$, so that $G_\alpha \subset G$ and $p \circ \varphi: D^n/G_\alpha \to D^n/G$ is the natural projection.

If $p: Y \to X$ is a continuous mapping between the topological spaces underlying two orbifolds, then p is said to be a *covering projection* if every point

$x \in X$ has an orbifold chart $\varphi: D^n/G \hookrightarrow X$ that is evenly covered by $p|p^{-1}(\varphi(D^n/G))$. In this case we also say that Y is a *covering orbifold* of X.

If $p: Y \to X$ is a covering projection, then the group of covering transformations $G_X(Y)$ acts properly discontinuously on the orbifold Y. The quotient orbifold $Y/G_X(Y)$ covers X via the map induced by p. Also, the quotient mapping $Y \to Y/G_X(Y)$ is a covering projection.

Conversely, if $H \times Y \to Y$ is a properly discontinuous action on the orbifold Y, then $Y \to Y/H$ is a covering projection whose group of covering transformations is H.

A covering of an orbifold $Y \to X$ is said to be *regular* if the induced covering $Y/G_X(Y) \to X$ is an isomorphism, i.e., if X is naturally isomorphic to the quotient orbifold of the group of covering transformations acting on Y.

The usual proof of the existence of universal coverings can be adapted to show that any orbifold X has a universal covering, $\tilde{X} \to X$. This is a regular covering, and the group of covering transformations is called the *orbifold fundamental group* of X, $\pi_1^{\text{orb}}(X)$. All connected coverings of X come, up to isomorphism, by dividing $\tilde{X}$ by a subgroup of $\pi_1^{\text{orb}}(X)$. A connected cover of X is regular if and only if it is isomorphic to the quotient of $\tilde{X}$ by a normal subgroup of $\pi_1^{\text{orb}}(X)$.

Suppose that X is an orbifold whose underlying space is Q. Let $p: P \to Q$ be a topological covering with group of covering transformations G. Since p is a local homeomorphism, we can induce the orbifold structure X on Q up to an orbifold structure Y on P. The projection $p: Y \to X$ becomes a covering of orbifolds. Clearly, G acts as the group of covering transformations for this covering of orbifolds. Thus if $P \to Q$ is a regular covering, then so is $Y \to X$, and the group of covering transformations is the same. Applying this to the universal topological covering of Q we see that X has a regular covering whose group of covering transformations is $\pi_1(Q)$. This proves Lemma 1.1.

LEMMA 1.1. *If Q is the space underlying an orbifold X, then there is a natural surjection $\pi_1^{\text{orb}}(X) \to \pi_1(Q) \to 1$.*

An orbifold is said to be *good* if its universal cover is a manifold (as an orbifold). There is a way to formulate this concept in terms of local groups. First, note that $\pi_1^{\text{orb}}(D^n/G) \cong G$. Thus if $\varphi: D^n/G \hookrightarrow X$ is an orbifold chart centered at x, then $\pi_1^{\text{orb}}(D^n/G)$ is identified with the local group at x. The map φ induces $\varphi_*: \pi_1^{\text{orb}}(D^n/G) \to \pi_1^{\text{orb}}(X)$.

The orbifold X is good if and only if for each chart $\varphi: D^n/G \hookrightarrow X$ the homomorphism $\varphi_*: \pi_1^{\text{orb}}(D^n/G) \to \pi_1^{\text{orb}}(X)$ is an injection. We denote this by saying that "the local groups inject."

This general discussion of orbifolds impinges on the problem of classifying actions of finite groups on homotopy 3-spheres. This is brought out clearly by the next theorem.

THEOREM 1.2. *Classifying finite group actions on homotopy 3-spheres, up to equivariant diffeomorphism, is the same as classifying closed, good, smooth, three-dimensional orbifolds with finite fundamental group, up to diffeomorphism.*

Proof. The correspondence between group actions and orbifolds is given as

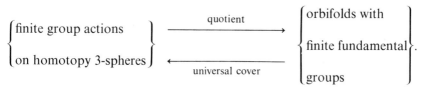

■

There is also a notion of locally smooth, topological orbifolds. One uses the same charts but requires that the overlap functions lift to homeomorphisms instead of diffeomorphisms. It turns out that this concept is the same as the underlying topological space with its stratification by local group type and with the local group type associated to each stratum. We shall describe in more detail these stratified spaces in dimensions two and three in the next two sections.

2. Two-Dimensional Orbifolds

The finite subgroups of $O(2)$, up to conjugacy, are (1) cyclic subgroups of $SO(2)$, (2) $O(1)$, and (3) dihedral groups $D_{2n} \subset O(2)$, $n \geq 2$. The resulting quotients of the 2-disk by these groups are shown in Fig. 2.1. From this it follows that if X is a two-dimensional orbifold without boundary, then its underlying topological space, Q, is a 2-manifold (possibly with boundary). There is a discrete set of points in the interior with nontrivial local groups.

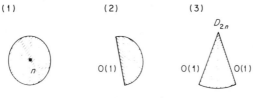

Figure 2.1

We label these points by the orders of their local groups. There is a discrete set of points on the boundary with dihedral local groups. The rest of the boundary consists of points with local groups O(1).

If we allow X to be an orbifold with boundary, then the situation is more complicated. There is a disjoint union of circles and closed intervals in the boundary of the underlying space which forms the space underlying ∂X. This space is unlabeled except for the end points of the intervals, which are labeled by O(1).

To each two-dimensional orbifold X we can associate its *underlying triple* (Q, K, ρ), where Q is the underlying space, $K \subset Q$ is the exceptional set, and ρ is the function on Q that assigns to each point its local group type. Given any such triple (where Q is a 2-manifold and K and ρ are as described above), there is an orbifold structure on Q that gives rise to this triple. There are many such. Any two are diffeomorphic by a diffeomorphism arbitrarily close to the identity.

One can check easily that there are four types of compact two-dimensional orbifolds that are not good:

(a) S^2 with one point labeled,
(b) S^2 with two points labeled by different integers,
(c) D^2 with one boundary point labeled by a dihedral group and the rest of ∂D^2 labeled by O(1), and
(d) D^2 with two boundary points labeled by dihedral groups of different order and the rest of ∂D^2 labeled by O(1).

Notice that all these orbifolds are closed.

All other compact two-dimensional orbifolds are good. The good orbifolds with finite fundamental group are

S(i) S^2, $\mathbf{RP}^2$; S^2 with 2 points labeled n, and $\mathbf{RP}^2$ with one point labeled.
S(ii) S^2 with 3 points labeled p, q, and r, where $1/p + 1/q + 1/r > 1$,
S(iii) D^2 with ∂D^2 labeled by O(1), D^2 with two boundary points labeled by D_{2n} and the rest of ∂D^2 labeled by O(1),
S(iv) D^2 with three boundary points labeled by dihedral groups of order $2p$, $2q$, and $2r$ with $1/p + 1/q + 1/r > 1$ and the rest of ∂D^2 labeled O(1);
S(v) D^2 with an interior point labeled by p and a boundary point labeled by a dihedral group of order $2n$, where $(2/p) + (1/n) > 1$ and the rest of ∂D^2 labeled by O(1); and
D orbifolds isomorphic to D^2/G for $G \subset O(2)$.

The orbifolds S(i)–S(v) are *spherical orbifolds* in the sense that they are diffeomorphic to S^2/G for some $G \subset O(3)$. Those of type D are called *2-disk orbifolds*.

X. Finite Group Actions on Homotopy 3-Spheres

All other compact two-dimensional smooth orbifolds have infinite fundamental group. It turns out that all of these are diffeomorphic to either flat or hyperbolic two-dimensional orbifolds. A *flat two-dimensional orbifold* is A/G, where $A \subset \mathbf{R}^2$ is a region bounded by straight lines and G is a discrete group of euclidean motions leaving A invariant. A *hyperbolic orbifold* is K/Γ, where $K \subset \mathbb{H}^2$ is a region bounded by geodesics (here $\mathbb{H}^2$ is the hyperbolic plane) and Γ is a discrete group of hyperbolic isometries which acts leaving K invariant.

Let X be a two-dimensional orbifold whose underlying space is Q and whose exceptional set is $K \subset Q$. Suppose that all local group of X are cyclic subgroups of SO(2). Then K is a discrete set of points in Q, and we can view ρ as a function from K to the natural numbers greater than one — ρ assigns to each $k \in K$ the order of the local group at k. An orbifold covering $Y \to X$ induces a ramified covering of Q, ramified over K with index of ramification over $k \in K$ dividing $\rho(k)$. Conversely, any ramified covering of this type corresponds to an orbifold cover of X.

3. Three-Dimensional Orbifolds

We shall consider three-dimensional orbifolds whose local groups are contained in SO(3). These are called *locally orientable* three-dimensional orbifolds. The finite subgroups of SO(3) are (1) the cyclic groups, (2) the dihedral groups, (3) the tetrahedral group, (4) the octahedral group, and (5) the icosahedral group. The quotients of D^3 by these groups are shown in Fig. 3.1. In each case the orbit space is homeomorphic to D^3. In case (a) the exceptional set is a line segment; the label n means that the local group is cyclic of order n. In cases (b)–(e), the exceptional set is a cone on three points. The central vertex is the image of the origin; it is the only point with noncyclic local group type.

If X is a locally orientable, three-dimensional orbifold, then we can extract the *underlying triple* (Q, K, ρ). The first element, Q, is the underlying space; the second, $K \subset Q$, is the exceptional set; and the third, ρ, is the function which assigns local group type to each point. In this case we can view ρ as a function from Q-{vertices of K} to the positive integers. Each point $q \in Q$ that is not a vertex has cyclic local group. We think of ρ as associating to that point the order of the local group. From this function one can recover the local group at each point of Q.

If (Q, K, ρ) is a triple, which locally near each point of Q is of one of the five types in Fig. 3.1, then there is an orbifold structure on Q whose underlying triple is (Q, K, ρ). There are many such structures. Any two are isomorphic by an isomorphism that is arbitrarily close to the identity on Q.

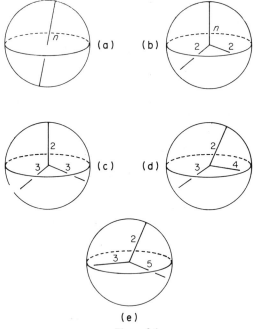

Figure 3.1

A three-dimensional orbifold is of *cyclic type* if all its local group types are represented by cyclic subgroups of SO(3). This simply means that the exceptional set K is a disjoint union of circles. An orbifold is of *dihedral type* if each of its vertices is dihedral, or equivalently, if each local group type is represented either by a cyclic or dihedral subgroup of SO(3).

Suppose that X is a three-dimensional, locally orientable orbifold whose underlying triple is (Q, K, ρ). Covering spaces of X are exactly ramified coverings of Q (with the total space being a topological manifold) which are ramified over K, so that above any $q \in (K\text{-vertices})$ the index of ramification divides $\rho(q)$. This means that any covering $Y \to X$ yields such a ramified covering on the underlying triples and, conversely, any ramified covering $P \to Q$ of the type specified above yields a covering of orbifolds $Y \to X$.

The universal cover of X corresponds to the universal ramified covering of (Q, K, ρ). The orbifold X is good if in the universal ramified covering the indices of ramification over any $q \in (K\text{-vertices})$ do not merely divide $\rho(q)$ but are equal to $\rho(q)$.

There is an explicit description of the fundamental group of the orbifold X in terms of the underlying triple (Q, K, ρ). Let $\Gamma = \pi_1(Q - K)$. Number the edges and circles of K as $e_1, \ldots, e_T$. Let $\mu_i \in \Gamma$ be the class of the *meridian*

X. Finite Group Actions on Homotopy 3-Spheres

around e_i. It is well-defined up to conjugation and taking inverses in Γ. The fundamental group $\pi_1^{\text{orb}}(X)$ is then $\Gamma/\{(\mu_1)^{\rho(e_1)}, \ldots, (\mu_T)^{\rho(e_T)}\}$. The kernel of $\Gamma \to \pi_1^{\text{orb}}(X) \to 1$ corresponds to an unramified cover of $Q - K$. This covering can be completed in only one way to form a ramified covering $\tilde{Q}_{\text{ram}} \to Q$, which is the universal covering of the orbifold.

REMARK 3.1. The order of μ_i in $\pi_1^{\text{orb}}(X)$ divides $\rho(e_i)$. The orbifold X is good if and only if the order of each μ_i is equal to $\rho(e_i)$. If X has finite fundamental group and is of dihedral type, then it is automatically good. Basically this follows from the fact that Q is a rational homology 3-sphere.

Let Y be a two-dimensional orbifold and X a three-dimensional orbifold. An embedding of Y in X is a mapping $f: Y \to X$ that is a homeomorphism onto its image such that for each $x \in X$ there is an orbifold chart $\varphi: D^3/G \hookrightarrow X$ centered at x so that either.

(1) $\varphi(D^3/G) \cap f(Y) = \emptyset$, or
(2) $G \subset O(2) \subset O(3)$; $f(Y) \cap \varphi(D^3/G) = \varphi((D^2 \times \{0\})/G)$; and $f^{-1} \circ \varphi$: $D^2/G \hookrightarrow Y$ is an orbifold chart (in the structure of Y) centered at $f^{-1}(x)$.

If $Y^2 \to X^3$ is an embedding, then we can cut X^3 open along Y. The result is a (smooth) orbifold with boundary. (If Y meets the boundary of X, then it is necessary to "round the corners" at $\partial Y = \partial X \cap Y$.)

If Y^2 is an orientable spherical orbifold, then $Y^2 \cong S^2/G$ for some $G \subset O(3)$. Thus Y is the boundary of D^3/G. A three-dimensional orbifold is *prime* if whenever $Y^2 \hookrightarrow X^3$ is an embedding of a spherical orbifold which locally separates X, then Y bounds an orbifold isomorphic to D^3/G in X. If $Y^2 \hookrightarrow X^3$ is an embedding of a spherical orbifold in X that separates, then we can write $X = X_1 \cup_Y X_2$, where X_1 and X_2 are orbifolds with Y as a boundary component. If $Y \cong S^2/G$, then we let $\hat{X}_i = X_i \cup_Y D^3/G$. We say that X is the *connected sum* of $\hat{X}_1$ and $\hat{X}_2$; $X \cong \hat{X}_1 \# \hat{X}_2$. If Y separates, but one side, say X_2, is isomorphic to D^3/G, then $\hat{X}_2 \cong S^3/G$ for some $G \subset SO(3) \subset SO(4)$. Otherwise, X is said to be a *nontrivial* connected sum.

PROPOSITION 3.2. *Let X be a closed three-dimensional orbifold of cyclic type with finite fundamental group. Then $X \cong X_1 \# \Sigma$, where X_1 is prime and Σ is a simply connected manifold.*

Proof. If X is a 3-manifold, then the result follows immediately from the existence of a prime decomposition and the Seifert–van Kampen theorem. Let us assume that X is not a manifold. Let Q be the underlying space for X and $K \subset Q$ the exceptional set. We know that Q is a 3-manifold. The prime decomposition theorem for manifolds allows us to write $Q \cong Q_1 \# \Sigma$, where

$K \subset (Q_1 - (3\text{-ball})) \subset Q'_1 \# \Sigma$ and where $Q_1 - K$ is prime. This induces a connected sum decomposition $X \cong X_1 \# \Sigma$, where Σ is a manifold.

We claim that Σ is simply connected. Since X is good and X is not a manifold, X_1 is good and X_1 is not a manifold. Thus, $\pi_1^{\text{orb}}(X_1) \neq \{e\}$. Since $\pi_1^{\text{orb}}(X) = \pi_1^{\text{orb}}(X_1) * \pi_1(\Sigma)$ is finite, $\pi_1(\Sigma) = \{e\}$.

Finally, it remains to show that X_1 is prime. If $Y^2 \hookrightarrow X_1$ is an embedded spherical orbifold which locally separates, then Y^2 is of cyclic type. We shall show that Y^2 is the boundary in X of an orbifold isomorphic to D^3/G for some $G \subset SO(3)$. The underlying space of Y is a closed 2-manifold. Since $\pi_1^{\text{orb}}(X_1)$, and hence $\pi_1(Q_1)$, is finite, this 2-manifold must separate Q_1. Thus Y^2 meets every component of K in an even number of points. If $Y^2 \cap K = \emptyset$, then Y^2 is diffeomorphic to a 2-sphere. Since $Q_1 - K$ is prime, Y^2 bounds a 3-ball in X_1. If $Y^2 \cap K \neq \emptyset$, then Y^2 must be a 2-sphere with two points labeled n. We have a decomposition $X_1 = X'_1 \cup_Y X''_2$. Since X_1 is good, $\pi_1^{\text{orb}}(Y) \to \pi_1^{\text{orb}}(X)$ injects. Thus we have a free product with amalgamation decomposition

$$\pi_1^{\text{orb}}(X_1) = \pi_1^{\text{orb}}(X'_1) *_{\mathbf{Z}/n\mathbf{Z}} \pi_1^{\text{orb}}(X''_1).$$

Since $\pi_1^{\text{orb}}(X_1)$ is finite, this decomposition must be trivial; i.e., either $Y \hookrightarrow X'_1$ or $Y \hookrightarrow X'_2$ must induce an isomorphism on orbifold fundamental groups. Suppose that it is $Y \hookrightarrow X'_1$. Let $\tilde{X}'_1$ be the universal cover of X'_1. It is a simply connected 3-manifold whose boundary is S^2. Thus, by the solution to the Smith conjecture, $X'_1 \cong (D^3/G) \# \Sigma$, where $G \subset O(3)$ and Σ is a homotopy 3-sphere. Since $Q_1 - K$ is prime, this homotopy 3-sphere must be standard. Thus $X'_1 \cong D^3/G$. This proves that X_1 is prime. ∎

4. Seifert-Fibered Orbifolds

In this section we shall introduce the notion of a Seifert fiber structure for a locally orientable three-dimensional orbifold. This will generalize the classical notion for 3-manifolds. The basic reason for introducing Seifert orbifolds is that they provide a bridge to quotients of linear group actions in S^3.

Let $S^1 \subset SO(4)$ be the subgroup of matrices

$$\begin{pmatrix} R(\zeta) & 0 \\ 0 & R(\zeta) \end{pmatrix}, \quad \zeta \in S^1,$$

where $R(\zeta)$ is the matrix for rotation by angle ζ in the plane. If we take the natural action of $SO(4)$ on S^3 and restrict it to this circle, then the result is a free action. The projection mapping to the quotient space, $p: S^3 \to S^2$, is the Hopf fibration. Let $N_{SO(4)}(S^1)$ be the normalizer of this circle in $SO(4)$.

X. Finite Group Actions on Homotopy 3-Spheres

It is an extension of the unitary group $U(2)$, by $\mathbf{Z}/2\mathbf{Z}$. There is also an extension

$$1 \to S^1 \to N_{SO(4)}(S^1) \xrightarrow{p} O(3) \to 1.$$

In this sequence the action of $O(3)$ on S^1 is given by $g\zeta g^{-1} = \zeta^{\det(p(g))}$ for all $g \in N_{SO(4)}(S^1)$.

Suppose that $G \subset N_{SO(4)}(S^1)$ is a finite subgroup. Since G normalizes the S^1, its action on S^3 sends fibers of the Hopf fibration to fibers. The induced action on the fibers is either complex linear or complex antilinear. There is induced on the quotient orbifold S^3/G a decomposition into one-dimensional sets; namely, the images of the fibers of the Hopf fibration. Most of these images are circles, but it is possible for an image to be an interval. This happens when there is an element $g \in G$ which leaves a fiber invariant but acts on this fiber in an orientation-reversing manner.

It is exactly these one-dimensional decompositions of three-dimensional linear orbifolds which serve as models for Seifert-fibered orbifolds.

DEFINITION 4.1. Let X be a locally orientable, three-dimensional orbifold. Let $\mathscr{F}$ be a decomposition of X into intervals and circles. We say that $\mathscr{F}$ is a (smooth) *Seifert fibration* of X if for each element $T \in \mathscr{F}$ there are

(1) an open set $U_T \subset X$, containing T, that is a union of elements of $\mathscr{F}$;
(2) a finite finite subgroup $G_T \subset N_{SO(4)}(S^1)$;
(3) a G_T-invariant open set $V_T \subset S^3$, which is a union of Hopf fibers; and
(4) a smooth isomorphism $\varphi_T : V_T/G_T \to U_T$ so that φ_T carries the decomposition of V_T/G_T by images of Hopf fibers to the decomposition which $\mathscr{F}$ induces on U_T.

We say that X is *Seifert-fibered* if it admits a Seifert fibration structure.

It is clear from this definition that if $G \subset N_{SO(4)}(S^1)$ is a finite subgroup, then the orbifold S^3/G has a natural Seifert fibration induced by the Hopf fibration in S^3. The base space of this natural Seifert fibration on S^3/G is the orbifold $S^2/p(G)$. The Hopf fibration induces a continuous map $\pi : S^3/G \to S^2/p(G)$, which is called the *projection* of the Seifert fibration. The fibers of π form the decomposition. Thus $\pi : S^3/G \to S^2/p(G)$, on the level of underlying spaces, is just the quotient mapping of the decomposition.

In general, if X is an orbifold with a smooth Seifert fibration $\mathscr{F}$, then there is a quotient space A for the decomposition and a continuous map $\pi : Q \to A$, where Q is the underlying space of X. Since X is locally isomorphic to S^3/G, A is locally isomorphic to $S^2/p(G)$. These local isomorphisms define on A the structure of a smooth two-dimensional orbifold. This orbifold is denoted B and is called the *base of the Seifert fibration*. The map $\pi : X \to B$ is called the *projection*.

Suppose that $\pi\colon X \to B$ is the projection of a Seifert fibration of an orbifold and that $\mu\colon C \to B$ is a covering of two-dimensional orbifolds. Form the "fiber product" of X and C over B, and call the result Y,

$$\begin{array}{ccc} Y & \xrightarrow{\pi'} & C \\ \mu' \downarrow & & \downarrow \mu \\ X & \xrightarrow{\pi} & B \end{array}$$

(Some care is needed in defining this "fiber product;" it is *not* the set theoretic fiber product.) The orbifold structure on X induces one on Y so that $\mu'\colon Y \to X$ is a covering of orbifolds. The fibers of $\pi'\colon Y \to C$ give a Seifert fibration on Y with base C. If $\mu\colon C \to B$ is a regular covering with group of covering transformations G, then the G-action on C induces a G-action on Y. The quotient orbifold is X. If C is a manifold, i.e., if the exceptional set is empty, then $\pi'\colon Y \to C$ is a smooth circle bundle.

In the next two lemmas we shall consider two consequences of this fiber product construction.

LEMMA 4.2. *Suppose that $\pi\colon X \to B$ is the projection of a Seifert fibration on an orbifold. Let $b \in B$ be a point with local group $G_b \subset O(2)$. Then there is a neighborhood N of $\pi^{-1}(b) \subset X$ of the form $(S^1 \times D^2)/G_b$, where G_b acts orthogonally on both factors and where $\pi|N$ is induced by projection onto the second factor.*

Proof. The point b has a neighborhood in B of the form D^2/G_b, where $G_b \subset O(2)$. Let N be the preimage of this neighborhood in X. Form the fiber product of N and D^2 over D^2/G_b. Call the result $\tilde{N}$. Then $\tilde{N}$ is a smooth circle bundle over D^2. Moreover, G_b acts on it by bundle maps. It follows that there is a product structure $\tilde{N} = S^1 \times D^2$, where G_b acts orthogonally on both factors. Hence $N \cong \tilde{N}/G_b \cong (S^1 \times D^2)/G_b$, and $\pi|N$ is induced by projection onto the second factor. ∎

REMARK. One consequence of Lemma 4.2 is that the orbifold X must be of dihedral type.

LEMMA 4.3. *Suppose that $\pi\colon X \to B$ is the projection of a Seifert fibration on an orbifold. Then π induces a surjection $\pi_1^{orb}(X) \to \pi_1^{orb}(B) \to 1$. The kernel of this homomorphism is a cyclic normal subgroup generated by the class of a generic fiber in the Seifert fibration. This kernel is nontrivial if $\partial X \neq \varnothing$.*

Proof. Form the fiber product of X and the universal cover $\tilde{B} \to B$. Call the result Y. Then $\pi_1^{orb}(B)$ is the group of covering transformations of Y and

X. Finite Group Actions on Homotopy 3-Spheres

the surjection $\pi_1^{\mathrm{orb}}(X) \to \pi_1^{\mathrm{orb}}(B) \to 1$ is induced by π. If B is good, then $\tilde{B}$ is either $\mathbf{R}^2$, S^2, or D^2, and $Y \to \tilde{B}$ is a smooth circle bundle. Hence $\pi_1^{\mathrm{orb}}(Y)$ is cyclic and generated by the class of the fiber. If $\pi_1^{\mathrm{orb}}(Y)$ is trivial and B is good, then $\tilde{B} = S^2$ and X is closed.

If $\tilde{B}$ is bad, then $\tilde{B}$ is either S^2 with one exceptional point or S^2 with two exceptional points with local groups having relatively prime order. It is easy to see that once again $\pi_1^{\mathrm{orb}}(Y)$ is generated by the class of the generic fiber. Also, Y, and hence X, is closed in this case. ∎

DEFINITION. If each local group of B is cyclic, then the Seifert fibration $\pi: X \to B$ is said to be of *restricted type*.

We next prove a technical proposition about Seifert fibrations of restricted type that will be useful later.

PROPOSITION 4.4. *Let X^3 be an orbifold of cyclic type and let Q be the underlying topological space. Suppose that Q is Seifert-fibered (in the classical sense) so that the exceptional set $K \subset Q$ is a union of fibers. Then X is a smooth Seifert-fibered orbifold.*

First we consider a simple relative version of the proposition.

LEMMA 4.5. *Let X be an orbifold with underlying topological space $S^1 \times D^2$ and exceptional set $S^1 \times \{0\}$. Suppose that there is given a Seifert fibration on ∂X that extends to a Seifert fibration (in the classical sense) on $S^1 \times D^2$. Then the Seifert fibration on ∂X extends to one for the orbifold X.*

Proof of Lemma 4.5. Suppose that $\rho(S^1 \times \{0\}) = n$. Let $Y \to X$ be the cyclic n-sheeted cover (branched along $S^1 \times \{0\}$). Then Y is a manifold. The fibration on ∂X lifts to one on ∂Y which is invariant under the $\mathbf{Z}/n\mathbf{Z}$-action. It is easy to extend this to a Seifert fibration of Y which is invariant under $\mathbf{Z}/n\mathbf{Z}$. Taking the quotient yields the Seifert fibration of X. ∎

Proof of Proposition 4.4. Let $v(K)$ be a disk bundle neighborhood of K in Q. By hypothesis, there is a smooth Seifert fibration of $Q - v(K)$. Use Lemma 4.5 on each component of $v(K)$ to extend this to a Seifert fibration for X. ∎

Finally, notice that if $\pi: X \to B$ is the projection of a Seifert fibration and $x \in X$ has local group G_x, then G_x is a subgroup of the local group $G_{\pi(x)}$. Thus if $K \subset X$ is a circle of points with local group cyclic of order >2, then $\pi(K)$ consists of points with local group of order larger than 2. This implies that $\pi(K)$ is a single point. This proves the following lemma.

LEMMA 4.6. *If $K \subset X$ is a circle of points whose local group has order greater than 2 and if $\mathscr{F}$ is a Seifert fibration of X, then K is a fiber of $\mathscr{F}$.*

5. Seifert-Fibered Orbifolds and Linear Actions

In [11], Seifert and Threlfall identified the Seifert fiber spaces with finite fundamental group with the orbit spaces of finite subgroups of SO(4) acting freely on S^3. In this section we shall generalize this to Seifert orbifolds and nonfree actions. Thus our goal is to identify Seifert orbifolds with finite fundamental group with linear orbifolds coming from the action of finite subgroups of $N_{SO(4)}(S^1)$ on S^3. As we saw in the last section, all such linear orbifolds are Seifert-fibered. Here we prove the converse.

THEOREM 5.1. *Let X^3 be a good, orientable three-dimensional orbifold with finite fundamental group. Suppose that we are given a Seifert fibration for X. There is a subgroup $G \subset N_{SO(4)}(S^1)$ so that X and S^3/G are isomorphic as orbifolds.*

By Remark 3.2, X is good, and thus $\tilde{X}$ is a simply connected 3-manifold. Since it is Seifert-fibered, it is diffeomorphic to S^3. Let $G = \pi_1^{orb}(X)$. Let $\mathscr{D}$ be the group of orientation-preserving diffeomorphisms of S^3. Identifying $\tilde{X}$ with S^3 gives a representation $G \hookrightarrow \mathscr{D}$. We wish to show that this representation is conjugate to one whose image is contained in $N_{SO(4)}(S^1)$.

Let B be the base of the given Seifert fiber structure on X, and let $\pi: X \to B$ be the projection. The argument is divided into two cases, depending on whether B is a good orbifold. Let $\tilde{B}$ denote the universal cover of B.

Case I. B is not good. Let $N = (O(2) \times O(2)) \cap SO(4)$. It is our plan to construct an effective action of N on $\tilde{X}$ so that the group of covering transformations G acts as a subgroup of N. Since it is easy to see that any such N-action on $\tilde{X}$ is equivariantly diffeomorphic to the standard linear N-action on S^3, this will prove that $X = \tilde{X}/G$ is isomorphic to $S^3/\rho(G)$, where ρ is the linear representation $G \hookrightarrow N \subset N_{SO(4)}(S^1)$.

According to the list in Section 2, if B is not good, then either it is S^2 with at most two exceptional points or it is D^2 with at most two dihedral points on the boundary (See Fig. 5.1). Hence B decomposes as

$$B = B_1 \bigcup_{\partial B_1} (\partial B_1 \times I) \bigcup_{\partial B_2} B_2,$$

where $B_i \cong D^2/G_i$ for some $G_i \subset O(2)$. (Actually, the argument in this case will work for any such B.) Let $X_i = \pi^{-1}(B_i)$ and let $Y = \pi^{-1}(\partial B_1 \times I)$. Also let, $\partial_i Y = Y \cap X_i$. Clearly,

$$X = X_1 \bigcup_{\partial_1 Y} Y \bigcup_{\partial_2 Y} X_2.$$

X. Finite Group Actions on Homotopy 3-Spheres

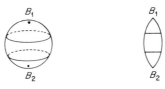

Figure 5.1

Let $\tilde{B}_i \subset \tilde{B}$ be the preimage of B_i. Each $\tilde{B}_i$ is connected. Let $\tilde{X}_i$ and $\tilde{Y}$ be the preimages of X_i and Y, respectively, in $\tilde{X}$. They are also connected. The group G acts on $\tilde{X}_i$, $\tilde{Y}$, and $\partial_i \tilde{Y}$ with quotients X_i, Y, and $\partial_i Y$.

There is a natural action of N on $S^1 \times D^2$ given by the standard linear action of $O(2)$ on each factor. This action has principal isotropy group $\mathbb{Z}/2\mathbb{Z}$ and has one singular orbit—the core. By Lemma 4.2, X_i is isomorphic to $(S^1 \times D^2)/G_i$, where G_i acts linearly on each factor. Such an identification induces one of $\tilde{X}_i$ with $S^1 \times D^2$, so that G acts as a subgroup of N. ($\tilde{X}_i$ is a manifold since X is good, and $\tilde{X}_i$ is a finite cover of X_i since G is finite.) Choose such identifications and let $\psi_i \colon N \times \tilde{X}_i \to \tilde{X}_i$ be the resulting actions. Also, let $\lambda_i \colon \partial_i Y \to T^2/G$ be the resulting identifications.

Since Y is Seifert-fibered with base $\partial B_1 \times I$, it is diffeomorphic to $\partial_1 Y \times I$. Choose an identification $\lambda \colon Y \to (T^2/G) \times I$ so that $\lambda | \partial_1 Y = \lambda_1$. Then λ defines an N-action on $\tilde{Y}$, so that $G \subset N$, and so that, when restricted to $\partial_1 \tilde{Y}$, this action agrees with ψ_1 restricted to $\partial \tilde{X}_1$. The problem is that, in general, the restriction of this action to $\partial_2 \tilde{Y}$ will differ from the action of ψ_2 restricted to $\partial \tilde{X}_2$. However, we shall show that after changing the action $\psi_2 \colon N \times \tilde{X}_2 \to \tilde{X}_2$ by an automorphism of N, these actions become equivalent. This will allow us to amalgamate these actions into an N-action on $\tilde{X}$.

The group N is a semidirect product $T^2 \times \mathbb{Z}/2\mathbb{Z}$. We denote the elements of N by pairs (t, ω^i), where $t \in T^2$ and ω is the generator of $\mathbb{Z}/2\mathbb{Z}$. The multiplication is defined by $(t, \omega^i)(t', \omega^j) = (t + (-1)^i t', \omega^{i+j})$. Consider the natural action on N of $S^1 \times S^1 \subset S^1 \times D^2$. A map $\alpha \colon S^1 \times S^1 \to S^1 \times S^1$ will be called an *affine map* if it is the composition of a map induced by $A \in GL(2, \mathbb{Z})$ acting linearly on $\mathbb{R}^2$ and a map which is translation by $\tau \in T^2$. If $\alpha \colon S^1 \times S^1 \to S^1 \times S^1$ is an affine map, then there is a Lie group automorphism $i_\alpha \colon N \to N$ so that

$$\alpha(n \cdot x) = i_\alpha(n) \cdot \alpha(x)$$

for all $n \in N$ and $x \in S^1 \times S^1$. The formulas for i_α are

$$i_\alpha(t, 1) = (At, 1),$$
$$i_\alpha(t, \omega) = (At + 2\tau, \omega).$$

LEMMA 5.2. *Let $G \subset N$ be a finite subgroup. Any isomorphism $\varphi: T^2/G \to T^2/G$ is isotopic (through orbifold diffeomorphisms) to one which lifts to an affine map on T^2.*

Proof. If $G \subset SO(2) \times SO(2)$, then T^2/G is again a torus. Any diffeomorphism of a torus is isotopic to a linear isomorphism, and any linear map on T^2/G lifts to an affine map on T^2.

If G is not contained in $SO(2) \times SO(2)$, then let $G_0 \subset G$ be the subgroup of index two given by $G_0 = G \cap (SO(2) \times SO(2))$. Then T^2/G_0 is a torus and is a double cover of T^2/G. The orbifold T^2/G is isomorphic to S^2, with four points labeled 2. The lemma will be established if we can show that φ can be deformed until it lifts to an affine map on T^2/G_0. First, we may assume that φ is orientation-preserving. Indeed, if $A \in GL(2, \mathbf{Z})$ has determinant -1, then it induces a linear map $L_A: T^2/G_0 \to T^2/G_0$ covering $\bar{L}_A: T^2/G \to T^2/G$; and either φ or $\varphi \circ \bar{L}_A$ is orientation-preserving. Let $[0] \in T^2/G$ be the image of $0 \in \mathbf{R}^2$ (this is one of the four distinguished points). Secondly, we may assume, by composing φ with the image of a translation of order two, that $\varphi([0]) = [0]$. We can view such a φ as an orientation-preserving homeomorphism of S^2 which fixes $[0]$ and which leaves invariant the other three distinguished points. Birman [1] showed that the group of isotopy classes of such φ is a braid group isomorphic to $PSL(2, \mathbf{Z})$. Furthermore, this isomorphism can be realized as follows. Begin with $A \in SL(2, \mathbf{Z})$, and let $L_A: T^2/G_0 \to T^2/G_0$ be its linear map. Then L_A induces an isomorphism $\bar{L}_A: T^2/G \to T^2/G$. The correspondence $A \to \bar{L}_A$ factors through $PSL(2, \mathbf{Z})$ to give the isomorphism. It follows that every $\varphi: T^2/G \to T^2/G$ is isotopic one which lifts to an affine map $T^2/G_0 \to T^2/G_0$. ∎

By using this lemma, we may assume that the identification $\lambda: Y \to (T^2/G) \times I$ is such that $\lambda|\partial_1 Y = \lambda_1$ and $\lambda_2 \circ (\lambda|\partial_2 Y)^{-1}$ lifts to an affine mapping α on T^2. This λ defines an action $\mu: N \times \tilde{Y} \to \tilde{Y}$, so that $\mu|\partial_1\tilde{Y} = \psi_1|\partial\tilde{X}_1$ and $\mu|\partial_2\tilde{Y} = (\psi_2 \circ i_\alpha)|\partial\tilde{X}_2$.

Thus we can amalgamate these three actions, ψ_1, μ, and $\psi_2 \circ i_\alpha$, to define an effective action $\psi: N \times \tilde{X} \to \tilde{X}$, so that $G \subset N$ and so that $\psi|G$ is the action of the group of covering transformations.

We can write $\tilde{X} \cong T_1 \cup T_2$, where each T_i is an N-invariant solid torus. The standard linear N-action on S^3 has a similar decomposition. By using these decompositions, one can easily construct an N-equivariant diffeomorphism from $\tilde{X}$ to S^3.

Case II. B is a good orbifold. In this case $\tilde{B}$ is isomorphic to S^2 and B is isomorphic to S^2/H for some $H \subset O(3)$. Let $Y \to X$ be the induced covering of X. We know that Y is Seifert-fibered with base $\tilde{B}$. Moreover, since $\tilde{B}$ has

X. Finite Group Actions on Homotopy 3-Spheres

no exceptional points, $Y \to \tilde{B}$ is a smooth circle bundle. As a result, the induced Seifert fibration on $\tilde{X}$ is also a smooth circle bundle over S^2.

LEMMA 5.3. *There is a smooth, free S^1-action on $\tilde{X}$ so that the orbits are the fibers of the induced Seifert fibration and so that $G = \pi_1^{\text{orb}}(X)$ normalizes the S^1-action.*

Proof. Each $b \in B$ has a neighborhood U_b isomorphic to D^2/G_b for some $G_b \subset O(2)$. By Lemma 4.2, the preimage of U_b in X is of the form $(S^1 \times D^2)/G_b$, where G_b acts orthogonally on each factor. For each $p \in D^2$ choose a distance function on $S^1 \times \{p\}$, so that this family of distance functions is G_b-invariant and smooth in p and so that the total length of each circle is $2\pi/n$, where n is the order of the kernel of $\pi_1^{\text{orb}}(X) \to \pi_1^{\text{orb}}(B)$. These distance functions define ones on the fibers of $(S^1 \times D^2)/G_b \to D^2/G_b$. Cover B by a finite collection of such open sets U_b and choose product structures and distance functions as above. Choose a smooth partition of unity subordinate to this cover and use this to average the distance functions on the base.

If we pull back these distance functions to the Seifert fibration $Y \to \tilde{B}$, then we have an S^1-fibration over S^2 with each fiber having total length $2\pi/n$. Pulling back to $\tilde{X}$, the total length of each fiber is 2π. Such a smooth family of distance functions, together with a choice of orientation, gives a free circle action on $\tilde{X}$ whose orbits are the fibers. Since the distance functions are G-invariant, this circle action is normalized by G. This completes the proof of the lemma. ∎

Any two smooth, free S^1-actions on S^3 are equivalent (since they give principal S^1-bundles over S^2). In particular, any such action is equivalent to the standard linear action of $S^1 \subset U(2)$, $\zeta \to \left(\begin{smallmatrix} \zeta & 0 \\ 0 & \zeta \end{smallmatrix}\right)$. Hence there is an identification of $\tilde{X}$ with S^3 in such a way that the circle action on $\tilde{X}$ normalized by G becomes $S^1 \subset U(2)$. We fix such an identification.

Recall that $\mathscr{D}$ is the group of orientation-preserving diffeomorphisms of S^3, with group multiplication defined by composition. Let $N_{\mathscr{D}}(S^1)$ be the subgroup of $\mathscr{D}$ which normalizes the standard S^1, and let $C_{\mathscr{D}}(S^1)$ be the centralizer (it is a subgroup of index 2 in $N_{\mathscr{D}}(S^1)$). The above identification of $\tilde{X}$ with S^3 gives a representation $G \hookrightarrow N_{\mathscr{D}}(S^1)$, where $G = \pi_1^{\text{orb}}(X)$. We are trying to show that G is actually conjugate to a subgroup of $N_{SO(4)}(S^1)$.

If $f \in N_{\mathscr{D}}(S^1)$, then define a diffeomorphism $\bar{f}: S^2 \to S^2$ by

$$\bar{f}(\pi(z)) = \pi(f(z)),$$

where $\pi: S^3 \to S^2$ is the Hopf fibration. Let $\varepsilon: N_{\mathscr{D}}(S^1) \to \{\pm 1\} = \text{Aut}(S^1)$ be the natural map. If f flips S^1, then, since f is orientation-preserving, $\bar{f}$ must be orientation-reversing. Hence $\varepsilon(f) = \deg(\bar{f})$.

Let $C^\infty(S^2, S^1)$ be the group of smooth mappings from S^2 to S^1. (The group multiplication is pointwise multiplication.) If $\psi \in C^\infty(S^2, S^1)$, then define a diffeomorphism $\hat\psi \in C_\mathscr{D}(S^1)$ by

$$\hat\psi(z) = \psi(\pi(z)) \cdot z,$$

where the dot denotes the standard S^1-action on S^3. There are exact sequences

$$1 \to C^\infty(S^2, S^1) \xrightarrow{i} N_\mathscr{D}(S^1) \xrightarrow{p} \operatorname{diff}(S^2) \to 1,$$
$$1 \to C^\infty(S^2, S^1) \xrightarrow{i} C_\mathscr{D}(S^1) \xrightarrow{p} \operatorname{diff}_+(S^2) \to 1,$$

where p is the homomorphism $f \to \bar f$ and i is $\psi \to \hat\psi$.

LEMMA 5.4. *If $f \in N_\mathscr{D}(S^1)$ and $\psi \in C^\infty(S^2, S^1)$, then $f^{-1} \circ \hat\psi \circ f = \hat\lambda$, where $\lambda = (\psi \circ \bar f)^{\varepsilon(f)}$.*

Proof. If $\alpha \in S^1$ and $z \in S^3$, then $f(\alpha \cdot z) = \alpha^{\varepsilon(f)} \cdot f(z)$. Thus

$$\begin{aligned} f \circ \hat\lambda(z) &= f(\psi(\bar f \circ \pi(z))^{\varepsilon(f)} \cdot z) \\ &= \psi(\bar f \pi(z)) \cdot f(z) \\ &= \psi(\pi(f(z))) \cdot f(z) \\ &= \hat\psi(f(z)). \quad \blacksquare \end{aligned}$$

LEMMA 5.5. *Any homomorphism $\mu: G \to \operatorname{diff}(S^2)$, with G a finite group, is conjugate to a homomorphism with image in $O(3)$.*

Proof. The orbifold $S^2/\mu(G)$ is good and has finite fundamental group. Hence it is smoothly isomorphic to S^2/H for some $H \subset O(3)$. An isomorphism $S^2/\mu(G) \to S^2/H$ lifts to a diffeomorphism of S^2 which conjugates $\mu(G)$ to H. $\blacksquare$

Let $\mathscr{G} \subset N_\mathscr{D}(S^1)$ be the full preimage of $O(3) \subset \operatorname{diff}(S^2)$ and let $\mathscr{G}^+ \subset \mathscr{G}$ be the full preimage of $SO(3)$. Thus $\mathscr{G}^+$ is the intersection of $\mathscr{G}$ and $C_\mathscr{D}(S^1)$.

COROLLARY 5.6. *Let $G \subset N_\mathscr{D}(S^1)$ be a finite group. There is an element $f \in N_\mathscr{D}(S^1)$ that conjugates G into $\mathscr{G}$. If $G \subset C_\mathscr{D}(S^1)$, then f conjugates G into $\mathscr{G}^+$.*

Proof. By the above lemma, there is $\bar f \in \operatorname{diff}(S^2)$ that conjugates $p(G)$ into $O(3)$. If $p(G) \subset \operatorname{diff}_+(S^2)$, then this conjugation sends $p(G)$ into $SO(3)$. Lifting $\bar f$ to $f \in N_\mathscr{D}(S^1)$ gives the required element. $\blacksquare$

X. Finite Group Actions on Homotopy 3-Spheres

Consider the ladder of groups

$$\begin{array}{ccccccccc} 1 & \longrightarrow & C^\infty(S^2, S^1) & \xrightarrow{i} & \mathscr{G} & \xrightarrow{p} & O(3) & \longrightarrow & 1 \\ & & \uparrow & & \uparrow & & \| & & \\ 1 & \longrightarrow & S^1 & \longrightarrow & N_{SO(4)}(S^1) & \xrightarrow{p} & O(3) & \longrightarrow & 1. \end{array}$$

By applying Corollary 5.6, we conjugate $G = \pi_1^{\mathrm{orb}}(X)$ into $\mathscr{G}$. We wish to do a further conjugation by an element of $C^\infty(S^2, S^1)$ to move G into $N_{SO(4)}(S^1)$. We shall first treat the special case in which $G \subset \mathscr{G}^+$.

Case IIA. $G \subset \mathscr{G}^+$. The centralizer of S^1 in $SO(4)$ is $U(2)$. So in this case our ladder becomes

$$\begin{array}{ccccccccc} 1 & \longrightarrow & C^\infty(S^2, S^1) & \xrightarrow{i} & \mathscr{G}^+ & \xrightarrow{p} & SO(3) & \longrightarrow & 1 \\ & & \uparrow & & \uparrow & & \| & & \\ 1 & \longrightarrow & S^1 & \longrightarrow & U(2) & \longrightarrow & SO(3) & \longrightarrow & 1. \end{array}$$

First, we shall define a homomorphism $\rho: G \hookrightarrow U(2)$ so that $p \circ \rho = p|G$. Then we shall use a group cohomology argument to construct an element $\psi \in C^\infty(S^2, S^1)$, so that $\hat{\psi}$ conjugates G to $\rho(G)$.

Suppose that $c \in SO(3)$ and that n is an integer. For any $\varphi \in C^\infty(S^2, S^1)$, define a new function $\prod_n (\varphi, c) \in C^\infty(S^2, S^1)$ by

$$\prod_n (\varphi, c) = \prod_{i=0}^{n-1} (\varphi \circ c^i)$$

Similarly, if $\theta \in C^\infty(S^2, \mathbf{R}^1)$, define $\sum_n (\theta, c)$ by

$$\sum_n (\theta, c) = \sum_{i=0}^{n-1} (\theta \circ c^i).$$

Let $\exp: \mathbf{R}^1 \to S^1$ be the universal cover. By a *lifting* $\tilde{\varphi}$ of $\varphi \in C^\infty(S^2, S^1)$, we mean that $\tilde{\varphi} \in C^\infty(S^2, \mathbf{R}^1)$ and that $\exp \circ \tilde{\varphi} = \varphi$.

LEMMA 5.7. *Suppose that $f \in \mathscr{G}^+$ has order n. Then there is a $j \in U(2)$ of order n and a $\psi \in C^\infty(S^2, S^1)$ such that*

(i) $f = \hat{\psi} \circ j$ *and*
(ii) ψ *lifts to* $\tilde{\psi}: S^2 \to \mathbf{R}^1$ *with* $\sum_n (\tilde{\psi}, \bar{f}) = 0$.

Moreover, this decomposition of f is unique.

Proof. We have $\bar{f} \in SO(3)$ and $\bar{f}^n = 1$. Let l be any lift of $\bar{f}$. Then $l^n \in S^1$. Hence there is an element $\zeta \in S^1$ with $l^n = \zeta^{-n}$. Set $k = l\zeta$. Clearly, k is also a lift of $\bar{f}$ and $k^n = 1$. Since $\bar{k} = \bar{f}$, we have that $f = \hat{\varphi} \circ k$ for some $\varphi \in C^{\infty}(S^3, S^1)$. We have

$$\text{id} = (\hat{\varphi} \circ k)^n = \prod_{i=0}^{n-1} (k^i \circ \hat{\varphi} \circ k^{-i}) \circ k^n = \prod_{i=0}^{n-1} k^i \circ \hat{\varphi} \circ k^{-i}.$$

Lemma 5.4 identifies this last product with the image of $\Pi_n(\varphi, k) = \Pi_n(\varphi, \bar{f})$. Hence $\Pi_n(\varphi, \bar{f}) = 1$. Let $\tilde{\varphi} \colon S^2 \to \mathbf{R}^1$ be any lifting of φ. Then $\sum_n(\tilde{\varphi}, \bar{f}) = m$ for some integer m. Define $\tilde{\psi} \in C^{\infty}(S^2, \mathbf{R}^1)$ by $\tilde{\psi}(x) = \tilde{\varphi}(x) - m/n$. Let $\omega \in S^1$ be defined by $\omega = \exp(m/n)$. Finally, let $j = \omega \cdot k$ and let $\psi = \exp \circ \tilde{\psi}$. Clearly, $f = \hat{\psi} \circ j$ is a decomposition as required.

Suppose that $f = \hat{\psi}' \circ j'$ is another such decomposition. Since $\bar{j}' = \bar{j} = \bar{f} \in SO(3)$, we have that $j' = \zeta \cdot j$ for some $\zeta \in S^1$. Thus $\tilde{\psi}'(x) = \tilde{\psi}(x) + d$ for some $d \in \mathbf{R}^1$ that projects onto ζ. Since $\sum_n(\tilde{\psi}', \bar{f}) = \sum_n(\tilde{\psi}, \bar{f}) = 0$ and $nd = \sum_n(\tilde{\psi}', \bar{f}) - \sum_n(\tilde{\psi}, \bar{f})$, it follows that $d = 0$ and, hence, that $\zeta = 1$. ∎

For each $f \in \mathscr{G}^+$ of finite order, let $\rho(f) = j$ and $\psi_f = \psi$ be the above unique decomposition. Also, let $\tilde{\psi}_f \in C^{\infty}(S^2, \mathbf{R}^1)$ be the lifting of ψ_f that sums to zero over the $\bar{f}$-orbits.

LEMMA 5.8. *Let $G \subset \mathscr{G}^+$ be a finite group. The map $f \to \rho(f)$ defines a homomorphism $\rho \colon G \to U(2)$. Furthermore,*

$$\tilde{\psi}_{f \circ g} = \tilde{\psi}_f + \tilde{\psi}_g \circ \bar{f}^{-1}.$$

Proof. We must show that $\rho(e) = e$ and that $\rho(f \circ g) = \rho(f) \circ \rho(g)$. The first is obvious. As for the second, we have $f = \hat{\psi}_f \circ \rho(f)$ and $g = \hat{\psi}_g \circ \rho(g)$. Hence $f \circ g = \hat{\psi}_f \circ \rho(f) \circ \hat{\psi}_g \circ \rho(g)$. Since $\rho(f)$ covers $\bar{f} \in SO(3)$, Lemma 5.4 implies that $\rho(f) \circ \hat{\psi}_g = (\psi_g \circ \bar{f}^{-1})\hat{} \circ \rho(f)$. Thus

$$(*) \qquad \hat{\psi}_{f \circ g} \circ \rho(f \circ g) = \hat{\psi}_f \circ (\psi_g \circ \bar{f}^{-1})\hat{} \circ \rho(f) \circ \rho(g)$$

$$= [\psi_f \cdot (\psi_g \circ \bar{f}^{-1})]\hat{} \circ \rho(f) \circ \rho(g).$$

The elements $\rho(f \circ g)$ and $\rho(f) \circ \rho(g)$ are in $U(2)$ and have the same image in $SO(3)$. Thus there is an element $\zeta \in S^1$, so that

$$\zeta \cdot \psi_{f \circ g} = \psi_f \cdot (\psi_g \circ \bar{f}^{-1}).$$

This means that there is $d \in \mathbf{R}^1$ with $\exp(d) = \zeta$ so that

$$d + \tilde{\psi}_{f \circ g} = \tilde{\psi}_f + (\tilde{\psi}_g \circ \bar{f}^{-1}).$$

Summing over G-orbits gives

$$\text{order}(G) \cdot d + \sum_{\alpha \in G} \tilde{\psi}_{f \circ g} \circ \bar{\alpha} = \sum_{\alpha \in G} \tilde{\psi}_f \circ \bar{\alpha} + \sum_{\alpha \in G} \tilde{\psi}_g \circ \bar{f}^{-1} \circ \bar{\alpha}.$$

X. Finite Group Actions on Homotopy 3-Spheres 203

Since $\tilde{\psi}_f$ sums over any f-orbit to zero and since a G-orbit is a disjoint union of f-orbits, $\sum_{\alpha \in G} \tilde{\psi}_f \circ \bar{\alpha} = 0$. Similarly, $\sum_{\alpha \in G} \tilde{\psi}_{f \circ g} \circ \bar{\alpha} = 0$ and

$$\sum_{\alpha \in G} \tilde{\psi}_g \circ \bar{f}^{-1} \circ \bar{\alpha} = 0.$$

Thus $d = 0$ and $\zeta = 1$. This proves that $\zeta \cdot \tilde{\psi}_{f \circ g} = \tilde{\psi}_f \cdot (\tilde{\psi}_g \circ \bar{f}^{-1})$. It follows immediately from (∗) that $\rho(f \circ g) = \rho(f) \circ \rho(g)$. ∎

Let $G \subset \mathcal{G}^+$ be a finite group and let $\rho \colon G \to N_{SO(4)}(S^1)$ be the homomorphism constructed above. Let $\tilde{\Psi} \colon G \to C^\infty(S^2, \mathbf{R}^1)$ be the function that assigns to $g \in G$ the element $\tilde{\psi}_g$. The last part of Lemma 5.8 says that $\tilde{\Psi}$ satisfies the cocycle condition. If we project $\tilde{\Psi}$ to $\Psi \colon G \to C^\infty(S^2, S^1)$, then the resulting cocycle is the "difference cocycle" for the two mappings of G into $\mathcal{G}^+$. Lemma 5.9 shows that if Ψ is a coboundary, then G and $\rho(G)$ are conjugate. We shall complete the proof of Case IIA by showing that $\tilde{\Psi}$ (and hence Ψ) is a coboundary.

LEMMA 5.9. *With notation as above, suppose that $\Psi \colon G \to C^\infty(S^2, S^1)$ is a coboundary, that is, suppose that there exists $\mu \in C^\infty(S^2, S^1)$, so that*

$$(\mu \circ \bar{g}^{-1}) \circ \mu^{-1} = \Psi(g)$$

for all $g \in G$. Then $\hat{\mu}$ conjugates G to $\rho(G)$.

Proof.

$$(\hat{\mu} \circ g \circ \hat{\mu}^{-1}) \circ \rho(g)^{-1} = \hat{\mu} \circ (g \circ \hat{\mu}^{-1} \circ g^{-1}) \circ \hat{\psi}_g$$
$$= [\mu \cdot (\mu^{-1} \circ \bar{g}^{-1}) \cdot \Psi(g)]\hat{\,}.$$

Thus $\hat{\mu} \circ g \circ \hat{\mu}^{-1} = \rho(g)$ if and only if $\Psi(g) = (\mu \circ \bar{g}^{-1}) \cdot \mu^{-1}$. ∎

The next two lemmas establish that $\tilde{\Psi}$ is indeed a coboundary.

LEMMA 5.10. *Let ω be a generator of $\mathbf{Z}/n\mathbf{Z}$ and suppose that ω acts on D^2 by rotation through $2\pi/n$ radians. Let $\theta \in C^\infty(D^2, \mathbf{R}^1)$ be such that $\sum_n(\theta, \omega) = 0$. Then there is $\beta \in C^\infty(D^2, \mathbf{R}^1)$ such that $\beta(\omega x) - \beta(x) = \theta(x)$ for all $x \in D^2$.*

Proof. Define

$$\beta(x) = -\sum_{i=0}^{n-2} \frac{n-i-1}{n} \theta(\omega^i x).$$

Since $\sum_n(\theta, \omega) = 0$, we have $-(1/n)\theta(\omega^{n-1}x) = \sum_{i=0}^{n-2}(1/n)\theta(\omega^i x)$. Hence

$$\beta(\omega x) - \beta(x) = -\sum_{i=0}^{n-3}\frac{n-i-1}{n}\theta(\omega^{i+1}x) + \sum_{i=0}^{n-2}\frac{1}{n}\theta(\omega^i x)$$

$$+ \sum_{i=0}^{n-2}\frac{n-i-1}{n}\theta(\omega^i x)$$

$$= -\sum_{i=1}^{n-2}\frac{n-i}{n}\theta(\omega^i x) + \sum_{i=0}^{n-2}\frac{n-i}{n}\theta(\omega^i x)$$

$$= \theta(x).$$

LEMMA 5.11. *Let* $\bar{\rho}: G \to SO(3)$ *be a representation of a finite group and suppose that* $\tilde{\Psi}: G \to C^\infty(S^2, \mathbf{R}^1)$ *satisfies the cocycle condition*

$$\tilde{\Psi}(f \circ g) = \tilde{\Psi}(f) + \tilde{\Psi}(g) \circ \bar{\rho}(f)^{-1}.$$

Then $\tilde{\Psi}$ *is a coboundary, i.e., there is a function* $\tilde{\mu} \in C^\infty(S^2, \mathbf{R}^1)$, *so that* $\tilde{\mu} \circ \bar{\rho}(g)^{-1} - \tilde{\mu} = \tilde{\Psi}(g)$ *for all* $g \in G$.

Proof. Suppose that $\bar{\rho}(f) = 1$. Then

(*) $$\tilde{\Psi}(f \circ g) = \tilde{\Psi}(f) + \tilde{\Psi}(g).$$

This means that, restricted to the kernel of $\bar{\rho}$, $\tilde{\Psi}$ is a homomorphism. Since ker $\bar{\rho}$ is a finite group and $C^\infty(S^2, \mathbf{R}^1)$ has no elements of finite order, $\tilde{\Psi}|\ker \bar{\rho} \equiv 0$. By invoking (*) once again, we see that $\tilde{\Psi}$ factors through $\bar{\rho}(G)$ to define a cocycle on that group. In view of this it suffices to solve the problem for the group $\bar{\rho}(G)$. We may therefore assume that $\bar{\rho}$ in injective.

We can cover S^2 by G-invariant open sets $U_1, \ldots, U_k$, where each U_i has the form

$$U_i \cong G \times_{H_i} D^2.$$

where $H_i \subset G$ is a cyclic isotropy group. By the previous lemma we can find $\tilde{\mu}_i \in C^\infty(D^2, \mathbf{R}^1)$, so that $\tilde{\Psi}(g) = \tilde{\mu}_i \circ \bar{\rho}(g)^{-1} - \tilde{\mu}_i$ for all $g \in H_i$. Extend $\tilde{\mu}_i$ to U_i by using the same formula for all $g \in G$.

Let $\{\lambda_i\}$ be a G-invariant smooth partitition of unity subordinate to $\{U_i\}$ and define

$$\tilde{\mu} = \sum_{i=1}^{k} \lambda_i \tilde{\mu}_i.$$

Clearly, $\tilde{\Psi}(g) = \tilde{\mu} \circ \rho(g)^{-1} - \tilde{\mu}$. ∎

X. Finite Group Actions on Homotopy 3-Spheres

Setting $\mu = \exp \circ \tilde{\mu}$, we have that $\hat{\mu}$ conjugates $G \subset \mathscr{G}^+$ to $\rho(G) \subset U(2)$. This completes the proof of Case IIA.

Case IIB. $G \subset \mathscr{G}$, but G is not contained in $\mathscr{G}^+$. Let $G' = G \cap \mathscr{G}^+$. Then G' is a subgroup of index 2 in G. By case IIA, we can conjugate G by an element of $C^\infty(S^2, S^1)$ so that G' is contained in $U(2)$. Choose arbitrarily an element $h \in G - G'$ and an element $k \in N_{SO(4)}(S^1)$ with $\bar{k} = \bar{h} \in O(3)$. There is an element $\varphi \in C^\infty(S^2, S^1)$, so that $k = \hat{\varphi} \circ h$.

LEMMA 5.12. $\varphi(x) = \varphi(\bar{h}^{-1}x)$ for all $x \in S^2$.

Proof. Both k^2 and h^2 are elements in $U(2)$ with the same image in $SO(3)$. Thus there is $\zeta \in S^1$ with $\hat{\zeta} \circ h^2 = k^2$. Since $k = \hat{\varphi} \circ h$, by using Lemma 5.4, we see that $\hat{\zeta} = \hat{\varphi} \circ \hat{\lambda}$, where $\lambda = (\varphi \circ \bar{h}^{-1})^{-1}$. Hence $\zeta = \varphi(x) \cdot \varphi(\bar{h}^{-1}x)^{-1}$ for all x. Therefore, it suffices to show that $\zeta = 1$. Let $\tilde{\varphi} \in C^\infty(S^2, \mathbf{R}^1)$ be a lifting of φ. Then $\tilde{\zeta} = \tilde{\varphi}(x) - \tilde{\varphi}(\bar{h}^{-1}x)$ is a constant that projects to ζ. By summing over an $\bar{h}$-orbit, we see that $\tilde{\zeta} = 0$ and hence that $\zeta = 1$. ∎

LEMMA 5.13. *If $g \in G' = G \cap U(2)$, then $hgh^{-1} = kgk^{-1}$, where h and k are as above.*

Proof. Again, hgh^{-1} and kgk^{-1} are both elements of $U(2)$ with the same projection in $SO(3)$. Hence, their difference is an element $\zeta \in S^1$. Since $k = \hat{\varphi} \circ h$, $k \circ g \circ k^{-1} = \hat{\varphi} \circ (h \circ g \circ h^{-1}) \circ \hat{\varphi}^{-1}$. Set $f = h \circ g \circ h^{-1}$. Then

$$\zeta = f \circ (\hat{\varphi} \circ f \circ \hat{\varphi}^{-1})^{-1} = (f \circ \hat{\varphi} \circ f^{-1}) \circ \hat{\varphi}^{-1} = \hat{\lambda} \circ \hat{\varphi}^{-1},$$

where $\hat{\lambda} = \varphi \circ \bar{f}^{-1}$. Hence $\zeta = (\varphi \circ \bar{f}^{-1}) \cdot \varphi^{-1}$. By applying this equality at $x = \bar{h}y$, we find that $\zeta = \varphi(\bar{f}^{-1}\bar{h}y) \cdot \varphi(\bar{h}y)^{-1} = \varphi(\bar{h}\bar{g}^{-1}y) \cdot \varphi(\bar{h}y)^{-1}$. Since $\varphi \circ \bar{h} = \varphi$, we have $\zeta = (\varphi \circ \bar{g}^{-1})^{-1} \cdot \varphi$. Choose a lifting $\tilde{\varphi} \in C^\infty(S^2, \mathbf{R}^1)$ for φ. Then $\tilde{\zeta} = \tilde{\varphi} - \tilde{\varphi} \circ \bar{g}^{-1}$ is a lifting for ζ. Summing over a $\bar{g}$-orbit shows that $\tilde{\zeta} = 0$ and hence that $\zeta = 1$. ∎

COROLLARY 5.14. *With notation as above, φ and its lifting $\tilde{\varphi}$ are G-invariant.*

Proof. The last step in the proof of Lemma 5.13 shows that $\tilde{\varphi}$ is invariant under $G' \subset G$. Lemma 5.12 shows that $\tilde{\varphi}$ is invariant under h. Since G' and h together generate G, it follows that $\tilde{\varphi}$, and hence φ, is G-invariant. ∎

Now we define $\rho: G \to N_{SO(4)}(S^1)$. Restricted to G' it is the identity. It sends an element of the form $g \circ h$, $g \in G'$, to $g \circ k$. By using Lemmas 5.12 and 5.13, the following becomes a straightforward calculation.

LEMMA 5.15. *The map $\rho: G \to N_{SO(4)}(S^1)$ is a homomorphism.*

LEMMA 5.16. *There is an element $\mu \in C^\infty(S^2, S^1)$ that conjugates G to $\rho(G) \subset N_{SO(4)}(S^1)$.*

Proof. According to Corollary 5.14, the element $\varphi \in C^\infty(S^2, S^1)$ has a G-invariant lifting $\tilde{\varphi} \in C^\infty(S^2, \mathbf{R}^1)$. Define μ by $\mu(x) = \exp(1/2\ \tilde{\varphi}(x))$. We claim that $\hat{\mu} \circ g \circ \hat{\mu}^{-1} = \rho(g)$ for all $g \in G$. Notice that $\hat{\mu}$ is also G-invariant. Thus

$$\hat{\mu} \circ g \circ \hat{\mu}^{-1} = \hat{\mu} \circ \hat{\mu}^{-\varepsilon(\alpha)} \circ g = (\hat{\mu}^{1-\varepsilon(\alpha)})\hat{\ } \circ g.$$

If $\varepsilon(g) = 1$, then $\hat{\mu} \circ g \circ \hat{\mu}^{-1} = g = \rho(g)$. If $\varepsilon(g) = -1$, then

$$\hat{\mu} \circ g \circ \hat{\mu}^{-1} = (\mu^2)\hat{\ } \circ g = \hat{\varphi} \circ g = \hat{\varphi} \circ (g \circ h^{-1}) \circ h = (g \circ h^{-1}) \circ \hat{\varphi} \circ h$$
$$= (g \circ h^{-1}) \circ k = \rho(g). \blacksquare$$

This completes the proof of Case IIB.

6. Statement of the Main Result

We shall establish the following result.

THEOREM 6.1. *Let X be a closed, orientable, three-dimensional orbifold with finite fundamental group. Suppose that X is of cyclic type and that at least one point in X has a local group of order >5. Then there is a Seifert-fibered orbifold X_1 and a homotopy 3-sphere Σ, so that X is diffeomorphic to $X_1 \# \Sigma$.*

This theorem, combined with Theorem 5.1, yields the following corollary.

COROLLARY 6.2. *Let $G \times \tilde{\Sigma} \to \tilde{\Sigma}$ be a finite, smooth, orientation-preserving group action on a homotopy 3-sphere. Suppose that all isotropy groups for this action are cyclic and at least one has order >5. Then the action is essentially linear.*

Proof of the Corollary Assuming the Theorem. Let X be the quotient orbifold $\tilde{\Sigma}/G$. Since G is an orientation-preserving action X is orientable. The local group of X at x is isomorphic to the isotropy group $G_{\tilde{x}}$ at any $\tilde{x} \in \tilde{\Sigma}$ that projects onto x. Thus X is of cyclic type and at least one local group of X has order >5. The fundamental group of X is isomorphic to G. Thus Theorem 6.1 implies that $X \cong X_1 \# \Sigma$, where X_1 is a Seifert-fibered orbifold and Σ is a homotopy 3-sphere. According to Theorem 5.1, X_1 is diffeomorphic

X. Finite Group Actions on Homotopy 3-Spheres

to S^3/H for some $H \subset SO(4)$. Thus, $\tilde{\Sigma}/G$ is diffeomorphic to $(S^3/H) \# \Sigma$. This means that the action of G on $\tilde{\Sigma}$ is essentially linear. (See Chapter I for a discussion of essentially linear actions.) ∎

According to Proposition 3.3, if X is a closed, orientable, three-dimensional orbifold of cyclic type with finite fundamental group, then it is isomorphic to $X_1 \# \Sigma$, where X_1 is prime and Σ is a homotopy 3-sphere. Thus we can reformulate Theorem 6.1 as follows.

THEOREM 6.3. *Let X be a closed, orientable, prime, three-dimensional orbifold of cyclic type with finite fundamental group. Suppose that there is a point $x \in X$ whose local group has order >5. Then X is a Seifert-fibered orbifold.*

The above remarks show that Theorem 6.3 implies Theorem 6.1. The rest of this chapter is devoted to proving Theorem 6.3.

7. A Special Case

Let X be an orbifold as in the hypothesis of Theorem 6.3. Let (Q, K, ρ) be the underlying triple of X, In this section we shall prove Theorem 6.3 under the additional hypotheses that

(i) Q is simply connected, and
(ii) $K \subset Q$ is connected.

The hypothesis that some local group has order >5 means that $\rho(K) = n > 5$. The hypothesis that $\pi_1^{\text{orb}}(X)$ is finite means that the cyclic n-sheeted branch cover of Q branched over K has finite fundamental group. We shall show that under all these assumptions Q is Seifert-fibered (in the classical sense), with K being a union of fibers. According to 4.4, this means that the smooth orbifold X is Seifert-fibered of restricted type.

Our argument begins in exactly the same way as that of the solution to the Smith conjecture. Let $v(K)$ be the interior of a disk bundle neighborhood of K in Q.

LEMMA 7.1. *With notation and assumptions as above, one of the following is true:*

(a) $Q - v(K)$ *is Seifert-fibered.*
(b) $Q - v(K)$ *has an incompressible torus which is not peripheral (i.e., not parallel to the boundary).*
(c) $\text{int}(Q - v(K))$ *has a complete metric of finite total volume all of whose sectional curvatures are -1. (Such a structure is called hyperbolic.)*

This lemma follows from the Jaco–Shalen, Johannson theorem and Thurston's uniformization theorem exactly as in the case of the Smith conjecture (see Chapter IV).

LEMMA 7.2. $Q - v(K)$ *does not have a closed, incompressible surface that is not peripheral.*

This follows from the Meeks–Yau result exactly as in the case of the Smith conjecture (see Chapter VII).

COROLLARY 7.3. *Either case* (a) *or case* (c) *of Lemma 7.1 obtains. If case* (c) *obtains, then the representation* $\pi_1(Q - v(K)) = \Gamma \hookrightarrow \mathrm{PSL}_2(\mathbf{C})$ *coming from a hyperbolic structure on* $Q - v(K)$ *is conjugate to a representation* $\Gamma \to \mathrm{PSL}_2(A) \subset \mathrm{PSL}_2(\mathbf{C})$, *where* A *is a ring of algebraic integers in a number field.*

Proof. The first statement follows immediately from Lemmas 7.1 and 7.2. The second follows from an application of Bass's theorem on subgroups of $\mathrm{PSL}_2(\mathbf{C})$ [Chapter VI]. The reasoning is the same as that which occurs in the case of the Smith conjecture [Chapter IV]. ∎

To prove Theorem 6.3 in the special case under consideration here, we first show that $Q - v(K)$ is Seifert-fibered. To do this it suffices to assume that $Q - K$ has a complete hyperbolic structure of finite volume and deduce a contradiction.

We make this assumption. By Corollary 7.3 we can choose the holonomy representation for the hyperbolic structure, so that $\pi_1(Q - K) = \Gamma$ is represented in $\mathrm{PSL}_2(A)$.

Let $\mu \in \Gamma$ be the class of the meridian about K. Since $\pi_1(Q) = \{e\}$, $\Gamma/\{\mu\} = \{e\}$, i.e., μ is a normal generator for Γ. The group $\pi_1^{\mathrm{orb}}(X) = G$ is $\Gamma/\{\mu^n\}$.

The situation which we have can be summarized as follows:

(i) There is a group $\Gamma \subset \mathrm{PSL}_2(\mathbf{C})$ that is discrete and torsion-free and acts on hyperbolic 3-space so that the quotient has finite volume.

(ii) Γ is actually contained in $\mathrm{PSL}_2(A)$, where A is a ring of algebraic integers.

(iii) Γ is normally generated by an element μ of trace ± 2.

(iv) $\Gamma/\{\mu^n\}$ is a finite group for some $n > 5$.

We shall show that the only torsion-free, discrete group in $\mathrm{PSL}_2(\mathbf{C})$, satisfies conditions (ii)–(iv), is a cyclic group. Since no cyclic group acts with quotient having finite volume, this will yield a contradiction and will establish that $Q - K$ cannot be hyperbolic.

X. Finite Group Actions on Homotopy 3-Spheres

Choose a prime ideal $\mathfrak{p} \subset A$ so that $n \in \mathfrak{p}$. Let p be the rational prime below $\mathfrak{p}$, i.e., $\mathfrak{p} \cap \mathbf{Z} = (p)$. Consider $\Gamma \subset \mathrm{PSL}_2(A) \to \mathrm{PSL}_2(A/\mathfrak{p})$. Let $H \subset \mathrm{PSL}_2(A/\mathfrak{p})$ be the image of Γ.

LEMMA 7.4. (a) *H is normally generated (over itself) by $[\mu]$ and*
(b) *μ becomes an element of order p (or 1) in $\mathrm{PSL}_2(A/\mathfrak{p})$.*

The first fact follows immediately, since μ normally generates Γ. The second follows from the fact that any matrix in $\mathrm{PSL}_2(A/\mathfrak{p})$ of trace ± 2 is conjugate to a matrix of the form $\pm \begin{pmatrix} 1 & \lambda \\ 0 & 1 \end{pmatrix}$ for some $\lambda \in A/\mathfrak{p}$. Clearly, any such upper triangular matrix is of order p (or 1) since $A/\mathfrak{p}$ is a finite field of characteristic p. As a consequence of Lemma 7.4(b), the element μ^n is sent to 1 in $\mathrm{PSL}_2(A/\mathfrak{p})$ and hence $\Gamma \to H \subset \mathrm{PSL}_2(A/\mathfrak{p})$ factors through $\Gamma/\{\mu^n\} = G$.

Let $\tilde{X}$ be the universal cover of X, and let $\tilde{X}_H$ be the covering corresponding to the quotient group H:

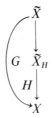

Let P be the underlying space of $\tilde{X}_H$. Also, let $\tilde{\Gamma}$ be the kernel of $\Gamma \to H$,

$$1 \to \tilde{\Gamma} \to \Gamma \to H \to 1.$$

By Dickson's theorem [see the appendix] on subgroups of PSL_2 of finite fields, the only possibilities for H are

(7.5) (α) H is cyclic (of order p of 1).
(β) H is conjugate to $\mathrm{PSL}_2(F)$, where F is a subfield of $A/\mathfrak{p}$.
(γ) $p = 2$ and H is a dihedral group of order $(2k+1) \cdot 2$.
(δ) $p = 3$ and H is isomorphic to the icosahedral group A_5.

Case α. H is cyclic. If H is cyclic, then it is generated by $[\mu]$, which has order p or 1. Thus $\tilde{X}_H \to X$ is either the trivial covering or the p-sheeted cover of Q branched along K. The group $\tilde{\Gamma}$ is isomorphic to $\pi_1(P - K)$. Clearly, under the representation $\Gamma \subset \mathrm{PSL}_2(A)$, the group $\tilde{\Gamma}$ is the subgroup of all matrices in Γ congruent to $\pm \begin{pmatrix} 1 & 0 \\ 0 & 1 \end{pmatrix}$ modulo $\mathfrak{p}$.

LEMMA 7.6. $\tilde{\Gamma}/[\tilde{\Gamma}, \tilde{\Gamma}] \otimes \mathbf{Z}/p\mathbf{Z} \cong \mathbf{Z}/p\mathbf{Z}.$

Proof. The group $\tilde{\Gamma}/[\tilde{\Gamma}, \tilde{\Gamma}]$ is isomorphic to $H_1(P - K)$. Therefore, the lemma states that $H_1(P - K; \mathbf{Z}/p\mathbf{Z}) \cong \mathbf{Z}/p\mathbf{Z}$. If H is trivial, then $P = Q$ and the result is immediate. If $H = \mathbf{Z}/p\mathbf{Z}$, then $P - K$ is a cyclic, p-fold covering of $Q - K$. We know that $H_1(Q - K) \cong \mathbf{Z}$. Let Z be the infinite cyclic cover of $Q - K$ and let $T: Z \to Z$ be the covering transformation corresponding to a generator of $\mathbf{Z}$. Then $P - K \cong Z/T^p$. We have an exact sequence

$$H_2(Q - K) \longrightarrow H_1(Z) \xrightarrow{1 - T_*} H_1(Z)$$
$$\longrightarrow H_1(Q - K) \longrightarrow \mathbf{Z} \longrightarrow 0.$$

Hence $1 - T_*: H_1(Z) \to H_1(Z)$ is an isomorphism. But $1 - T_*^p \equiv (1 - T_*)^p$ mod p. Thus $1 - T_*^p = (1 - T_*)^p: H_1(Z; \mathbf{Z}/p\mathbf{Z}) \to H_1(Z; \mathbf{Z}/p\mathbf{Z})$. Hence $1 - T_*^p$ is an isomorphism modulo p. It follows that $H_1(P - K; \mathbf{Z}/p\mathbf{Z}) \cong \mathbf{Z}/p\mathbf{Z}$. This completes the proof of Lemma 7.6. ∎

Let $H_n \subset \mathrm{PSL}_2(A/\mathfrak{p}^n)$ be the image of $\tilde{\Gamma}$ under reduction modulo $\mathfrak{p}^n$. By the previous lemma, $(H_n/[H_n, H_n]) \otimes \mathbf{Z}/p\mathbf{Z}$ is either 0 or $\mathbf{Z}/p\mathbf{Z}$. On the other hand, $H_n \subset \mathrm{PSL}_2(A/\mathfrak{p}^n)$ consists of a group of matrices congruent to $\pm \begin{pmatrix} 1 & 0 \\ 0 & 1 \end{pmatrix}$ mod $\mathfrak{p}$.

LEMMA 7.7. *The group of matrices in* $\mathrm{PSL}_2(A/\mathfrak{p}^n)$ *congruent to* $\pm \begin{pmatrix} 1 & 0 \\ 0 & 1 \end{pmatrix}$ *mod* $\mathfrak{p}$ *is a nilpotent p-group.*

Proof. Let $C_1 \subset \mathrm{PSL}_2(A/\mathfrak{p}^n)$ be the group of these matrices. Let $C_{n-1} \subset C_1$ be the subgroup of matrices congruent to $+\begin{pmatrix} 1 & 0 \\ 0 & 1 \end{pmatrix}$ mod $\mathfrak{p}^{n-1}$. We claim that C_{n-1} is contained in the center of C_1. The proof is a simple computation:

$$\begin{pmatrix} a & b \\ c & d \end{pmatrix} \begin{pmatrix} \alpha & \beta \\ \gamma & \delta \end{pmatrix} \begin{pmatrix} d & -b \\ -c & a \end{pmatrix} \equiv \begin{pmatrix} \alpha & \beta \\ \gamma & \delta \end{pmatrix} \mod \mathfrak{p}^n,$$

if $a \equiv d \equiv \pm 1(\mathfrak{p})$, $b \equiv c \equiv 0(\mathfrak{p})$, $\alpha \equiv \delta \equiv 1(\mathfrak{p}^{n-1})$, and $\beta \equiv \gamma \equiv 0(\mathfrak{p}^{n-1})$. By induction on n, one easily establishes that C_1 is a nilpotent group and, in fact, has order a power of p. ∎

It follows immediately from Lemma 7.7 that $H_n \subset \mathrm{PSL}_2(A/\mathfrak{p}^n)$ is a nilpotent p-group.

Since the abelianization of H_n tensored with $\mathbf{Z}/p\mathbf{Z}$ is cyclic, it follows that H_n itself is a cyclic group. Thus if we reduce $\tilde{\Gamma}$ modulo any power of $\mathfrak{p}$, then the result is a cyclic group. But every nontrivial element in $\tilde{\Gamma}$ is nontrivial modulo some power of $\mathfrak{p}$. Hence every nontrivial element of $\tilde{\Gamma}$ is detected in a cyclic image of $\tilde{\Gamma}$. This means that the commutator subgroup $[\tilde{\Gamma}, \tilde{\Gamma}]$ is trivial, i.e., that $\tilde{\Gamma}$ is abelian. This is a contradiction since by hypothesis $\tilde{\Gamma}$ is the fundamental group of a complete hyperbolic manifold of finite volume. This shows that Case (α) never occurs.

X. Finite Group Actions on Homotopy 3-Spheres

Case β (for $p > 5$). H is conjugate to $\mathrm{PSL}_2(F)$, F a field of characteristic $p > 5$. Let μ_H and μ_G denote the images of $\mu \in \Gamma$ in H and G, respectively. Consider the action of G on $\tilde{X}$. The element μ_G has fixed point set a circle lying above $K \subset Q$. Consequently, the action of $\mu_H \in H$ on $\tilde{X}_H$ has as fixed point set a union of circles $S_1, \ldots, S_t$, each projecting onto K. (Here $t \geq 1$.) Case β for $p > 5$ is ruled out by the following lemma applied to P, the underlying space of $\tilde{X}_H$.

LEMMA 7.8. *Let M be a closed, orientable 3-manifold with finite fundamental group. Let F be a finite field of characteristic $p > 5$. There is no action of $\mathrm{PSL}_2(F)$ on M, so that*

(i) *all isotropy groups are cyclic and*
(ii) *an element of order p, $g \in \mathrm{PSL}_2(F)$, has fixed points.*

Proof. Suppose there were an action of $\mathrm{PSL}_2(F)$ on M satisfying (i) and (ii). Since M has a finite fundamental group and $p \neq 2$, the dimension of $H_1(M; \mathbf{Z}/p)$ is ≤ 1. By Smith theory [2, p. 126], if $\mathbf{Z}/p$ acts on M with fixed point set W in M, then the dimension of $H_1(W; \mathbf{Z}/p)$ is ≤ 2. Thus the fixed point set of g, denoted by $W(g)$, is either one circle or two. The normalizer $N\langle g\rangle \subset \mathrm{PSL}_2(F)$ of $\langle g\rangle$ acts on $W(g)$. If $\alpha \in N\langle g\rangle$ fixes a component S of $W(g)$, then the group $\langle \alpha, g\rangle$ generated by α and g acts on S and hence on some tubular neighborhood of S. Since all isotropy groups are cyclic, it is possible to choose this neighborhood v so that $\langle \alpha, g\rangle$ acts freely on $v - S$. In particular, $\langle \alpha, g\rangle$ acts freely and in an orientation-preserving manner on ∂v. This implies that $\langle \alpha, g\rangle$ is an abelian group. Thus α is in the centralizer $Z\langle g\rangle$ of $\langle g\rangle$. This proves that the order of $N\langle g\rangle/Z\langle g\rangle$ is bounded above by the number of components of $W(g)$. Thus the order of $N\langle g\rangle/Z\langle g\rangle$ is ≤ 2.

Consider the element $h_\lambda = \pm \begin{pmatrix} 1 & \lambda \\ 0 & 1 \end{pmatrix}$, $\lambda \neq 0$ in F. Its normalizer contains all elements of the form $\pm \begin{pmatrix} \alpha & 0 \\ 0 & \alpha^{-1} \end{pmatrix}$, where α is a nonzero element of the prime field of characteristic p. None of these elements, save $\pm$identity, commute with h_λ. Thus $N\langle h_\lambda\rangle/Z\langle h_\lambda\rangle$ has order divisible by $(p-1)/2$. On the other hand, every element of order p in $\mathrm{PSL}_2(F)$ is conjugate to an element h_λ for some λ. Thus for any element of order p in $\mathrm{PSL}_2(F)$, g, $N\langle g\rangle/Z\langle g\rangle$ has has order divisible by $(p-1)/2$. If $p > 5$, this contradicts the fact, established above, that $N\langle g\rangle/Z\langle g\rangle$ has order at most 2. ∎

Case β (for $p = 2, 3,$ or 5), Case γ, and Case δ. We shall need the following lemma.

LEMMA 7.9. *Let X be a three-dimensional orbifold of cyclic type with finite fundamental group. It is impossible for the exceptional set to have four or more components labeled 2 or three or more components labeled by n, with $n > 2$.*

Proof. Let us consider the second case. Suppose there are three components of the exceptional set of X all labeled $n > 2$. Take the orbifold cover Y of X corresponding to the universal topological cover of the underlying space of X. In Y there are at least three components of the exceptional set labeled n—say $\tilde{K}_1$, $\tilde{K}_2$, and $\tilde{K}_3$. Let $\tilde{Q}$ be the topological space underlying Y. There is a regular branched covering of $\tilde{Q}$ branched over $\tilde{K}_1 \sqcup \tilde{K}_2$ with group of covering transformations $\mathbf{Z}/n\mathbf{Z} \times \mathbf{Z}/n\mathbf{Z}$. This corresponds to an orbifold covering of Y. The preimage of $\tilde{K}_3$ in this covering is at least n-circles. We can repeat this process ad infinitum. This contradicts the fact that $\pi_1^{\text{orb}}(X)$ is finite.

A similar argument works in the case of 4-circles labeled 2. Details are left to the reader. ∎

We have $p < 5$, F is a finite field of characteristic p, and $H \subset \text{PSL}_2(F)$ is noncyclic and normally generated by an element μ_H of order p. Consider the regular covering $\tilde{X}_H \to X$. Let (P, J, τ) be the underlying triple for $\tilde{X}_H$, and let $\pi: P \to Q$ be the ramified covering of underlying spaces. The map π is ramified over K, with index of ramification p. Since $p \leq 5$ and $\rho(K) = n > 5$, we see that J is $\pi^{-1}(K)$ and that $\tau(J) = n/p$.

LEMMA 7.10. *The number of components of J is divisible by the number of cosets of $Z\langle \mu_H \rangle$ in H.*

Proof. H acts on P leaving J invariant. Since $J/H = K$, H acts transitively on the components of J. Let J_1 be one of these components, and let $\text{stab}(J_1)$ be its stabilizer. Thus the number of components is the order of $H/\text{stab}(J_1)$. As we have seen in the proof of Lemma 7.8, $\text{stab}(J_1) \subset Z\langle \mu_H \rangle$. Therefore, the number of components of J is divisible by the order of $H/Z\langle \mu_H \rangle$. ∎

If $p = 3$ or 5 and H is isomorphic to $\text{PSL}_2(F)$, $\text{order}(F) = p^s$, then the centralizer of an element μ_H of order p has order p^s and thus J has at least $(p^{2s} - 1)/2 \geq 4$ components; $\tau(J) = n/p$.

If $p = 3$ and H is isomorphic to A_5, then the centralizer of μ_H has order 3, and J has at least 20 components; $\tau(J) = n/3$.

If $p = 2$ and H is isomorphic to $\text{PSL}_2(F)$, $\text{order}(F) = 2^s$, then J has at least $2^{2s} - 1 \geq 3$ components; $\tau(J) = n/2 \geq 3$.

If $p = 2$ and H is isomorphic to a dihedral group of order $(2k + 1) \cdot 2$, then J has at least $(2k + 1) \geq 3$ components; $\tau(J) = n/2 \geq 3$.

All these possibilities are ruled out by Lemma 7.9.

X. Finite Group Actions on Homotopy 3-Spheres

At this point we have ruled out the possibility that $Q - K$ has a complete hyperbolic structure of finite volume. Thus by Lemma 7.1, $Q - v(K)$ must be Seifert-fibered. To complete the proof of Theorem 6.3 in the special case under consideration we need only show that this Seifert fibration structure extends over $\overline{v(K)}$ with K being a fiber. The next proposition guarantees this.

PROPOSITION 7.11. *Let X be a three-dimensional orbifold with finite fundamental group. Let K_1 be a component of the exceptional set of X. Let $v(K_1) \subset X$ be a neighborhood of K_1. Suppose that the orbifold $X - v(K_1)$ is Seifert-fibered. Then X is Seifert-fibered.*

Proof. Let B be the base orbifold for the Seifert fibration on $X - v(K_1)$. There is a boundary circle C for B corresponding to $\partial v(K_1)$. Let μ_1 be the meridian in $\partial v(K_1)$. Then μ_1 projects to some multiple of C in B. Unless this multiple is 0, Lemma 4.4 shows that the Seifert fibration extends over X.

If the multiple is zero, then $\pi_1^{\text{orb}}(X) = \pi_1^{\text{orb}}(X - v(K_1))/\{\mu^{\rho(K_1)}\}$ has $\pi_1^{\text{orb}}(B)$ as a quotient. Hence $\pi_1^{\text{orb}}(B)$ is finite. This means that B is diffeomorphic to D^2/G for some $G \subset O(2)$. Since ∂B has a component that is a circle, G is actually a subgroup of $SO(2)$. As a result, the underlying topological space of $X - v(K_1)$ is a solid torus, and the exceptional set of $X - v(K_1)$ is either empty or the core of this solid torus. This means that the topological space underlying X is the union of two solid tori, and the exceptional set of X is either the union of the two cores or one of the cores. In these cases it is easy to construct a Seifert fibration of the underlying topological space, so that the exceptional set is a union of fibers. By Lemma 4.4 this implies that X is a Seifert-fibered orbifold. ∎

8. Completion of the Proof

In this section we deduce Theorem 6.3 from the special case proved in Section 7. Basically, the argument is by induction on the order of the group. It is based on a result for orbifolds (Theorem 8.1) that generalizes a theorem for 3-manifolds proved by Waldhausen [10] and Gordon and Heil [5].

Recall that if X^3 is Seifert-fibered, then the class of a generic fiber generates a normal, cyclic subgroup N of $\pi_1^{\text{orb}}(X)$. This group is nontrivial unless X is diffeomorphic to S^3/G. The following theorem shows that often the existence of such a normal subgroup is also sufficient for X to be Seifert-fibered when $\partial X \neq \emptyset$.

THEOREM 8.1. *Let X^3 be a prime, good, orientable three-dimensional orbifold with nonempty boundary. X is Seifert-fibered if and only if $\pi_1^{\text{orb}}(X)$ contains a nontrivial, normal, cyclic subgroup.*

The proof of this theorem is contained in [6]. Basically, one follows the original Waldhausen argument in [10].

A subgroup $N \subset \Gamma$ is said to be *characteristic* if any automorphism of Γ leaves N invariant.

Another result of [6] that is needed here follows.

PROPOSITION 8.2. *Let $N \subset \pi_1^{\text{orb}}(X)$ be the normal, cyclic subgroup generated by the fiber of a Seifert fibration of X. Unless $X \cong T^2 \times I$, N is characteristic.*

This proposition is not difficult. The main idea is that if N is not characteristic, then there is another normal, cyclic group in $\pi_1^{\text{orb}}(X)$. This group projects to a nontrivial, normal cyclic subgroup of the base of the Seifert fibration. This limits the base severely. An examination of the possible base spaces completes the argument.

By using these two results, we complete the proof of Theorem 6.3. The argument is by induction on the order of the fundamental group.

Let X be an orbifold that satisfies the hypothesis of Theorem 6.3. This means that X is a closed, orientable, prime orbifold of cyclic type with finite fundamental group. Let Q be the underlying space of X, and let $K \subset Q$ be the exceptional set. We know that $Q - K$ is irreducible. Let K_1 be a component of K for which $\rho(K_1) > 5$.

Case 1. Q is not simply connected. Let $\tilde{Q} \to Q$ be the universal topological cover. There is an orbifold covering $Y \to X$, so that on the level of spaces the projection is $\tilde{Q} \to Q$. The exceptional set of Y is the preimage of K in $\tilde{Q}$. We call it $\tilde{K}$. Let $\tilde{K}_1$ be the preimage of K_1. Since $\tilde{Q} - \tilde{K}$ is a covering of $Q - K$, it follows that $\tilde{Q} - \tilde{K}$ is irreducible. This means that Y satisfies the hypothesis of Theorem 6.3. By induction, Y is Seifert-fibered. By Lemma 4.6, $\tilde{K}_1$ must be a union of fibers. Let $v(K_1)$ be a disk bundle neighborhood of K_1 in Q and let $v(\tilde{K}_1)$ be its preimage in $\tilde{Q}$. We can easily arrange for $Y - v(\tilde{K}_1)$ to be Seifert-fibered.

Case 1a. $Y - v(\tilde{K}_1)$ is not isomorphic (as an orbifold) to $T^2 \times I$. In this case the normal, cyclic group $N \subset \pi_1^{\text{orb}}(Y - v(\tilde{K}_1))$ generated by the class of a generic fiber is nontrivial (by Lemma 4.3) and characteristic (by Proposition 8.2). Hence it forms a nontrivial, normal, cyclic subgroup of $\pi_1^{\text{orb}}(X - v(K_1))$. Since X is prime, $X - v(K_1)$ is also prime. Thus, according to Theorem 8.1, $X - v(K_1)$ is Seifert-fibered. By Proposition 7.11, we can choose this Seifert fibration so that it extends to one on all of X.

Case 1b. $Y - v(\tilde{K}_1)$ is isomorphic to $T^2 \times I$. In this case $\tilde{K} = \tilde{K}_1$, and hence $K = K_1$ in Q. Thus $Q - v(K_1)$ is finitely covered by $T^2 \times I$. The only

X. Finite Group Actions on Homotopy 3-Spheres

orientable manifold with one boundary component that is covered by $T^2 \times I$ is the twisted I-bundle over the Klein bottle. This manifold has two S^1-fibrations. At least one of them extends to a Seifert fibration of Q with K being a fiber. According to Proposition 4.4, this implies that X is a Seifert-fibered orbifold.

Case 2. Q is simply connected and K has at least two components. Let K_2 be a component of K distinct from K_1. Let p be a prime dividing $\rho(K_2)$. Since Q is simply connected, there is a p-sheeted, branched, cyclic covering of Q branched over K_2. Let $P \to Q$ be this branched, cyclic covering. Since $p | \rho(K_2)$, there is a corresponding covering of orbifolds $Y \to X$. The exceptional set is the preimage of $K - K_2$ in P if $p = \rho(K_2)$. Otherwise, it is the preimage of all of K in P. Let $\tilde{K}$ be the exceptional set, and let $\tilde{K}_1$ be the preimage of K_1 in P.

The orbifold Y satisfies all the hypotheses of Theorem 6.3, except possibly the condition that Y is prime. By Proposition 3.3 and induction, Y is isomorphic to $Y_1 \# \Sigma$ where Y_1 is Seifert-fibered and Σ is a homotopy 3-sphere. By Lemma 4.6, $\tilde{K}_1$ must be a union of fibers. Let $v(K_1)$ be a disk bundle neighborhood of $K_1 \subset Q$, and let $v(\tilde{K}_1)$ be its preimage in P. The orbifold $Y_1 - v(\tilde{K}_1)$ is Seifert fibered.

Case 2a. $Y_1 - v(\tilde{K}_1)$ is not isomorphic to $T^2 \times I$. In this case, as in Case 1a, $\pi_1^{\text{orb}}(Y - v(\tilde{K}_1)) = \pi_1^{\text{orb}}(Y_1 - v(\tilde{K}_1))$ has a nontrivial, characteristic, cyclic subgroup N. This subgroup N is a normal subgroup in $\pi_1^{\text{orb}}(X - v(K_1))$. Since X is prime, so is $X - v(K_1)$. Hence Theorem 8.1 says that $X - v(K_1)$ is Seifert-fibered. By Proposition 7.11, this means that X itself is Seifert-fibered.

Case 2b. $Y_1 - v(\tilde{K}_1)$ is isomorphic to $T^2 \times I$. Since $Y - v(\tilde{K}_1)$ has two boundary components and $X - v(K_1)$ has one, the group of covering transformations of Y over X has even order. By construction this group is a cyclic group of prime order. Hence the group is isomorphic to $\mathbf{Z}/2\mathbf{Z}$. It acts by interchanging the boundary components. Thus we have an extension

$$1 \to \mathbf{Z} \times \mathbf{Z} \to \pi_1^{\text{orb}}(X - v(K_1)) \to \mathbf{Z}/2\mathbf{Z} \to 1.$$

It follows that $\pi_1^{\text{orb}}(X - v(K_1))$ has a nontrivial, cyclic, normal subgroup. Since X is prime, so is $X - v(K_1)$. Thus Theorem 8.1 implies that $X - v(K_1)$ is Seifert-fibered. As before, Proposition 7.11 allows us to find a Seifert fibration on all of X.

Case 3. Q is simply connected and K has only one component. This is exactly the special case dealt with in Section 7.

This completes the proof of Theorem 6.3 and hence of Theorem A.

Appendix

Let F be a field with p^n elements. Let $\mu \in \mathrm{PSL}_2(F)$ be the element $\pm\begin{pmatrix}1 & 1\\0 & 1\end{pmatrix}$. We shall give Dickson's argument [3] classifying groups $G \subset \mathrm{PSL}_2(F)$ that contain μ. Two such groups are considered equivalent if there are a finite extension $\tilde{F}$ of F and an element $\alpha \in \mathrm{PSL}_2(\tilde{F})$ that normalizes $\mathrm{PSL}_2(F)$, commutes with μ, and conjugates one to the other. This leads to a classification of groups $G \subset \mathrm{PSL}_2(F)$ that contain a nontrivial element of trace ± 2 (or, equivalently, that contain an element of order p). Once again the classification is up to conjugation by elements in $\mathrm{PSL}_2(\tilde{F})$, $\tilde{F}$ a finite extension of F, that normalize $\mathrm{PSL}_2(F)$. The reason is that any nontrivial element of trace ± 2 in $\mathrm{PSL}_2(F)$ is conjugate to $\pm\begin{pmatrix}1 & \lambda\\0 & 1\end{pmatrix}$, $\lambda \neq 0$. We let $\tilde{F}$ be the extension of F obtained by adjoining $x = \sqrt{\lambda}$ to F. In $\mathrm{PSL}_2(\tilde{F})$ the matrix $\pm\begin{pmatrix}x & 0\\0 & x^{-1}\end{pmatrix}$ normalizes $\mathrm{PSL}_2(F)$ and conjugates $\pm\begin{pmatrix}1 & \lambda\\0 & 1\end{pmatrix}$ to $\pm\begin{pmatrix}1 & 1\\0 & 1\end{pmatrix}$.

Let us begin then with $G \subset \mathrm{PSL}_2(F)$ containing μ. There are three classes of such groups, as we shall see

Class I. Subgroups of upper triangular matrices in $\mathrm{PSL}_2(F)$;

Class II. groups conjugate to $\mathrm{PSL}_2(F')$ or a $\mathbf{Z}/2\mathbf{Z}$-extension of $\mathrm{PSL}_2(F')$ for some subfield $F' \subset F$; and

Class III. exceptional groups for F a field of order a power of 2 or 3.

We shall discuss in more detail later the various groups in these classes.

Let us set up some notation. *A maximal unipotent subgroup*, or MU subgroup for short, is any subgroup conjugate in $\mathrm{PSL}_2(F)$ to the group of strictly upper triangular matrices

$$B_\infty = \left\{ \begin{pmatrix}1 & x\\0 & 1\end{pmatrix} \text{ for } x \in F \right\},$$

Let $G \subset \mathrm{PSL}_2(F)$, and let $B \subset \mathrm{PSL}_2(F)$ be a MU subgroup. If $G \cap B$ is nontrivial, then we say that it is a MU *subgroup of* G. Any MU subgroup of G has order p^l for some $1 \leq l \leq n$ (where l might depend on the MU subgroup).

LEMMA A1. *Any MU subgroup of G is a p-Sylow subgroup of G.*

Proof. A subgroup is a p-Sylow subgroup if it is a maximal p-group (i.e., if it is not contained in any larger p-group). Any MU subgroup of $\mathrm{PSL}_2(F)$ is a p-Sylow subgroup, and any two distinct MUs in $\mathrm{PSL}_2(F)$ have trivial intersection. From this it follows easily that any maximal p-group in G must be of the form $G \cap B$, and, conversely, that if $G \cap B$ is nontrivial,

X. Finite Group Actions on Homotopy 3-Spheres

then it is a maximal p-group of G. Thus the p-Sylow subgroups of G are the groups $G \cap B$ that are nontrivial. ∎

COROLLARY A2. *Any two MU subgroups of G are conjugate.*

Case I. *G has only one MU subgroup. Then G is a group of upper triangular matrices.* If G has only one MU subgroup, then that subgroup must be B_∞ (since $\mu \in G$) and B_∞ must be normal in G. This implies that every element of G normalizes B_∞, i.e., that G is a group of upper triangular matrices. This completes Case I.

For the rest of this appendix we assume that G has more than one MU subgroup.

LEMMA A3. *The number of MU subgroups of G is $fp^r + 1$, where p^r is the order of $G \cap B_\infty$ and $f \geq 1$.*

Proof. B_∞, acting by conjugation on the MU subgroups of $\mathrm{PSL}_2(F)$, fixes B_∞ and acts freely and transitively on the others. Thus $G \cap B_\infty$, acting by conjugation on the MU subgroups of G, has one fixed point — $G \cap B_\infty$ — and acts freely on the remaining ones. Thus the number of MU subgroups of G is $fp^r + 1$, where p^r is the order of $G \cap B_\infty$. ∎

We adhere to the following notation:

$\operatorname{order}(G \cap B_\infty) = p^r$.

number of MU subgroups of $G = fp^r + 1$, $f \geq 1$.

$N_\infty = $ normalizer in G of $G \cap B_\infty$.

$\operatorname{order}(N_\infty) = dp^r$.

order $G = dp^r(fp^r + 1)$.

Of course, N_∞ consists of all upper triangular matrices in G. The MU subgroup $G \cap B_\infty$ is naturally identified with a subgroup of the additive group of F. This identification sends $\pm \begin{pmatrix} 1 & x \\ 0 & 1 \end{pmatrix}$ in $G \cap B_\infty$ to $x \in F$. Call the image of $G \cap B_\infty$ under this identification V. It is automatically a vector space over the prime field $F_p \subset F$. Let $F' \subset F$ be the subset of all $a \in F$ such that $a \cdot V \subset V$. One sees easily that $F' \subset F$ is a subfield. Clearly, V is a vector space over F'. Let the order of F' be p^l, $l \mid n$. Since the order of V is p^r, we see that $l \leq r$. Since $1 \in V$, $F' \subset V$.

Consider the sequence

$$1 \to G \cap B_\infty \to N_\infty \xrightarrow{\psi} F^*/\{\pm 1\}, \qquad \text{where } \psi(\pm \begin{pmatrix} \alpha & x \\ 0 & \alpha^{-1} \end{pmatrix}) = \pm \alpha.$$

The image of ψ is a subgroup of $F^*/\{\pm 1\}$ of order d. Such a group is cyclic and generated by $\pm\eta_0$, where η_0 is a primitive dth root of -1. The action of $\pm\alpha \in F^*/\{\pm 1\}$ on $G \cap B_\infty = V$ is by multiplication by α^2

$$\begin{pmatrix} \alpha & Y \\ 0 & \alpha^{-1} \end{pmatrix} \begin{pmatrix} 1 & x \\ 0 & 1 \end{pmatrix} \begin{pmatrix} \alpha^{-1} & -Y \\ 0 & \alpha \end{pmatrix} = \begin{pmatrix} 1 & \alpha^2 x \\ 0 & 1 \end{pmatrix}.$$

Thus $\eta_0^2 \in (F')^*$. This shows that $d | (p^l - 1)$.

We wish to arrange that N_∞ meets the diagonal matrices, $\{\pm(\begin{smallmatrix}\alpha & 0 \\ 0 & \alpha^{-1}\end{smallmatrix})\}$, in exactly the cyclic group generated by

$$\pm \begin{pmatrix} \eta_0 & 0 \\ 0 & \eta_0^{-1} \end{pmatrix}.$$

LEMMA A4. *There is an element of B_∞ that conjugates G to $G' \subset \mathrm{PSL}_2(F)$, so that $G' \cap \{\pm(\begin{smallmatrix}\alpha & 0 \\ 0 & \alpha^{-1}\end{smallmatrix})\}$ is the cyclic group generated by*

$$\pm \begin{pmatrix} \eta_0 & 0 \\ 0 & \eta_0^{-1} \end{pmatrix}.$$

(Notice that such conjugation leaves fixed $G \cap B_\infty$.)

Proof. If $\eta_0 = \pm 1$, then no conjugation is necessary since $N_\infty = G \cap B_\infty$. Suppose that $\eta_0 \neq \pm 1$. Let $\pm(\begin{smallmatrix}\eta_0 & x \\ 0 & \eta_0\end{smallmatrix})$ be an element of N_∞. Since $\eta_0 \neq \pm 1$, we see that $\eta_0 \neq \eta_0^{-1}$. Conjugate G by

$$\pm \begin{pmatrix} 1 & x/(\eta_0 - \eta_0^{-1}) \\ 0 & 1 \end{pmatrix}.$$

The conjugate group contains

$$\pm \begin{pmatrix} 1 & \frac{x}{\eta_0 - \eta_0^{-1}} \\ 0 & 1 \end{pmatrix} \begin{pmatrix} \eta_0 & x \\ 0 & \eta_0^{-1} \end{pmatrix} \begin{pmatrix} 1 & \frac{x}{\eta_0 - \eta_0^{-1}} \\ 0 & 1 \end{pmatrix} = \pm \begin{pmatrix} \eta_0 & 0 \\ 0 & \eta_0^{-1} \end{pmatrix}. \quad \blacksquare$$

We assume that we have made the required conjugation and have renamed the new group G. At this point we have

(A5) (a) a subfield $F' \subset F$ with p^l elements,
     (b) an F'-vector subspace $V \subset F$ of order p^r containing F', and
     (c) an integer d such that $d | (p - 1)$, and such that
     (d) $N_\infty \subset G$ consists of all products

$$\left\{ \pm \begin{pmatrix} 1 & v \\ 0 & 1 \end{pmatrix} \begin{pmatrix} \eta_0^t & 0 \\ 0 & \eta_0^{-t} \end{pmatrix} \quad \text{for } v \in V \text{ and } \eta_0 \text{ a primitive } d\text{th-root of } -1 \right\}.$$

X. Finite Group Actions on Homotopy 3-Spheres

LEMMA A6. *There are* $(fp^r + 1)$ *right cosets of* N_∞ *in* G. *Each nonidentity coset contains at most* $2d$ *elements of order* p (*at most* d *if* $p = 2$).

Proof. Since the order of N_∞ is $p^r d$ and the order of G is $(1 + fp^r) dp^r$, the first statement follows immediately.

Let $V_j = \begin{pmatrix} \alpha_j & \beta_j \\ \gamma_j & \delta_j \end{pmatrix}; j = 1, \ldots, fp^r$, be coset representatives for the nonidentity cosets (thus $\gamma_j \neq 0$). The condition that a product

$$\pm \begin{pmatrix} \alpha_j & \beta_j \\ \gamma_j & \delta_j \end{pmatrix} \begin{pmatrix} \eta & \eta v \\ 0 & \eta^{-1} \end{pmatrix},$$

with $v \in V$ and $\eta^2 \in (F')^*$, be an element of order p is that its trace $\eta \alpha_j + \gamma_j \eta v + \eta^{-1} \delta$ be ± 2. By fixing j and η, there are at most two solutions (one if $p = 2$) to

$$\eta \alpha_j + \gamma_j \eta v + \eta^{-1} \delta = \pm 2. \blacksquare$$

On the other hand, G has $(1 + fp^r)$ MU subgroups. These groups have trivial intersection and each contains $(p^r - 1)$ nontrivial elements of trace ± 2. Thus $G - N_\infty$ has $fp^r(p^r - 1)$ elements of trace ± 2. In light of Lemma A6, this implies that

$$fp^r(p^r - 1) \leq 2dfp^r \quad \text{if} \quad p \text{ is odd}$$

and

$$fp^r(p^r - 1) \leq dfp^r \quad \text{if} \quad p = 2.$$

Since $d | (p^l - 1)$ and $l \leq r$, this yields

(A7) $l = r$ and $d = (p^r - 1)/2$ or $d = p^r - 1$, if p is odd.
        $l = r$ and $d = p^r - 1$, if $p = 2$.
        $F' = V$.

(A8)   If $d = (p^r - 1)/2$, then $N_\infty = B_\infty \cap \text{PSL}_2(F')$. This is also true for $p = 2$. If $d = p^r - 1$ and $p > 2$, then N_∞ equals $\{\pm \begin{pmatrix} \eta & \eta \lambda \\ 0 & \eta^{-1} \end{pmatrix} | \eta^2 \in (F')^* \text{ and } \lambda \in F'\}$. Furthermore, if $d = p^r - 1$ and $p > 2$, then the degree $[F : F']$ is even.

Case IIA. If p is odd and $d = (p^r - 1)/2$, then $G = \text{PSL}_2(F')$, unless $p^r = 3$, in which case G is conjugate to $\text{PSL}_2(F')$. As we have seen (A6) each nonidentity coset of N_∞ contains at most $2d = (p^r - 1)$ elements of order p. This gives a maximum of $fp^r(p^r - 1)$. On the other hand, this is exactly the number of elements of order p outside N_∞. We conclude that for each j and each $\eta \in (F')^*$ there are two solutions to

$$\alpha_j \eta + \gamma_j \eta v + \delta_j \eta^{-1} = \pm 2.$$

Since each coset of N_∞ contains an element of order p, and hence of trace 2, we can assume that we have chosen $v_j = \begin{pmatrix} \alpha_j & \beta_j \\ \gamma_j & \delta_j \end{pmatrix}$ so that $\alpha_j + \delta_j = 2$. By substituting, we see that

$$\alpha_j(\eta - \eta^{-1}) + \gamma_j \eta v = \pm 2 - 2\eta^{-1}$$

has two solutions for each $\eta \in (F')^*$. (Here, α_j and γ_j are fixed in F, and "solution" means $v \in F'$ satisfying the equation.) If we take $\eta = \pm 1$, then above equation reduces to $\gamma_j v = \pm 4$. For there to be two $v \in F'$ solving this equation, it is necessary that $\gamma_j \in (F')^*$. If the order of F' is not 3, then choose $\eta \neq \pm 1, \eta \in (F')^*$. There are solutions $v \in F'$ for $\alpha_j(\eta - \eta^{-1}) + \gamma_j \eta v = \pm 2 - 2\eta^{-1}$. Since $\eta - \eta^{-1} \in (F')^*$ and $\gamma_j \in (F')^*$, it follows that $\alpha_j \in F'$. Consequently, $\delta_j = 2 - \alpha_j$ is in F'. Finally, $\beta_j = (\alpha_j \delta_j - 1)/\gamma_j$ is also in F'. This proves that except in the case for which F' has order 3, $G \subset \mathrm{PSL}_2(F')$.

If F' has order 3, then the argument above shows that each nonidentity coset of N_∞ contains an element of the form

$$\pm \begin{pmatrix} \alpha_j & \beta_j \\ \pm 1 & 2 - \alpha_j \end{pmatrix}.$$

Multiplying by $\pm \begin{pmatrix} 1 & \pm 1 \\ 0 & 1 \end{pmatrix} \in N_\infty$, we change this coset representative to

$$\pm \begin{pmatrix} \langle \alpha_j \rangle & \mp(1 + \alpha_j^2) \\ \pm 1 & -\alpha_j \end{pmatrix}.$$

Conjugate G by $\pm \begin{pmatrix} 1 & \pm \alpha \\ 0 & 1 \end{pmatrix}$. As the reader can easily check, the resulting group (which we continue to call G) has the following properties.

(a) The lower left-hand entry of every element of G is contained in F' and
(b) $\pm \begin{pmatrix} 0 & -1 \\ 1 & 0 \end{pmatrix} \in G$.

Any group with these two properties is easily seen to be contained in $\mathrm{PSL}_2(F')$.

We have shown that in all cases under Case IIA, G is contained in $\mathrm{PSL}_2(F')$ (after conjugation). The order of G is $(1 + fp^r)p^r(p^r - 1)/2$, and the order of $\mathrm{PSL}_2(F')$ is $(p^{2r} - 1)p^r/2$. Thus $f = 1$ and $G = \mathrm{PSL}_2(F')$. This completes Case IIA.

Case IIB. If $p > 3$ and $d = (p^r - 1)$, then n/r is even and G consists of all products $\pm \begin{pmatrix} \alpha & \beta \\ \gamma & \delta \end{pmatrix} \begin{pmatrix} x & 0 \\ 0 & x^{-1} \end{pmatrix}$ where $\pm \begin{pmatrix} \alpha & \beta \\ \gamma & \delta \end{pmatrix}$ is in $\mathrm{PSL}_2(F')$ and $x \in F^*$ with $x^2 \in F'$. Thus G is a $\mathbf{Z}/2\mathbf{Z}$-extension of $\mathrm{PSL}_2(F')$. We saw in (A8) that if $d = (p^r - 1)$ with p odd, then n/r is even. We also saw that N_∞ is the group of all products

$$\left\{ \pm \begin{pmatrix} \eta & 0 \\ 0 & \eta^{-1} \end{pmatrix} \begin{pmatrix} 1 & \lambda \\ 0 & 1 \end{pmatrix} \middle| \lambda \in F' \text{ and } \eta^2 \in (F')^* \right\}.$$

X. Finite Group Actions on Homotopy 3-Spheres

Since n/r is even, every element of $(F')^*$ has a square root in F. In particular, $\eta = \sqrt{-1}$ is in F. Thus

$$t = \pm \begin{pmatrix} \sqrt{-1} & 0 \\ 0 & -\sqrt{-1} \end{pmatrix}$$

is an element of $N_\infty \subset G$. Its normalizer in G is all diagonal matrices $\{\pm \begin{pmatrix} \eta & 0 \\ 0 & \eta^{-1} \end{pmatrix} | \eta^2 \in (F')^*\}$ unless G contains an element of the form $+\begin{pmatrix} 0 & -\tau^{-1} \\ \tau & 0 \end{pmatrix}$. In the latter case, the normalizer of t is the $\mathbf{Z}/2\mathbf{Z}$-extension of the group of diagonal matrices generated by any such element. Depending on which of the two cases obtains, there are either $(1 + fp)p^r$ or $(1 + fp^r)p^r/2$ elements in G conjugate to t. (In the second case, note that f must be odd.) Of these conjugates, fp^{2r} or $(fp^r - 1)p^r/2$ are outside N_∞. All conjugates of t have trace 0. If an element in the coset $V_j N_\infty$ has trace 0, then it is a product

$$\begin{pmatrix} \alpha_j & \beta_j \\ \gamma_j & \delta_j \end{pmatrix} \begin{pmatrix} \eta & \eta\lambda \\ 0 & \eta^{-1} \end{pmatrix},$$

where $\alpha_j \eta + \gamma_j \eta \lambda + \delta_j \eta^{-1} = 0$. For j fixed and η fixed, there is at most one solution $\lambda \in F'$. Thus $G - N_\infty$ contains at most $(fp^r)(p^r - 1)$ elements of trace 0. This means t has $(fp^r - 1)p^r/2$ conjugates in G, and hence in G there is an element of the form $\pm \begin{pmatrix} 0 & -\tau^{-1} \\ \tau & 0 \end{pmatrix}$.

We now count the elements of trace 0 (i.e., the elements of order 2) in $G - N_\infty$. For each coset $V_j N_\infty$ that contains such an element we choose the coset representative V_j to be of trace 0:

$$V_j = \begin{pmatrix} \alpha_j & \beta_j \\ \gamma_j & -\alpha_j \end{pmatrix}, \qquad \gamma_j \neq 0.$$

Some of these cosets can contain more than one element of trace 0. For $V_j \cdot N_\infty$ to contain two such elements, there must be $\eta \in F^*$ with $\eta^2 \in F'$ and $\lambda \in F'$ with either $\lambda \neq 0$ or $\eta \neq \pm 1$ so that

(A9) $\qquad \alpha_j(\eta - \eta^{-1}) + \gamma_j \eta \lambda = 0.$

If there are such η and λ, then $\alpha_j/\gamma_j = \eta^2 \lambda(1 - \eta^2)$. Thus $\alpha_j/\gamma_j \in F'$. Conversely, if $\alpha_j/\gamma_j \in F'$ and $\eta \in F^*$ is any element with $\eta^2 \in F'$, then there is a unique solution to A9 with $\lambda \in F'$. This proves

(A10) *Each nonidentity right coset of N_∞ that contains two distinct elements of trace 0 contains exactly $(p^r - 1)$ such elements.*

Let A be the number of nonidentity cosets of N_∞ that contain $(p^r - 1)$ elements of tr 0, and let B be the number that contain exactly 1. Then

(A11) $\quad (fp^r - 1)p^r/2 \leq A(p^r - 1) + B \leq A(p^r - 1) + fp^r - A.$

LEMMA A12. *Let* $V = \begin{pmatrix} \alpha & \beta \\ \gamma & -\alpha \end{pmatrix}$ *and* $V' = \begin{pmatrix} \alpha' & \beta' \\ \gamma' & -\alpha' \end{pmatrix}$ *be elements of trace 0 in G. Suppose that* α/γ *and* α'/γ' *both belong to* F'. *The elements V and V' are in the same coset of* N_∞ *if and only if* $\alpha/\gamma = \alpha'/\gamma'$.

Proof. The only if part of the lemma is clear. Conversely, if V and V' belong to different cosets of N_∞, then $VV' = V(V')^{-1}$ is not an element of N_∞. Thus

$$\begin{pmatrix} \alpha & \beta \\ \gamma & -\alpha \end{pmatrix} \begin{pmatrix} \alpha' & \beta' \\ \gamma' & -\alpha' \end{pmatrix} = \begin{pmatrix} \alpha\alpha' + \beta\gamma' & \beta'\alpha - \beta\alpha' \\ \alpha'\gamma - \alpha\gamma' & \gamma\beta' + \alpha\alpha' \end{pmatrix}$$

is not upper triangular. Hence $\alpha'\gamma \neq \alpha\gamma'$. This implies that $\alpha'/\gamma' \neq \alpha/\gamma$. ∎

Applying Lemma A12 we see that $A \leq p^r$, and hence

$$A(p^r - 1) + fp^r - A \leq p^{2r} - p^r + fp^r - p^r = p^r(p^r + f - 2).$$

In light of inequality A11 we have

$$(fp^r - 1)p^r/2 \leq p^r(p^r + f - 2)$$

or

$$(fp^r - 1)/2 \leq (p^r + f - 2)$$

Since f is odd, there are only two possibilities: either $f = 1$, or $f = 3$ and $p^r = 3$. In the hypothesis of Case IIB we assume $p > 3$. Thus we conclude that in this case $f = 1$.

Recall that there is an element of the form $+\begin{pmatrix} 0 & -\tau^{-1} \\ \tau & 0 \end{pmatrix}$ in G. Let us consider its p^r conjugates by all elements of $G \cap B_\infty$:

$$\left\{ \pm \begin{pmatrix} x\tau & -x^2\tau - \tau^{-1} \\ \tau & -x\tau \end{pmatrix} \middle| x \in F' \right\}.$$

These are all elements of trace 0. The ratios α_j/γ_j run through the p^r elements of F'. By Lemma A12, this implies that these p^r elements are in p^r distinct right cosets of N_∞. Since $f = 1$, there are exactly p^r such cosets. Thus we can use these elements as the coset representatives for all the nonidentity cosets of N_∞. It follows that G is generated by N_∞ and $\pm\begin{pmatrix} 0 & -\tau^{-1} \\ \tau & 0 \end{pmatrix}$. Also notice that each nonidentity coset of N_∞ in G has a representative of the form $\pm\begin{pmatrix} \alpha_j & \beta_j \\ \gamma_j & \delta_j \end{pmatrix}$, where α_j/γ_j is in F'. Consequently, all elements of $G - N_\infty$ are of the form $\pm\begin{pmatrix} \alpha & \beta \\ \gamma & \delta \end{pmatrix}$, where α/γ is in F'. In particular,

$$\pm\begin{pmatrix} 0 & -\tau^{-1} \\ \tau & 0 \end{pmatrix}\begin{pmatrix} \tau & -\tau - \tau^{-1} \\ \tau & -\tau \end{pmatrix} = \begin{pmatrix} -1 & 1 \\ \tau^2 & -1 - \tau^2 \end{pmatrix}$$

has this property. This proves that $\tau^2 \in F'$. Clearly, then, both N_∞ and $\pm\begin{pmatrix} 0 & -\tau^{-1} \\ \tau & 0 \end{pmatrix}$ are contained in the $\mathbf{Z}/2\mathbf{Z}$-extension of $PSL_2(F')$ described in Case IIB. This implies that G is contained in this extension. The order of G

X. Finite Group Actions on Homotopy 3-Spheres

is twice that of $PSL_2(F')$. This implies that G is equal to this extension. This completes Case IIB.

Case IIIA. If $p = 2$ and $p^r > 2$, then $G = PSL_2(F')$. We know by A8 that N_∞ consists of all upper triangular matrices in $PSL_2(F')$, order $F' = 2^r$ for $r > 1$. We also know, by A6, that each coset of N_∞ contains at most $(p^r - 1)$ elements of order 2. Since there are a total of $fp^r(p^r - 1)$ such elements in $G - N_\infty$, each coset contains exactly $(p^r - 1)$ elements of order 2. For each nonidentity coset of N_∞ choose a representative of order 2:

$$V_j = \begin{pmatrix} \alpha_j & \beta_j \\ \gamma_j & \alpha_j \end{pmatrix}, \quad \gamma_j \neq 0, \quad j = 1, \ldots, fp^r.$$

If $V_j \cdot N_\infty$ contains $p^r - 1$ elements of order 2 (i.e., of trace 0), then for each $\eta \in (F')^*$ there is an element $x \in F'$, so that

$$\alpha_j \eta + \gamma_j \eta x + \alpha_j \eta^{-1} = 0.$$

Since F' has more than two elements, there is $\eta \in (F')^*$ with $\eta \neq \eta^{-1}$. For any such η, we have

$$\alpha_j / \gamma_j = \eta x / (\eta + \eta^{-1}).$$

This proves that $\alpha_j / \gamma_j \in F'$ for $j = 1, \ldots, fp^r$. Thus for every $\begin{pmatrix} \alpha & \beta \\ \gamma & \delta \end{pmatrix} \in G - N_\infty$ the quotient α/γ is in F'. If $i \neq j$, then $V_i V_j = V_i V_j^{-1}$ is in $G - N_\infty$. Hence

$$(\alpha_i \alpha_j + \beta_j \gamma_i)/(\gamma_j \alpha_u + \alpha_j \gamma_i)$$
$$= (\alpha_i / \gamma_i)(\alpha_j / \gamma_j) \cdot [(\alpha_i / \gamma_i) + (\alpha_j / \gamma_j)]^{-1} + [(\alpha_i / \gamma_i) + (\alpha_j / \gamma_j)]^{-1} \beta_j / \gamma_j$$

is in F'. It follows easily that β_j / γ_j is in F'. Since $(\alpha_j)^2 - \beta_j \gamma_j = 1$ and α_j / γ_j and β_j / γ_j belong to F', it results that $(\gamma_j)^2 \in F'$. Since F' is of characteristic 2, γ_j also belongs to F'. Thus α_j, β_j, and γ_j belong to F'. This proves that $G \subset PSL_2(F')$. On the other hand, the order of G is $(1 + fp^r)(p^r)(p^r - 1)$ and the order of $PSL_2(F)$ is $(1 + p^r)p^r(p^r - 1)$. Hence $G = PSL_2(F')$.

Case IIIB. If $p^r = 2$, then G is a dihedral group of order $2(1 + 2f)$. (If $f = 1$, then $G = PSL_2(F_2)$.) We know that $d = 1$, and hence that the order of G is $2(1 + 2f)$. The subgroup $G \cap B_\infty$ is cyclic of order 2 and is its own normalizer N_∞. Thus G contains $(1 + 2f)$ conjugates of $\begin{pmatrix} 1 & 1 \\ 0 & 1 \end{pmatrix}$:

$$\begin{pmatrix} 1 & 1 \\ 0 & 1 \end{pmatrix}, \quad V_1, \ldots, V_{2f}.$$

Let

$$V_j = \begin{pmatrix} \alpha_j & \beta_j \\ \gamma_j & \delta_j \end{pmatrix}, \text{ where } \gamma_j \neq 0.$$

Since V_j is of order 2, $\delta_j = \alpha_j$. We claim that each nonidentity coset of N_∞ contains exactly one element of order 2. If this is true, then the V_j are representatives for the nonidentity cosets of N_∞. The reason that each nonidentity coset of N_∞ contains at most one element of trace 0 is that

$$\begin{pmatrix} \alpha_j & \beta_j \\ \gamma_j & \alpha_j \end{pmatrix} \begin{pmatrix} 1 & 1 \\ 0 & 1 \end{pmatrix} = \begin{pmatrix} \alpha_j & \alpha_j + \beta_j \\ \gamma_j & \gamma_j + \alpha_j \end{pmatrix},$$

which does not have trace 0 if $\gamma_j \neq 0$. This proves that $U_j = V_j(\begin{smallmatrix}1 & 1\\0 & 1\end{smallmatrix})$ is not of order two.

Consider now a product $V_i \cdot V_j$, $i \neq j$. We claim that this product is not of order 2. If it is, say $V_i V_j = V_l$, then since all the V_i are conjugate to $V_0 = (\begin{smallmatrix}1 & 1\\0 & 1\end{smallmatrix})$, we have

$$(g_i V_0 g_i^{-1}) \cdot (g_j V_0 g_j^{-1}) = g_l V_0 g_l^{-1}$$

for some g_i, g_j, and g_l in G. Conjugating by g_l^{-1} gives

$$(g_l^{-1} g_i) V_0 (g_i^{-1} g_l)(g_l^{-1} g_j) V_0 (g_j^{-1} g_l) = V_0$$

or $V_i' \cdot V_j' = V_0$. As we have already seen, this is impossible. Thus we have a homomorphism from G onto $\mathbf{Z}/2\mathbf{Z}$ that sends each V_j nontrivially and each U_j to the identity. In particular, the $\{U_0, \ldots, U_{2f}\}$ form a normal subgroup of G. The action of V_0 on this subgroup sends U_i to U_i^{-1}. For the function $U_i \to U_i^{-1}$ to be a homomorphism of the group of the U_i, that group must be abelian. Thus the group of the U_i is an abelian subgroup of odd order in $PSL_2(F)$. Since F is of characteristic 2, the only such groups are cyclic. Thus G itself is dihedral.

Case IIIC. If $p = 3$ and $d = p^r - 1$, then G is either the $\mathbf{Z}/2\mathbf{Z}$-extension of $PSL_2(F')$ or an icosahedral group. As we saw in Case IIB, either $f = 1$, or $f = 3$ and $p^r = 3$. If $f = 1$, then the argument given in Case IIB is valid to show that G is the given $\mathbf{Z}/2\mathbf{Z}$-extension of $PSL_2(F)$. The remaining case is $p^r = 3$ and $f = 3$. The order of G is $(1 + 3 \cdot 3)(3)(2) = 60$. A straightforward argument shows that G is isomorphic to A_5. All such groups turn out to be conjugate by an upper triangular matrix of the form $\pm(\begin{smallmatrix}1 & \alpha\\0 & 1\end{smallmatrix})$ for $\alpha^2 \in F$. This completes all possible cases and finishes the classification.

Notice that if $G \subset PSL_2(F)$ is normally generated by $\mu = \pm(\begin{smallmatrix}1 & 1\\0 & 1\end{smallmatrix})$, then G is of one of the following types:

(I′) G is cyclic.
(II′) G is conjugate to $PSL_2(F')$ for F' a subfield of F.
(III′) $p = 3$ and G is isomorphic to A_5 or $p = 2$ and G is isomorphic to a dihedral group of order $(1 + 2f) \cdot 2$.

X. Finite Group Actions on Homotopy 3-Spheres

References

[1] Birman, J., Braids, links, and mapping class groups, *Ann. of Math. Studies*, Vol. 82, Princeton Univ. Press, Princeton, New Jersey, 1974.
[2] Bredon, G., "Introduction to Compact Transformation Groups." Academic Press, New York and London, 1972.
[3] Dickson, L., "Linear Groups." Dover, New York, 1958.
[4] Du Val, P., "Homographies, Quaternions and Rotations." Oxford Univ. Press (Clarendon), London and New York, 1964.
[5] Gordon, C. Mc A., and Heil, W., Cyclic normal subgroups of fundamental groups of 3-manifolds, *Topology* **14** (1975), 305–309.
[6] Morgan, J. W., 3-dimensional Seifert-fibered orbifolds, to appear.
[7] Orlik, P., Seifert manifolds, *Lecture Notes in Mathematics*, Vol. **291**, Springer-Verlag, Berlin and New York, 1972.
[8] Satake, I., On a generalization of the notion of manifold, *Proc. Nat. Acad. Sci. U.S.A.* **42** (1956), 359–363.
[9] Seifert, H., and W. Threlfall, Topologische Untersuchung der Diskontinuitätsbereiche endlicher Bewegungsgruppen des dreidimensionalen sphärischen Raumes I, *Math. Ann.* **104** (1930), 1–70; II, *Math. Ann.* **107** (1932), 543–586.
[10] Waldhausen Q. F., Gruppen mit zentrum und 3-dimensionale Mannigfaltigkeiten, *Topology* **6** (1967), 505–517.

CHAPTER XI

A Survey of Results in Higher Dimensions

Michael W. Davis

Department of Mathematics
Ohio State University
Columbus, Ohio

The field of compact transformation groups deals with actions of compact Lie groups on manifolds. Its origins can be traced back to the advent of the study of groups of linear transformations in the nineteenth century. The actions of such linear groups on euclidean spaces, disks, spheres, projective spaces, etc., provide a rich and natural class of basic examples. The guiding principle of the field has always been to compare arbitrary actions to linear actions, that is, to determine the extent to which arbitrary actions resemble the basic linear examples.

It is natural to begin by studying compact transformation groups of the simplest manifolds: euclidean spaces, disks, and spheres. The only examples that spring to mind are closed subgroups of the appropriate orthogonal group. Naïvely, one might conjecture that these linear actions are the only possibilities; that is, for $M^n = \mathbf{R}^n$, D^n, or S^n and for G a compact Lie group of homeomorphisms of M^n, one might conjecture that G is conjugate by a homeomorphism to a subgroup of $O(n)$ (or of $O(n + 1)$ when $M^n = S^n$). We shall hereafter refer to this as the naïve conjecture. Prior to 1950 it seems that people felt this might be true, although it was clearly regarded as an extremely difficult problem. The conjecture is true for $n = 2$. As stated, it is false for $n = 3$: Bing [2] produced a nonlinear involution on S^3 in 1952. However, Bing's example was essentially a local pathology in that it resulted from the nondifferentiability of the involution. There is mounting evidence (much of it in this volume) that the naïve conjecture holds for $n = 3$ provided that we stick to groups of *diffeomorphisms*. For $n \geq 4$, the conjecture is false in either category. Despite this, the determination of the exact extent to which actions of compact Lie groups on euclidean spaces, disks, or spheres resemble linear actions remains today one of the central problems in the field.

The purpose of this chapter is to survey some of the work on this problem and to describe some counterexamples to the naïve conjecture in dimensions greater than three. We shall be particularly concerned with the nature of the fixed point set. For another discussion of this subject see Bredon [4].

The earliest work in this area focused on periodic transformations, that is, on actions of finite cyclic groups. In the twenties Brouwer [5] and Kerekjarto [15] proved that any orientation-preserving periodic transformation of the 2-disk or 2-sphere was conjugate to a rotation. A gap in the proof was later filled by Eilenberg [10].

In dimension three there are substantial partial results. In the case for which the group G is compact and connected, the three-dimensional naïve conjecture was proved by Montgomery and Zippen [20]. For smooth actions of $G = \mathbf{Z}_2$ it is due to Livesay [16, 17] and to Waldhausen [31]. For smooth, orientation-preserving, periodic transformations with nonempty fixed point set the three-dimensional naïve conjecture is equivalent to the Smith conjecture solved in this volume. Rubenstein [25] has proved it for free actions of certain finite groups on S^3.

In order to attack the naïve conjecture in higher dimensions, one wants to find topological properties of linear compact transformation groups and determine the extent to which they are shared by arbitrary compact groups of homeomorphisms of euclidean spaces, disks, and spheres. Smith focused attention on one such property of linear actions: The fixed point set of any group of linear transformations of $\mathbf{R}^n$ is a linear subspace. Obviously, such a subspace intersects D^n in a linear subdisk and S^{n-1} in a linear subsphere. Thus a necessary condition for a G-action on euclidean space (respectively, disk or sphere) to be equivalent to a linear G-action is that the fixed point set must be homeomorphic to a lower-dimensional euclidean space (respectively, disk or sphere). Moreover, the embedding of the fixed point set in the ambient space must be standard.

The first positive results toward establishing this necessary condition were proved by Smith in a series of papers [26–28] published in the 1930s and 1940s. He proved that the condition holds for cyclic groups of prime power order provided that we are concerned only with homology with coefficients in $\mathbf{Z}_p$ (where p is the prime in question). For example, Smith showed that if X is a reasonably nice (e.g., compact) space with the mod p homology of a sphere, then the fixed point set of a periodic transformation of period p^m also has the mod p homology of a sphere. An easy induction can be used to extend this to actions of finite p-groups (p a prime). The results were extended to actions of tori (using integral coefficients) by Conner [6] and Floyd [12] and they can easily be extended to the case in which G is a p-group extended by a torus. Here is a precise statement.

THEOREM (Smith). *Suppose that G is a p-group and that X is either a finite dimensional or a compact G-space that has the mod p Cech cohomology of an n-sphere. The fixed point set X^G has the mod p Cech cohomology of an r-sphere for some $-1 \leq r \leq n$.*

A similar result holds for pair (X, A) with the mod p Cech cohomology of (D^n, S^{r-1}). For further information see [3, Chapter III].

Initially, Smith believed that it was merely a defect in his methods that forced him to restrict to p-groups and to homology with $\mathbf{Z}_p$-coefficients. He clearly believed that these results should extend to coefficients in $\mathbf{Z}$, and at least to all finite cyclic groups. We shall give a simple counterexample to this conjecture in Section 2.

It has now become clear that Smith's results, together with some conditions on the Euler characteristic, are essentially the only homological conditions that must be satisfied by fixed point sets of group actions on euclidean spaces, disks, and spheres. As an illustration of this principle, we shall now state some results that precisely give the homotopy types that can occur as fixed point sets of smooth G-actions on disks. Let us first remark that it follows from the differentiable slice theorem that any such fixed point set is a smooth submanifold and hence has the homotopy type of a finite complex. For finite groups there is the following result.

THEOREM. *Let F be a finite complex.* (1) *If G is a finite group of p-power order, p a prime, then F is homotopy equivalent to the fixed point set of a smooth G-action on some disk if and only if F is $\mathbf{Z}_p$-acyclic.*

(2) *If G is a finite group not of prime power order, then there is an integer n_G such that F is homotopy equivalent to the fixed point set of some smooth G-action on a disk if and only if $\chi(F) \equiv 1 \pmod{n_G}$.*

The integer n_G can be explicitly computed (see Section 7). The answer is also known when G is a compact Lie group of positive dimension. If the identity component G_0 is not a torus, then there is no condition on F: Any finite complex (including the empty set) can occur. If G_0 is a torus, then F

must satisfy either condition (1) or (2) of the above theorem, depending on whether G/G_0 has prime power order. Statement (1) in this theorem is essentially due to Jones [14]. In all other cases it is the work of Oliver [21].

One consequence of this result is that most groups can act smoothly on a disk with empty fixed point set. The existence of such actions runs contrary to one's intuition. It becomes even more counterintuitive when we combine this fact with Mostow's embedding theorem. This theorem states that we can embed any compact, smooth G-manifold in some linear action on some large dimensional euclidean space. Thus we can embed a G-action on a disk in a linear action and take an invariant regular neighborhood to obtain a smooth G-action on another disk (of larger dimension) with homotopy equivalent fixed point set. Therefore, for any sufficiently complicated G, we can find a linear action of G on $\mathbf{R}^N$ and a smoothly embedded G-invariant disk $D^N \subset \mathbf{R}^N$ such that G fixes no points of D^N.

1. The Montgomery–Samelson Example[1]

Suppose that B^m is a compact, contractible m-manifold with boundary. The boundary of B is, of course, a homology $(m - 1)$-sphere. However, it is well known that for $m \geq 4$, ∂B can be nonsimply connected. Let us suppose this. Let G be a closed subgroup of $O(n)$ such that G fixes only the origin of $\mathbf{R}^n$. Consider the manifold $M^{n+m} = D^n \times B^m$ with G-action defined by $g(x, b) = (gx, b)$. If $m + n \geq 6$, then by the h-cobordism theorem, ∂M is a sphere. The G-action on ∂M is nonlinear since its fixed point set ∂B is not simply connected, hence not a sphere. The point of this example is that at most we can expect to prove results about the homology of the fixed point set: We cannot hope to control its fundamental group.

2. G-Complexes

A *G-complex* is a G-space built by successively adjoining "equivariant cells" of the form $G/H \times D^n$. The following general principle is important in the construction of examples: every finite contractible G-complex X gives rise to a smooth G-action on a disk with fixed point set homotopy equivalent to X^G. Similarly, every finite dimensional contractible G-complex gives rise to a smooth G-action on some euclidean space. The point is that provided we are willing to be generous about dimensions, we may replace our equivariant cells by equivariant handles of the form $(G \times_H D(V)) \times D^n$, where

---

[1] See [19].

XI. A Survey of Results in Higher Dimensions

$D(V)$ is the unit disk in some linear H-space V (see [21]). Alternatively, in the case in which G is finite and X is a simplicial complex, we can embed X in some triangulation of a linear action on $\mathbf{R}^N$, N large, and then take a regular neighborhood. (See p. 57 in [3].)

3. The Brieskorn Examples[2]

Consider the polynomial $f: \mathbf{C}^{n+1} \to \mathbf{C}$ defined by

$$f(z) = \sum_{i=0}^{n} (z_i)^{a_i},$$

where the a_is are integers greater than one. The hypersurface $f^{-1}(0)$ has an isolated singularity at the origin. Its *link* $\Sigma^{2n-1}(a_0, \ldots, a_n)$ is defined as the intersection of the hypersurface with a sufficiently small sphere centered at the origin. Each such link is an $(n-2)$-connected smooth manifold. Formulas for computing $H_{n-1}(\Sigma^{2n-1}(a_0, \ldots, a_n))$ in terms of the exponents can be found in Milnor [18]. If this group is trivial (and if $2n - 1 > 3$), then the link is homeomorphic to S^{2n-1}. (This happens quite frequently.)

These links have many interesting symmetries. For example, for each $0 \leq i \leq n$, $\Sigma^{2n-1}(a_0, \ldots, a_n)$ admits a periodic diffeomorphism of period a_i defined by

$$\omega(z_0, \ldots, z_i, \ldots, z_n) = (z_0, \ldots, \omega z_i, \ldots, z_n),$$

where ω is a primitive a_ith root of unity. The fixed point set is clearly $\Sigma^{2n-3}(a_0, \ldots, \hat{a}_i, \ldots, a_n)$. More specifically, consider $\Sigma^5(3, 2, 2, 2)$. This manifold is well known to be diffeomorphic to S^5. It admits an action of $\mathbf{Z}_6 = \mathbf{Z}_3 \oplus \mathbf{Z}_2$ defined as above by acting on the first coordinate by a third root of unity and on the second coordinate by a second root of unity. The fixed point sets of $\mathbf{Z}_3$, $\mathbf{Z}_2$, and $\mathbf{Z}_6$ are, respectively, $\Sigma^3(2, 2, 2)$, $\Sigma^3(3, 2, 2)$, and $\Sigma^1(2, 2)$. These manifolds are, respectively, real projective 3-space, the lens space $L(3, 1)$, and the disjoint union of two circles. Thus S^5 admits a smooth action of $\mathbf{Z}_6$ such that (1) the fixed point set of $\mathbf{Z}_6$ is not a homology sphere with any coefficients and (2) the fixed point sets of the subgroups of order two and three are not homology spheres with integer coefficients. (Note, however, that $\mathbf{R}P^3$ is a mod 3 homology sphere and that $L(3, 1)$ is a mod 2 homology sphere.) A similar example was constructed earlier by Floyd [11] on the 41-sphere. Such examples provide rather convincing evidence that Smith's results cannot be improved.

---

[2] See [13].

By slight modifications of the above example one can construct similar examples in every odd dimension $2n - 1$, with $n \geq 3$. It is also easy to modify these examples to obtain, in higher dimensions, periodic transformations of the sphere such that the fixed point set is a knotted sphere of codimension two. For example, an involution on $\Sigma^{2n-1}(p, q, 2, \ldots, 2)$ will do the job for suitably chosen p and q.

4. Oliver's Example[3]

In this section we shall discuss Oliver's beautiful construction of a smooth SO(3)-action on the 8-disk with no fixed points.

Let $G = \mathrm{SO}(3)$. As a first step we shall construct a contractible finite G-complex X with no fixed points. Let X be any G-complex with orbit space as pictured in Fig. 1. Thus X is made up of five equivariant cells of the

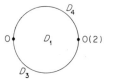

Figure 1

form $G/H \times D^n$, with $n \leq 2$. In Fig. 1 we have labeled each cell by its isotropy group H. Here D_n denotes the dihedral group of order $2n$, $O(2)$ is the orthogonal group, and O is the octahedral group. It is easy to see that there is a G-complex with this orbit space. By construction there are no fixed points.

We next want to show that X is contractible. Divide the orbit space into three pieces A', B', C' as shown in Fig. 2. Let $A = \pi^{-1}(A')$, $B = \pi^{-1}(B')$,

Figure 2

and $C = \pi^{-1}(C')$ be their inverse images in X. The proof that X is contractible is a straightforward exercise using van Kampen's theorem and the Mayer–Vietoris sequence for $X = A \cup_C B$. We shall sketch the details.

If H is a finite subgroup of $G = \mathrm{SO}(3)$, denote its inverse image in S^3 by $\tilde{H}$, so that $\pi_1(G/H) = \tilde{H}$. Since A deformation retracts to G/O and B deformation retracts to $G/O(2) = \mathbf{R}P^2$, we have that $\pi_1(A) = \tilde{O}$ and $\pi_1(B) = \mathbf{Z}_2$.

---

[3] See [23].

XI. A Survey of Results in High Dimensions

Also, $\pi_1(C) = \tilde{D}_4 *_{\tilde{D}_1} \tilde{D}_3$. The kernel of $\pi_1(C) \to \pi_1(B)$ is clearly $\mathbf{Z}_8 *_{\mathbf{Z}_2} \mathbf{Z}_6$. Since the binary octahedral group $\tilde{O}$ is generated by the cyclic groups of order eight and six, this kernel maps onto $\pi_1(A)$. Hence, X is simply connected.

The computation of the homology of X runs as follows:

$$\bar{H}_i(A) = \bar{H}_i(G/O) = \begin{cases} \mathbf{Z}_2, & i = 1, \\ \mathbf{Z}, & i = 3, \\ 0, & \text{otherwise.} \end{cases}$$

$$\bar{H}_i(B) = \bar{H}(\mathbf{R}P^2) = \begin{cases} \mathbf{Z}_2, & i = 1, \\ 0, & \text{otherwise.} \end{cases}$$

$$\bar{H}_i(C) = \begin{cases} \mathbf{Z}_2 \oplus \mathbf{Z}_2, & i = 1, \\ \mathbf{Z}, & i = 3, \\ 0, & \text{otherwise.} \end{cases}$$

The Mayer–Vietoris sequence shows that X is acyclic and hence contractible.

Next we want to "thicken" X to be an 8-disk. The idea is to thicken A and B to 8-manifolds with boundary M_A and M_B and to thicken C by a 7-manifold with boundary M_C in such a way that M_C is embedded in both ∂M_A and ∂M_B. The 8-disk will then be formed by gluing M_A and M_B along M_C. The manifolds M_A and M_B will each be unit disk bundles associated to certain G-vector bundles of the form $G \times_O V$ and $G \times_{O(2)} W$, respectively, where V and W are linear representations of O and $O(2)$ respectively. Let V_1 be the two-dimensional representation of O defined by $O \to D_3 \subset O(2)$. Note that V_1 contains points with isotropy subgroups $D_4 \subset O$. Let V_2 be the three-dimensional representation of O defined by regarding O as the full group of symmetries of the tetrahedron. Then V_2 has points with isotropy group D_3. Let $V = V_1 \oplus V_2$.

For $m \in \mathbf{Z}_+$, let W_m be the two-dimensional representation of $O(2)$ with kernel the cyclic group of order m. Set $W = W_3 \oplus W_4 \oplus W_k$, where $k \equiv \pm 2(12)$. Then W has points with isotropy subgroup $D_3 \subset O(2)$ and $D_4 \subset O(2)$, since W_3 and W_4 do.

Next we want to find a copy of C in both ∂M_A and ∂M_B. This is easy. For example to find it in ∂M_A, pick points x and y in $\partial M_A/G$ corresponding to orbits of type G/D_3 and G/D_4, respectively. Join x and y by an arc of orbits of type G/D. This gives a copy of C' in $\partial M_A/G$ and hence an embedding of

C in ∂M_A; similarly for ∂M_B. Let Y_A and Y_B be closed invariant regular neighborhoods of C in ∂M_A and ∂M_B, respectively. We want to show that Y_A is equivariantly diffeomorphic to Y_B. It clearly suffices to check this in a neighborhood of the G/D_3 and G/D_4 orbits. And this is just a matter of checking whether the tangential representations of the isotropy groups D_3 and D_4 in Y_A agree with those of Y_B. The condition that $k \equiv \pm 2(12)$ precisely guarantees this. Thus, we set $M_C = Y_A = Y_B$ and $M = M_A \cup_{M_C} M_B$. M is a compact contractible 8-manifold since it is homotopy equivalent to X. Since ∂M is the union of two sphere bundles along M_C, the calculation of $\pi_1(\partial M)$ is the same as for $\pi_1(X)$. Hence ∂M is simply connected and M is therefore an 8-disk. Thus SO(3) *can act smoothly on D^8 without fixed points.*

Let $I \subset \mathrm{SO}(3)$ be the icosahedral group. Since no isotropy group of M contains I, Oliver's example gives a bonus: *There is a smooth action of I on D^8 without fixed points.*

Actually, Floyd and Richardson had earlier constructed a fixed point free action of I on a disk. Their construction starts with the action of I on the Poincaré homology sphere G/I. This action has one fixed point. Removing a neighborhood gives an action of I on a compact acyclic 3-manifold Q. The diagonal action of I on the join $I * Q$ gives a finite contractible I-complex with no fixed points. This can be thickened to a disk. For further details see [3, p. 55]

5. Local Properties: Groups of Homeomorphisms versus Groups of Diffeomorphisms

The differentiable slice theorem asserts that any orbit of a smooth G-manifold M has an equivariant tubular neighborhood, i.e., an orbit has an invariant neighborhood in M that is G-diffeomorphic to a G-vector bundle over it. In particular, a fixed point has an invariant neighborhood on which the action is linear. It follows that the fixed point set is a smoothly embedded submanifold. Therefore, fixed point sets of smooth actions are locally indistinguishable from those of linear actions.

If the action is not smooth, we can have local pathologies. Fixed point sets need not be manifolds and even if they are manifolds, they need not be embedded in a locally flat fashion. The first example of this type was Bing's [2] involution of the 3-sphere with fixed point set an Alexander's horned sphere. Examples of actions on spheres where the fixed point set is not a manifold can be obtained by suspending any of the examples in Sections 1 or 3. (There is, however, a local version of Smith's theorem which asserts that if G is a p-group acting on a manifold, then the fixed point set is a mod p homology manifold.)

6. Work of Lowell Jones[4]

In this section we shall discuss some of the work of Jones [14] on fixed point sets of certain periodic transformation of disks.

A G-action on a space X is said to be *semifree* if for each $x \in X$ the isotropy group G_x is equal to either the trivial group or to G. If G is cyclic of order n and acts semifreely on a disk, then it follows Smith's theorem that the fixed point set is $\mathbf{Z}_p$-acyclic for every prime p dividing n. Hence, the fixed point set is actually $\mathbf{Z}_n$-acyclic. Conversely, Jones [14] proved the following result.

THEOREM (Jones). *Let G be the cyclic group of order n. A finite complex F is homotopy equivalent to the fixed point set of a smooth G-action on some disk if and only if F is $\mathbf{Z}_n$-acyclic.*

As we pointed out in Section 2, this theorem is equivalent to the statement that any such F occurs as the fixed point set of a semifree G-action on some finite contractible G-complex. Thus we want to start with F and successively adjoin a finite number of cells of free orbits to kill its fundamental group and its homology, in this way obtaining a finite contractible G-complex X with $X^G = F$. If we have succeeded in building an $(m-1)$-connected G-complex X, then there is no problem in equivariantly attaching cells to kill $H_m(X)$. Simply choose a set of generators for $H_m(X)$ as a $\mathbf{Z}[G]$-module, represent these generators by spheres, and attach a free orbit of cells for each generator. The trouble is that we will introduce new homology in dimension $m+1$. For an arbitrary F there is no guarantee that this process will terminate after a finite number of steps. Jones's key observation is that the hypothesis that F is $\mathbf{Z}_n$-acyclic ensures that the "extra" homology we have introduced in dimension $m+1$ is a free module over $\mathbf{Z}[G]$. Thus we can kill this extra homology at the next stage without introducing any further homology. This observation essentially comes down to the following algebraic lemma.

LEMMA. *Suppose that G is cyclic of order n and that A is a $\mathbf{Z}[G]$-module such that (1) A is finite of order prime to n and (2) G acts trivially on A. Let $\psi: R \to A$ be an epimorphism with R a free $\mathbf{Z}[G]$-module. Then $\ker \psi$ is a free $\mathbf{Z}[G]$-module.*

A proof can be found in Jones [14].

Using similar ideas it is easy to extend Jones's result to actions of finite groups of prime power order.

---

[4] See [14].

THEOREM. *Let G be a p-group (p a prime). A finite complex F is homotopy equivalent to the fixed point set of a smooth G-action on some disk if and only if F is $\mathbf{Z}_p$-acyclic.*

To prove this, we choose $H \triangleleft G$ with G/H cyclic. Attach cells of G/H-orbits to F to obtain a finite contractible G-complex X. Then attach cells of free orbits in adjacent dimensions so as to make the action effective without introducing any homology.

7. Actions on Disks

Having dealt with p-groups in the previous section, we shall in this section summarize the definitive work of Oliver [21, 22, 24] on fixed point sets of finite group actions on disks for groups not of prime power order. He shows that in this case the only restriction on the homotopy type of the fixed point set is that its Euler characteristic must sometimes satisfy a certain congruence relation.

THEOREM (Oliver [21]). *For any finite group G not of prime power order there is an integer n_G such that a finite complex F is homotopy equivalent to the fixed point set of some smooth G-action on a disk if and only if $\chi(F) \equiv 1 \pmod{n_G}$.*

Oliver then goes on to explicitly calculate n_G. In order to give this calculation, we shall first define an integer $m(G)$. Let $\mathscr{G}^1$ be the class of finite groups G with G/P cyclic for some $P \triangleleft G$, with P of prime power order. For each prime q, let $\mathscr{G}^q$ be the class of groups G with G/H of q-power order for $H \triangleleft G$ and $H \in \mathscr{G}^1$. Thus $\mathscr{G}^1$ consists of those groups that are "cyclic mod p" and $\mathscr{G}^q$ consists of those that are "q-hyperelementary mod p" for some prime p. By definition, $m(G)$ will be either 0, 1, or a product of distinct primes. It is 0 if and only if $G \in \mathscr{G}^1$, and $q \mid m(G)$ if and only if $G \in \mathscr{G}^q$.

The result of Oliver's calculation is that almost always $n_G = m(G)$. However, for a certain class of 2-hyperelementary groups, $n_G = 4$, while $m(G) = 2$. Since the definition of this class of groups is fairly complicated we shall simply refer to it as the "exceptional case" and direct the reader to Oliver [22, p. 345] for the precise definition.

THEOREM (Oliver [24]). *In the exceptional case, $n_G = 4$. Otherwise, $n_G = m(G)$.*

If $n_G = 1$, then the congruence in the theorem is automatically satisfied. Hence, for such groups G any finite complex (including the empty set!) can

XI. A Survey of Results in Higher Dimensions

occur as the homotopy type of the fixed point set of a smooth G-action on a disk. Notice that it is easy for a group to satisfy the condition $m(G) = 1$ (and hence $n_G = 1$). Any nonsolvable group has this property, as does any sufficiently complicated solvable group. The smallest abelian group with $n_G = 1$ is $Z_{30} \oplus Z_{30}$, while the smallest solvable groups with this property have order 72. There are two such: $S_4 \oplus Z_3$ and $A_4 \oplus S_3$ [21, p 175].

Before discussing the proofs of the above theorems, let us warm up by proving the following elementary result which makes the main results possible.

PROPOSITION (Oliver), *Suppose that F is the fixed point set of a smooth G-action on a disk, then $\chi(F) \equiv 1 \pmod{m(G)}$.*

Let M be any compact smooth G-manifold. The proof of this is based on the following three facts:

(1) If G is a p-group, then $\chi(M^G) \equiv \chi(M) \pmod{p}$.
(2) If G is a p-group and M is Z_p-acyclic, then M^G is Z_p-acyclic.
(3) If G is cyclic and M is rationally acyclic, then $\chi(M^G) = 1$.

Statement (1) is a standard result in Smith theory and can be found, for example, in Bredon [3, p. 145]. Statement (2) is Smith's theorem. Statement (3) follows from the more general fact that if G is cyclic, then $\chi(M^G)$ is equal to the Lefschetz number of a generator of G.

Proof of the Proposition. Let D be a disk with smooth G-action. If $G \in \mathscr{G}^1$, then G/P is cyclic for some $P \triangleleft G$, with P a p-group. By (2), D^P is Z_p-acyclic, hence rationally acyclic. Therefore, by (3), $\chi(D^G) = \chi((D^P)^{G/P}) = 1$. If $G \in \mathscr{G}^q$, then G/H is a q-group for some $H \in \mathscr{G}^1$. We have just shown that $\chi(D^H) = 1$. Hence by (1), $\chi(D^G) = \chi((D^H)^{G/H}) \equiv 1 \pmod{q}$.

A *G-resolution* of a finite complex F is an n-dimensional finite G-complex Y, such that Y is $(n-1)$-connected, $H_n(Y)$ is a projective $Z[G]$-module, and $Y^G = F$. If $H_n(Y)$ is a free $Z[G]$-module (or even stably free), then we can add cells of free orbits to kill it without introducing any further homology, obtaining in this way a finite contractible G-complex. Let $\gamma_G(F, Y)$ denote the class of $H_n(Y)$ in $\tilde{K}_0(Z[G])$. Given F, Oliver breaks the problem of constructing a finite contractible G-complex X with $X^G = F$ into two parts: (a) building a G-resolution Y and (b) analyzing the obstruction $\gamma_G(F, Y)$. He first proves the following result.

PROPOSITION (Oliver). *Suppose that G is a finite group not of prime power order. A finite complex F has a G-resolution if and only if $\chi(F) \equiv 1 \pmod{m(G)}$.*

In order to construct a finite contractible G-complex with fixed point set F, we are free to vary the G-resolution Y. Thus the obstruction we are interested in has indeterminacy in the subgroup $\beta(G) = \{\gamma_G(\text{point}, Y)\} \subset \tilde{K}_0(\mathbf{Z}[G])$. There results an obstruction $\gamma_G(F) \in \tilde{K}_0(\mathbf{Z}[G])/\beta(G)$. Oliver [21] shows that if G is not hyperelementary, then $\gamma_G(F) = 0$ and hence $n_G = m(G)$. He also shows that even when G is hyperelementary $m(G)$ divides n_G. Finally, in a later paper [24] he analyzes the hyperelementary case and shows that $n_G = m(G)$ precisely in the nonexceptional case.

8. Actions on Spheres

We have already given examples of nonlinear actions on spheres in Sections 1 and 3. Although there is a large body of literature on this subject we shall mention only one further example. Stein [30] constructed a smooth action of the binary icosahedral group on S^7 with only one fixed point. Further work along these lines can be found in Assadi [1] and Doverman and Petrie [8].

9. Actions on Euclidean Spaces

As we pointed out in Section 2, the construction of such actions is related to the construction of finite dimensional contractible G-complexes. Since we are free to add infinitely many cells, the obstructions in the projective class group are no longer relevant. Also, since $\mathbf{R}^n$ is not compact, the Lefschetz fixed point theorem no longer imposes constraints on the Euler characteristics of fixed point sets of cyclic subgroups. Hence it is generally much easier to construct actions with exotic fixed point sets on euclidean spaces than on either disks or spheres.

For example, exotic fixed point sets occur even for actions of finite cyclic groups which are not of prime power order. The first example of this type was constructed by Connor and Floyd [7]. If p and q are relatively prime integers, then they showed that a certain linear action of the cyclic group $\mathbf{Z}_{pq}$ on S^3 admits an equivariant self-map of degree 0. By taking the infinite mapping telescope of this map they obtain a $\mathbf{Z}_{pq}$-action on a contractible 4-complex with empty fixed point set and hence, a smooth $\mathbf{Z}_{pq}$-action on euclidean space with empty fixed point set. As Bredon observes [3, p. 62], a slight modification of this construction shows that any finite complex is homotopy equivalent to the fixed point set of some $\mathbf{Z}_{pq}$-action on a euclidean space.

The fixed point set of a linear action on euclidean space is a linear subspace; hence it is either a point or noncompact. Smith [29] asked if a compact manifold other than a point could occur as fixed point set. This was answered by Edmonds and Lee [9]. They showed that a smooth closed manifold occurs as the fixed point set of some finite cyclic group action on euclidean space if and only if its tangent bundle admits a complex structure.

References

[1] Assadi, A., Finite group actions on simply connected manifolds and CW complexes. *Mem. Amer. Math. Soc.* **35** (257) (1982).

[2] Bing, R. H., A homeomorphism between the 3-sphere and the sum of two solid horned spheres, *Ann. of Math.* **56**, 354–362.

[3] Bredon, G., "Introduction to Compact Transformation Groups." Academic Press, New York, 1972.

[4] Bredon, G., Book review, *Bull. Amer. Math. Soc.* **83** (1977), 711–718.

[5] Brouwer, L. E. J., Über die periodischen Transformationen der Kugel, *Math. Ann.* **80** (1919), 262–280.

[6] Conner, P. E., On the action of the circle group, *Michigan Math. J.* **4** (1957), 241–247.

[7] Conner, P. E., and Floyd, E. E., On the construction of periodic maps without fixed points. *Proc. Amer. Math. Soc.* **10** (1959), 354–360.

[8] Doverman, K. H., and Petrie, T., G-surgery II. *Mem. Amer. Math. Soc.* No. 260 **37** (1982).

[9] Edmonds, A. L., and Lee, R., Fixed point sets of group actions on Euclidean space, *Topology* **14** (1975), 339–345.

[10] Eilenberg, S., Sur les transformations périodiques de la surface du sphère, *Fund. Math.* **22** (1934), 28–41.

[11] Floyd, E. E., Examples of fixed point sets of periodic maps, *Ann. of Math.* **55** (1952), 167–171.

[12] Floyd, E. E., Fixed point sets of compact abelian groups of transformations, *Ann. of Math.* **65** (1957), 30–35.

[13] Hirzebruch, F., Singularities and exotic spheres, *Sem. Bourbaki* No. 314 **19**, (1966/7).

[14] Jones, L., A converse to the fixed point theory of P. A. Smith, I, *Ann. of Math.* **94** (1971), 52–68.

[15] Kerekjarto, B., Über die periodischen Transformationen de Kreisschube and Kugel-flache, *Math. Ann.* **80** (1919), 36–38.

[16] Livesay, G. R. Involutions with two fixed points of the 3-sphere, *Ann. of Math.* **78** (1963), 582–593.

[17] Livesay, G. R., Fixed point free involutions on the 3-sphere, *Ann. of Math.* **78** (1960), 603–611.

[18] Milnor, J. W., Singular points of complex hypersurfaces. *Ann. of Math. Studies*, Vol. 61. Princeton Univ. Press, Princeton, New Jersey, 1969.

[19] Montgomery, D., and Samelson, H., Examples for differentiable group actions on spheres, *Proc. Nat. Acad. Sci.* **47** (1961), 1202–1205.

[20] Montgomery, D., and Zippen, L., "Topological Transformation Groups." Wiley (Interscience), New York, 1955.

[21] Oliver, R., Fixed point sets of group actions of finite acyclic complexes, *Comment. Math. Helv.* **50**, (1975), 155–177.

[22] Oliver, R., Group actions on disks, integral permutation representations, and the Burnside ring, "Algebraic and Geometric Topology." *Proc. Sympos. Pure Math.*, Vol. 32, 339–346. Amer. Math. Soc., Providence, Rhode Island, 1978.

[23] Oliver, R., Weight systems for SO(3)-actions, *Ann. of Math.* **110** (1979), 227–241.

[24] Oliver, R., G-actions on disks and permutation representations II, *Math. Z.* **157** (1977), 237–263.

[25] Rubinstein, J. H. Free actions of some finite groups on S^3 I, *Math. Ann.* **240** (1979), 165–175.

[26] Smith, P. A., Transformations of finite period, *Ann. of Math.* **39** (1938), 127–164.

[27] Smith, P. A., Transformations of finite period, II, *Ann. of Math.* **40** (1939), 690–711.

[28] Smith, P. A., Transformations of finite period, III, *Ann. of Math.* **42** (1941), 446–458.

[29] Smith, P. A., New results and old problems in finite transformation groups, *Bull. Amer. Math. Soc.* **66** (1960), 401–415.

[30] Stein, E., Surgery on products with finite fundamental groups, *Topology* **16** (1977), 473–493.

[31] Waldhausen, F., Über Involutionen der 3-Sphäre, *Topology* **8** (1969), 81–91.

INDEX

A

Accidental parabolic, *see* Parabolic, accidental
Acylindrical, three-manifold, 89
Ahlfors finiteness theorem, 64, 68
Algebraic integers, 15, 16, 24, 25, 34, 127, 128, 134, 135, 208, *see also* Hyperbolic geometry and algebraic integers
Amalgamated free product, *see* Free product, amalgamated
Andre'ev's theorem, 43, 117, 118
Arboreal group theory, 130–132, 133
Arboreal presentation theorem, 128, 131–132

B

Bass's theorem, 15, 22, 34, 36, 208, *see also* GL_2-subgroup theorem
Beltrami differential, 86, 92
Beltrami equation, 82, 86
Bers's theorem, 82–83, 92
Boundary pattern, 107–109, 110–111, 115, 119, 120–124
  geometric realization, 109, 113–115, 122–124
Bounded image theorem, 40, 42, 85, 87–105
Bounded surface, 90
  maximal, 41, 91

Branched covering, 139
  cyclic, 12–13, 139, 141, 182, 207
  regular, 139, 140, 141, 146, 178
Brieskorn examples, 231–232

C

Characteristic submanifold, *see* Characteristic subpair
Characteristic subpair, 41, 88, 102
Combination theorem, *see* Maskit combination theorem
Complete hyperbolic manifold, 47, *see also* Hyperbolic structure
Conformal transformation, of S^2, 44, 47
Coxeter group, generalized, 108, 112
Cyclic branched covering, *see* Branched covering, cyclic

D

Deformation theory, *see* Hyperbolic structure, deformation theory
Dehn lemma and loop theorem, 24–25, 34, 57
  equivariant version, 14, 142
Dickson, 182, 209, 216–224

E

Energy of a map, 155, 156, 158, 161
Essentially linear action, *see* Linear action, essentially

F

Fibering over S^1, 42, 97, 106–107
Fixed point theorem, 40, 42, 84–87
Free product, amalgamated, 16, 31–32, 34, 127, 129, 130, 131

G

GL_2-subgroup theorem, 127–136, *see also* Bass's theorem
Gluing theorem, 39, 42, 70–71
Graph of groups, 132

H

Hierarchy, 55–60, 119
Homogeneously regular, 154, 155, 159
Horoball, 46–47, 64
Horosphere, 46–47
Hyperbolic element, of $PGL_2(\mathbf{C})$, 45–46, 52, 64
Hyperbolic geometry
  and algebraic integers, 23–28
  introduction, 43–47
Hyperbolic structure, 14, 24, 28–31, 51, 52, 71, 76, 83, 106, 114, 142, 147, 207, 208
  convex, 60, 61–70, 117, 118, 123
  convex core of, 63, 64, 65
  deformation theory, 78–84
  geometrically finite, 63, 65, 71, 73, 79, 114, *see also* Kleinian group, geometrically finite, 63, 64, 81
  on pared manifold, 60

I

Icosahedral group, 168, 177, 189, 209, 234
Incompressible splitting surface, *see* Splitting surface
Incompressible surface, 14, 16, 21, 31–35, 139, 140, 141, 143, 208
  in branched coverings, 139–152
  case of no, 21–36
Incompressible torus, *see* Torus, incompressible

J

Jones, work of, 235–236

K

Kleinian group, 38, 40, 42, 47–51, 76, 78
  elementary, 48
  geometrically finite, 63, 64, 81
  limit set, 47–48, 66–67
  Nielsen convex region, 63
  region of discontinuity, 48, 97, 104–105
  surfaces at infinity, 48
  with torsion, 107–116
    geometrically finite, 110

L

Linear action, 181–182, 196–206, 227–230
  essentially, 4, 5, 12, 181, 182, 206
Loop theorem, *see also* Dehn lemma and loop theorem
  equivariant, 139, 144, 153–163, 168–169, 170, 172
  for involutions, 140, 147–151
  geodesic, 168, 170, 172

M

Margulis's theorem, 39, 54, 64
Maskit combination theorem, 71, 74–78
Minimal surface, 14, 154, 158, 168
  existence theorems, 153–163
    for closed, 160–162
    free boundary value problem, 162–163
    for manifolds with boundary, 159–160
Morrey, 154–156, 158
Mostow's theorem, 52, 54, 79, 81, 82, 106

O

Oliver's example, 232–234
Orbifold, 71, 182, 183–187, 206, 212–215
  covering, 185–186
  fundamental group, 186
  good, 186
  Seifert fibered, *see* Seifert fibered, orbifold
  three-dimensional, 189–192
  two-dimensional, 187–189

P

Parabolic, accidental, 97, 98, 101–102, 112, 113, 115, 116, 123

Index

Parabolic element, of $PGL_2(\mathbf{C})$, 45, 46, 52, 53
Parabolic locus, pared manifold, 59
  with boundary pattern, 108
Pared manifold, 38, 58–59, 69, 113–116, 119–124
Periodic diffeomorphism, 4, 7, 21, 24, 55, 228
Periodic homeomorphism, 3, 7–8, 27
Peripheral subgroup, 14, 29, 31, 32, 33, 35, 52, 58, 109, 112, 113, 121
Peripheral surface, 14
$PGL_2(\mathbf{C})$, *see* Projective general linear group
Plateau problem, 154–158
Poincaré conjecture, 5–6
Poincaré metric, 43–44
Projective general linear group, 45
Projective special linear group, 23–24, *see also* Projective general linear group
$PSL_2(\mathbf{C})$, *see* Projective special linear group

Q

Quasi-conformal homeomorphism, 80, 81, 86, 92, 94, 99, 107
Quasi-fuchsian, 73, 76, 77, 78, 81, 96–97, 104–105, 123
Quasi-isometry, 79, 80, 81, 98–99

S

Seifert fibered, 14, 16, 23, 27–28, 30–31, 34, 88, 147, 182, 183, 207, 213
  orbifold, 182, 183, 192–206, 207, 213, 215
  and linear actions, 196–206
Seifert fibration, *see* Seifert fibered
Seifert manifold, *see* Seifert fibered
Skinning map, 39, 83, 96
Smith conjecture, 4, 21, 24, 27, 141, 146, 153, 169, 173, 181, 182, 208, 228

algebraic approach, 22–23
early progress, 8–9
even period, 8–9
generalization, 5, 167, 172, 183
higher dimension, 9
history, 7–8
outline of proof, 11–16
reformulation, 5
solution, 4
Splitting surface, 96, 97, 98, 99
Superincompressible surface, 38, 39, 57, 58, 59, 61, 73, 75, 114
Surgery, along a disk, 143, 144–146

T

Teichmüller, 85
Teichmüller distance, 82, 85, 86
Teichmüller map, 40, 82, 86
Teichmüller space, 39, 49, 158
Thurston's theorem, *see* Thurston's uniformization theorem
Thurston's uniformization theorem, 14, 22, 29, 30, 37–125, 140, 142, 208
Torus, incompressible, 16, 29, 30, 31, 34, 141, 207
Torus cusp, 54, 64, 65
Torus theorem, 30, 147

U

Unipotent element, 34, 127, 130, 134
Unipotent subgroup, 34, 35, 127, 133, 134, 216

W

Window, of three-manifold, 88, 90
  involution, 94

Z

Z-cusp, 64, 66, 69
  finite, 68, 69, 80

Pure and Applied Mathematics

A Series of Monographs and Textbooks

Editors **Samuel Eilenberg and Hyman Bass**

Columbia University, New York

RECENT TITLES

CARL L. DEVITO. Functional Analysis

MICHIEL HAZEWINKEL. Formal Groups and Applications

SIGURDUR HELGASON. Differential Geometry, Lie Groups, and Symmetric Spaces

ROBERT B. BURCKEL. An Introduction to Classical Complex Analysis: Volume 1

JOSEPH J. ROTMAN. An Introduction to Homological Algebra

C. TRUESDELL AND R. G. MUNCASTER. Fundamentals of Maxwell's Kinetic Theory of a Simple Monatomic Gas: Treated as a Branch of Rational Mechanics

BARRY SIMON. Functional Integration and Quantum Physics

GRZEGORZ ROZENBERG AND ARTO SALOMAA. The Mathematical Theory of L Systems

DAVID KINDERLEHRER and GUIDO STAMPACCHIA. An Introduction to Variational Inequalities and Their Applications

H. SEIFERT AND W. THRELFALL. A Textbook of Topology; H. SEIFERT. Topology of 3-Dimensional Fibered Spaces

LOUIS HALLE ROWEN. Polynominal Identities in Ring Theory

DONALD W. KAHN. Introduction to Global Analysis

DRAGOS M. CVETKOVIC, MICHAEL DOOB, AND HORST SACHS. Spectra of Graphs

ROBERT M. YOUNG. An Introduction to Nonharmonic Fourier Series

MICHAEL C. IRWIN. Smooth Dynamical Systems

JOHN B. GARNETT. Bounded Analytic Functions

EDUARD PRUGOVEČKI. Quantum Mechanics in Hilbert Space, Second Edition

M. SCOTT OSBORNE AND GARTH WARNER. The Theory of Eisenstein Systems

K. A. ZHEVLAKOV, A. M. SLIN'KO, I. P. SHESTAKOV, AND A. I. SHIRSHOV. Translated by HARRY SMITH. Rings That Are Nearly Associative

JEAN DIEUDONNÉ. A Panorama of Pure Mathematics; Translated by I. Macdonald

JOSEPH G. ROSENSTEIN. Linear Orderings

AVRAHAM FEINTUCH AND RICHARD SAEKS. System Theory: A Hilbert Space Approach

ULF GRENANDER. Mathematical Experiments on the Computer

HOWARD OSBORN. Vector Bundles: Volume 1, Foundations and Stiefel-Whitney Classes

K. P. S. BHASKARA RAO AND M. BHASKARA RAO. Theory of Charges

RICHARD V. KADISON AND JOHN R. RINGROSE. Fundamentals of the Theory of Operator Algebras, Volume I

EDWARD B. MANOUKIAN. Renormalization

BARRETT O'NEILL. Semi-Riemannian Geometry: With Applications to Relativity

LARRY C. GROVE. Algebra

E. J. MCSHANE. Unified Integration

STEVEN ROMAN. The Umbral Calculus

JOHN W. MORGAN AND HYMAN BASS (Eds.). The Smith Conjecture

IN PREPARATION

ROBERT B. BURCKEL. An Introduction to Classical Complex Analysis: Volume 2

RICHARD V. KADISON AND JOHN R. RINGROSE. Fundamentals of the Theory of Operator Algebras, Volume II

A. P. MORSE. A Theory of Sets, Second Edition

SIGURDUR HELGASON. Groups and Geometric Analysis, Volume I: Radon Transforms, Invariant Differential Operators, and Spherical Functions

E. R. KOLCHIN. Differential Algebraic Groups